THE COMPLETE BOOK OF PLANT PROPAGATION

GRAHAM CLARKE
and
ALAN TOOGOOD

WARD LOCK

ACKNOWLEDGEMENTS

The publishers are grateful to the following for granting permission to reproduce the colour photographs: the *Amateur Gardening* Library (pp. *i*, *ii* (upper and lower), *iii* (lower), *iv* (upper and lower), *vii* (lower), *xii*, *xviii*, *xix*, *xxv* (upper), *xxvii* and *xxix*); Photos Horticultural Picture Library (pp. *iii* (upper), *v* (upper), *vi*, *xi*, *xii*, *xxi*, *xxii*, *xxiii*, *xxvi*, *xxx* (upper and lower), *xxxi* (lower) and *xxxii*); Harry Smith Horticultural Photographic Collection (pp *v* (lower), *vii* (upper), *ix*, *x*, *xiv*, *xvi*, *xvii*, *xx*, *xxiv*, *xxv* (lower), *xxvii* and *xxxi* (upper).

All the line drawings with the exception of those in Chapter 1 were drawn by Nils Solberg.

First published in paperback 1992
by Ward Lock Limited, Villiers House, 41/47 Strand,
London WC2N 5JE, England
A Cassell Imprint

Reprinted 1991, 1992

Text filmset by Litho Link Limited, Welshpool, Wales

Printed in Hong Kong
by Colorcraft Ltd.

British Library Cataloguing in Publication Data
Clarke, Graham
The complete book of plant propagation.
1. Gardens. Plants. Propagation
I. Title II. Toogood, Alan R. *1941-*
635.04

ISBN 0-7063-7079-1

CONTENTS

Preface 4

PART I

Techniques 5

1 History of Plant Propagation 6
2 Equipment and Tools 13
3 Hygiene 25
4 Seed Collecting 35
5 Seed Sowing Outdoors 41
6 Seed Sowing Under Cover 54
7 Raising Plants from Spores 61
8 Stem Cuttings 64
9 Root cuttings 73
10 Leaf and Leaf-bud Cuttings 76
11 Division 82
12 Layering 91
13 Plantlets 99
14 Vegetative Propagation of Bulbs 102
15 Grafting 108
16 Budding 116
17 Micropropagation 120
18 Training Plants 122

PART II

Plants and Their Propagation 128

19 Hardy Ornamental and Fruiting Plants 129
20 Tender Ornamental and Fruiting Plants 205
21 Hardy and Tender Vegetables and Culinary Herbs 243

Appendix: Useful Addresses 250
Index 252

Preface

Raising your own plants from seed or cuttings, division, layering, or any of the other popular methods of propagation, has to be one of the most pleasurable and satisfying jobs a gardener can do. There is nothing quite like the feeling of achievement you get when you disappear down to the greenhouse on a crisp spring morning, to discover that your cherished tray of sown seed now takes on a faint shade of green. The first delicate seed leaves are raising themselves above the compost, and you realize that you have played the most vital role in initiating possibly hundreds of new, living organisms.

Similarly, when a row of semi-ripe shrub cuttings stands firm and refuses to wilt, one is often overwhelmed by a desire to venture out and take cuttings of everything in sight. Armed with a sharp knife, plastic bag and tie-on labels, there is no stopping the more adventurous among us. Plant propagation is infectious, and the more of it you do, the more fascinating and rewarding the hobby seems.

Certainly, any well-seasoned gardener will tell you that, no matter how many years you may have tended an allotment, or grown flowers for indoor arrangements, or fruit for pies and flans, the most satisfying element of all is that of being able to start the plants into growth. To originate. To conceive.

It is, therefore, the aim of this book to help you along the propagation path. In Part I, *Techniques*, you will discover all you need to know in terms of seed sowing and taking cuttings. You will see how some plants are raised by spores, while others grow vigorously from sections of root. Many plants may be split in two, three or more and the divisions will grow successfully on their own. You'll discover that certain plants produce their offspring on leaves or long arching stems. Many increase by means of underground stems, tubers and bulbs. All aspects of plant multiplication – including layering, grafting and budding, plus modern scientific techniques such as micropropagation – are discussed.

Part II, *Plants and Their Propagation*, is an extensive encyclopaedia split into three chapters, these covering the propagation of hardy ornamental and fruiting plants, tender ornamental and fruiting plants, and hardy and tender vegetables and culinary herbs. This is quite possibly the most comprehensive propagation guide found in any book on the subject – there are about 1800 entries.

The encyclopaedia embraces the major plants grown in Northern Europe and the United States, plus a good sprinkling of unusual kinds, and each entry includes details of how, when and where to propagate them.

Specially commissioned line drawings clearly show all the techniques of plant propagation and colour photographs illustrate many of the plants discussed.

We have written with the amateur gardener in mind so the text is easily followed but, it has to be said, professional gardeners and nurserymen, as well as horticultural students and academics, will find the information contained within the covers of this book to be of tremendous help to them in their work.

G. C. & A. T.

PART I

HISTORY OF PLANT PROPAGATION
EQUIPMENT AND TOOLS
HYGIENE
SEED COLLECTING
SEED SOWING OUTDOORS
SEED SOWING UNDER COVER
RAISING PLANTS FROM SPORES
STEM CUTTINGS
ROOT CUTTINGS
LEAF AND LEAF-BUD CUTTINGS
DIVISION
LAYERING
PLANTLETS
VEGETATIVE PROPAGATION OF BULBS
GRAFTING
BUDDING
MICROPROPAGATION
TRAINING PLANTS

CHAPTER 1

History of Plant Propagation

Man and Nature

Reproduction, in all its Earthly forms, is the most natural phenomenon of all. Early man realized that life comes and life goes, and for our race to continue its existence on this planet, we needed to understand the most important elements of survival, and to cherish them. Instinctively, we began to comprehend life and what it comprised; how it came, and how it went.

It soon became evident that all life, in all its forms, is in one way or another, dependent on the plant kingdom. We consumed plants, and we consumed meat, but the animals whose flesh we so keenly devoured also consumed plants. Many of our ancestors relied on plants to give them clothing and shelter – and the animals whose skins prehistoric man wore would not have existed without plants. Plants provided our absolute needs, to say nothing of our later luxuries, such as furniture and paper.

Most important was the realization that we needed plants for food. They formed a very healthy and pleasant supplement to the diet of hunted meat and fish, and it was in our best interests to learn about them; the way they grew, how they drew nourishment and moisture from the soil, and how they reproduced themselves so that there was always a plentiful supply for the insatiable human appetite. Food crops of the ancient civilizations consisted mainly of cereals; it was not until relatively modern times that the familiar vegetables and fruit of today were added to the diet.

We learnt well from Nature. We observed that plants would produce strange and sometimes attractive flowers and that, soon afterwards, the stems which once carried these blossoms would be holding even stranger pods, capsules, or clumps of dried hard seeds. Most plants would disperse their offspring. Some would throw them out and away from the plant, perhaps for an incredible distance. Others would produce such fine seeds that even the slightest breeze lifted them and carried them way out of sight. Some plants merely allowed their seeds to fall to the ground immedately under the canopy of leaves, so that beneath the shelter of the mother plant the next generation would begin its existence.

Prehistoric man also noticed how some plants would become damaged, by the elements, or by a foraging animal, including man himself. Pieces of the plant would break off and lay on the ground, or they would be half-submerged in the soft earth. In due course, many of these stems and twigs would show signs of growth and development. In time they would become plants in their own right and worthy of investigation for food! Without realizing it at the time, we had discovered the methods of plant procreation that we know today as **sowing seeds** (sexual) and **taking cuttings** (asexual).

Early experiments

As the centuries passed, Man became aware that certain plants required specific conditions if they were to give of their best. This meant that from the very earliest stage in the life of some plants, we had to take extra care of them, to ensure that they would not die, but instead thrive to provide us with sustenance, and even clothing as we were now discovering. We realized that some seeds had to be soaked, or chipped, or subjected to a period of freezing, *before* they were sown. We learnt that some types of seed germinated in a matter of hours, while others took weeks, months, years, and so long that perhaps they did not grow at all!

The Greeks

Some 2000 years ago the Greek philosopher Theophrastus, who is best remembered for his botanical works and the book *Enquiry into Plants*, wrote of the widely cultivated plants of the day. Olive trees and date palms were grown from seeds, and so were rather more familiar garden vegetables, such as beet, cabbage, celery, cress, cucumber, leeks, lettuce, onion, radish and turnip.

Parsley, thyme and hollyhocks also feature heavily in his ancient writings.

Cuttings from plants were also described by Theophrastus. The earliest types of cutting were, in fact, rooted suckers which had been unceremoniously pulled off the parent plant and set in the ground. Today we call these pieces of shoot with roots attached: 'Irishman's cuttings'. However, it had been observed that some plants could be propagated successfully by taking cuttings *without* roots already attached, and these included almond, apple, basil, bay, fig, marjoram, pear and pomegranate.

The Romans

Over the years Man became more knowledgeable and was able to use early technology to aid the propagation and growing of plants. The innovative Romans used natural minerals to progress the sowing of seeds. They used mica (transparent crystals from granite) as we use glass today, to retain within a small area the warmth from the day's sun, and to protect their tender young plants and seedlings from the elements whilst still allowing them to take advantage of full daylight.

Gradually we were able to associate the cultivation of plants with the medium in which they were so struggling for survival. Hormone rooting powders, of course, were not available in these ancient times, so we improvised. Again, it is a Roman gardener who is credited for realizing first that if we dipped the bases of cuttings into ox manure it was possible to encourage rooting and the development of a good root system. It is believed, too, that his contemporaries were the first to grow cuttings from pieces of plant root (believed to be from a thistle), although sound documentation on the subject, pre-dating the 17th century, is hard to find.

The Chinese

The ancient Chinese have long been accredited for their amazingly artistic culture. Over many centuries their dynasties have left us with wonderful legacies of fine arts: porcelain, pottery, painted screens, furniture, sculpture, and so on. Quite independently, the Chinese were also discovering their own methods of plant propagation. For example, the technique of rooting a piece of plant *above* ground level while still being attached to the parent plant is generally called 'air layering' (Fig. 1.1). It is also

Fig. 1.1

Fig. 1.1 Many weird and wonderful ways of propagating plants by cuttings and layering were tried, even as late as the 18th century.

often referred to as 'Chinese layering' because some believe that it was the Far Easterners who were responsible for discovering its merits. However, the ancient Greeks are also understood to have used this in the propagation of trees, and the Roman statesman and agriculturist Cato (234-149BC) described it in his writings. Today, air layering is used for those subjects which are normally diificult to raise by other means, and is one of the quickest ways of obtaining young, reasonably-sized plants.

The Chinese and Romans also knew about grafting and budding. The gardens of Imperial Rome were magnificent, with roof gardens, balconies and courtyards containing colourful and aromatic plants like violets, narcissi, hyacinths and roses. Indeed, roses belong to the earliest Roman traditions, and so were a common planting wherever they conquered. It was having realized that growing roses from seeds was a slow process that budding and grafting (Fig. 1.2) became a well-used skill. Today we are most familiar with these practices in the propagation and wholesale distribution of fruiting and ornamental trees. But it is thanks to the Roman plantsmen that Man discovered a rootstock can influence the size and vigour of a tree.

Chinese gardeners some 1200 years ago successfully raised many varieties of tree peony by hybridization, and propagated them vegetatively by budding on to stocks of the wild tree peony, or moutan.

Today we use rubber bands, air tapes, grafting wax or bituminous paint to protect our grafts and buds from drying out in the

Fig. 1.2 The art of grafting was realized many centuries ago. Medieval gardeners practised the act of taking the tip of a branch from one tree (scion), and grafting it on to the base of another (rootstock), to be severed when the two became united.

Fig. 1.2

sunlight, and from damage by wind or insects. However, ancient civilizations had no such luxuries, so they used mixtures of chalk, sand, clay, straw and cattle manure.

Europe and the UK

Relatively little of the horticultural scene of old England is recorded prior to the 16th century. The English herbalist Thomas Hill wrote, among other works, *The Gardener's Labyrinth*. He talks, briefly, of apples and 'yea Cions to be grafted', but his keenest interest was in the vegetable garden. Of sowing salad vegetables he recommends making a 'salad ball' by mixing seeds of half-a-dozen different vegetables, placing them in a pellet of dung and growing them intertwined. With our 20th century knowledge of seeds and their germination, one wonders about his success rate!

He seemed obsessed with the desire to increase the size of garden vegetables. He suggests that sowing conglomerations of seeds together, in the same hole, is a way of guaranteeing large specimens, and monster onions can be grown by sowing them *inside* gourd seeds. In addition, cucumbers could be 'stretched' by tantalizing watering and by growing the fruits inside hollow canes.

The Victorians

More recently, the age of the Victorian gardener brought about its own innovations and developments, and they are largely much as we use them today. Grafting had been developed into a fine art by this time, and there were rather more types of grafting and budding used then compared to now. Indeed, since the 18th century, growers and nurserymen have actively striven to decrease the number of different types of graft, as it was felt that there were too many, and that propagation of this nature needed to be simplified.

In other areas of propagation, the Victorians were developing their own styles. During these heady horticultural days, plant collectors were scouring the earth for new botanical specimens; they brought them back to Britain and the growers of the day were expected to be able to propagate from them; to increase quantities and to learn about the plants' uses. Just as war between nations will cause scientific advancement, so this period of plant importation hastened the developments in propagation. We had particularly advanced in the structure of our propagating houses and covered gardens generally, so that it could be truthfully said that almost anything from anywhere could be grown – provided it was in a viable state when received.

The principal method of plant introduction at this time was by seed, but so often in tropic humidity and rain forest conditions seed was hard to come by, difficult to dry properly and practically impossible to germinate in the artificial atmospheres created for them. Many eminent horticulturists gave close and reasoned thought to the problem, but it was an amateur naturalist who eventually hit upon the answer.

The Wardian case

Dr. Nathaniel Ward, an East London general practitioner, had used a wide-mouthed glass bottle covered with a lid to observe the metamorphosis of a Sphinx butterfly from its chrysalis state. He had buried the tiny insect organism in moist mould. After some days, emerging from the mould came a seedling grass and fern. He noticed that the mould had kept damp because during the heat of the day moisture had risen from the mould, had condensed on the side of the bottle and run down again, so maintaining a constant level of humidity. (The grass and fern, incidentally, thrived for four years, without disturbance, until the lid rusted and allowed the entry of rain-water.) This was the start of the now familiar 'bottle garden', examples of which can be found decorating millions of homes throughout the world. More importantly, however, it was the prototype for the Wardian case: a plant cabin with glass sides and tops, suitably strengthened at the corners and edges, and sealed hermetically (or as much as was possible then); in other words, practically an air-sealed miniature greenhouse.

The plant case had a dramatic effect on the plant hunter's and propagator's success rates, and allowed the introduction and development of thousands of new plants. In

1842 Conrad Loddiges and Sons, the great nurserymen of London, claimed to have used 500 cases for the transportation of plants from all over the world. 'Whereas', they reported, 'we used to lose 19 out of 20 cases during the voyage, 19 out of 20 is now the average that survive'.

Bottom heat

The Victorians may have discovered how to transport seeds and plants successfully, but what about the actual propagation of them and its associated difficulties? Some gardeners began to realize that to encourage and speed the good root development of some plant cuttings, it was desirable to have a degree of heating *underneath* the placed cutting in its growing medium. This was termed 'bottom heat', and the 19th century gardeners perfected it, albeit often using complicated apparatus (Fig. 1.3), including wood and glass cases and paraffin lamps! Today, of course, our bottom heat is achieved with the use of electrically heated propagating cases and soil-warming cables.

Fig. 1.3

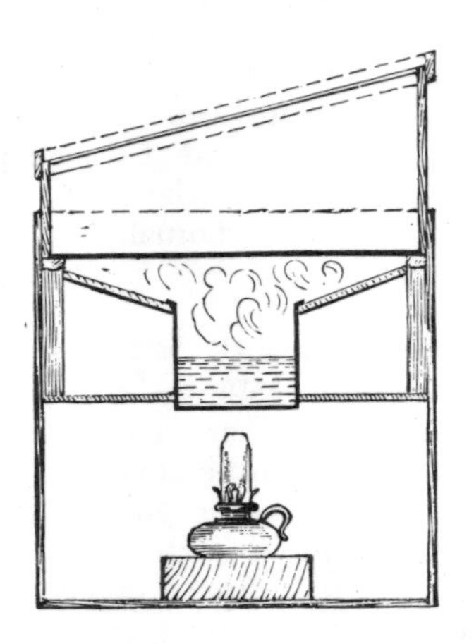

Fig. 1.3 Early propagating cases involved the use of wood frames, glass and paraffin lamps.

The Bell glass

One hundred years ago some gardeners also used a novel way of taking cuttings. They took two clay pots of different sizes, and placed one within the other. The space between was filled with compost, and the cuttings inserted into it. Water was poured into the centre pot, and this would moisten the compost enough for the cutting to root. A bell glass would then be placed over the whole lot, and a nicely humid atmosphere would be created and maintained (Fig. 1.4).

Bell glasses were, if you like, the first type of cloche. They were indeed bell-shaped, made of ordinary transparent thick glass, and had a bulbous knob at the top for handling. They were used for all kinds of propa-

Fig. 1.4

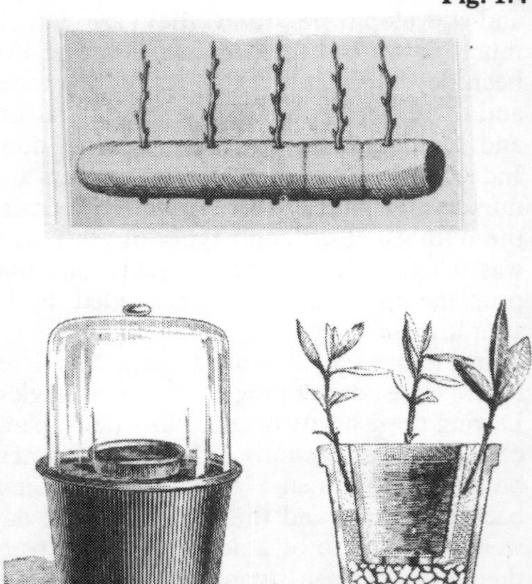

Fig. 1.4 *Top*: One of the early forms of taking cuttings involved boring holes into a branch of willow, or something similar, and burying this in the soil. *Below*: Some Victorian gardeners rooted cuttings in the space between two clay pots, which were covered with a bell glass.

Fig. 1.5

Fig. 1.5 Ornamental plant containers, including this Victorian fern case, developed from the need to propagate plants within an enclosed space.

gation and the forcing of young plants, and the original surviving glasses are today rare to find and considered highly collectable.

Indoors, of course, the Wardian Case was being used for encouraging young plant growth. A variation of this was the Fern Case (Fig. 1.5) and, although it invariably had a lid, it was not necessarily required to be air-tight. These highly ornate glass-sided cases were most often used for plant decoration – ferns being the best types of plant for prolonged siting within them. However, some of the more adventurous plant lovers also used them for propagation, and the cases afforded the observer a perfect view of the seedlings' or cuttings' development, all within the relative comfort of the drawing room!

Plant propagation today

So to the present day! Many aspects of plant propagation remain, in principle, as they did 10 000 years ago when Man first used agriculture for food, but there are still advancements being made. An example is micropropagation. This involves the production of plants from tiny plant parts, cells or tissues, grown clinically in a tube or phial where the nutrition, moisture level, and so on can be rigidly controlled.

Micropropagation

The development of tissue culture – a phrase that is sometimes used synonymously with micropropagation – must be credited to the German plant physiologist Haberlandt, who first disclosed his findings in 1902. Thirty years later scientists were able to grow tomatoes *in vitro* (using a laboratory-type facility and aseptic techniques similar to those used in culturing bacteria, fungi and other micro-organisms).

Of all the methods of propagation, it is with micropropagation that the amateur will find least success. The ordinary greenhouse, potting shed or kitchen window sill are not ideal conditions in which to experiment with the process. It is, however, an ideal form of plant increase for some ambitious nurserymen, who are able to invest in the modern technology required, as it enables them to obtain a great many (perhaps hundreds) of new plants from one single parent. Orchid and fruit tree producers, in particular, are finding this a very economic solution to their stock problems, and as conditions of complete hygiene are of paramount importance, scientists are using micropropagation to breed out viruses from certain plants.

Future progress

Other technological advancements connected to plant propagation include artificial lighting, heating, misting and humidifying. Research stations everywhere, particularly those working on the problems of the Third World, are constantly finding new and improved methods of propagating plants. Science advances every day, but as we near the 21st century, it is the sheer pleasure and satisfaction that the average gardener gains from succeeding in raising his own plants, that makes the subject of propagation so fascinating. Add to this the significant financial savings to be made from raising plants, as opposed to buying them, and you have another very good reason for investigating the options available.

CHAPTER 2

Equipment and Tools

As you progress through this book you will see that there are at least a dozen basic methods of propagating plants, such as seed sowing, taking leaf, stem, root or bud cuttings, layering, grafting, division, budding, micropropagation, and so on. There are, as well, many more minor or lesser forms, such as separating plantlets (as with some house plants, like chlorophytum) and bulblets, taking basal cuttings (as with many herbaceous plants), spore culture (with ferns), taking pipings (as with dianthus), and so on. For each of these, a certain tool, or piece of equipment will be required to aid the propagator, and to help ensure the success rate of his task.

Let us examine the requirements of each type of propagation in turn.

Seed sowing outdoors

Other than the normal gardening tools, relatively little specialist equipment is required for sowing seeds out in the open.

Spade, fork and rake

One each of these is necessary to obtain a good soil condition prior to sowing (Fig. 2.1). The rake should be light to handle, but at the same time have a fairly wide head, preferably metal, as this is longer lasting and should easily cope with the heavier, roughly dug soil.

Hoes

Vegetables and some annual flower seeds need to be sown in straight lines, so a drill has to be taken out. Before the soil is removed, it is important to obtain the straight line. For this, use either a garden line (string tied taut between two pegs), or a straight-edged board. The drill can be excavated by using the corner of a rake, or a swan-necked draw hoe. The handle of the latter is inserted in a socket, which is connected with the blade by a curved, solid neck. The width of the metal blade varies between 5 cm (2 in) and 23 cm (9 in). The draw hoe is used by chopping down into the soil, and pulling it towards you. It can enter the soil quite deeply and is also good for weeding and earthing up ridges for potatoes, celery, and so on.

Trowels

These can be used for spot planting individual seeds, and for planting out rooted

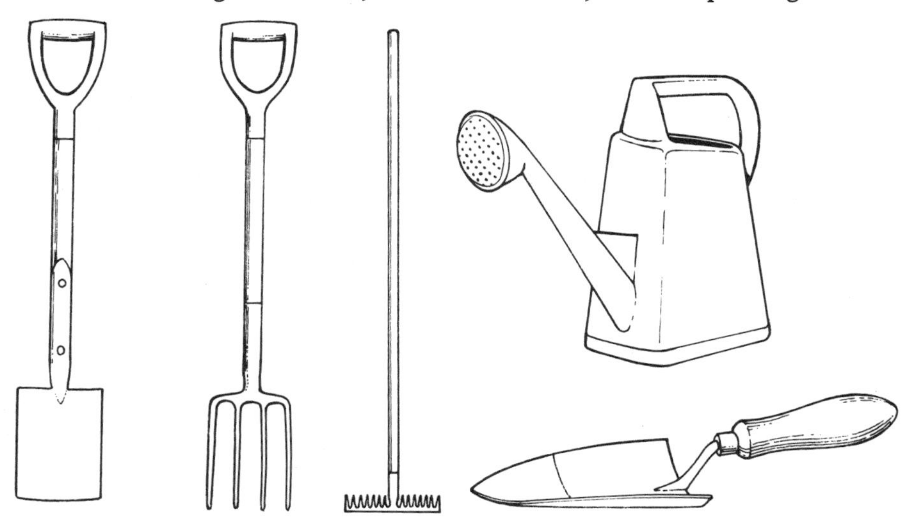

Fig. 2.1

Fig. 2.1 A basic set of hand tools required for plant propagation comprises a spade, fork, rake, trowel and watering can.

cuttings and bulbs, and transplanting seedlings. It is an indispensable tool for lifting small or young plants with a minimum of root disturbance. When choosing a trowel, it is essential to see that the blade is firmly fixed to the handle and that both are soundly made.

Large dibbers

Purpose-made dibbers may be purchased, or home-made ones may be had from using the upper handle from an old spade or fork, and making a dull point to the base.

The practice of dibbling is more or less the same as that of using the trowel, but with some differences. The dibber is used to create a hole, into which large individual seeds or small plants are set. To make the hole, the dibber is simply pushed into the soil and moved from side to side. However, this action compacts the side of the hole, and is something many gardeners feel does no good to the soil. Ideally, always use a trowel, especially if the soil is heavy, or not in a good working condition.

Of course, unlike a trowel, a dibber cannot be used for lifting seedlings (with the soil around their roots).

Watering cans

After sowing or planting outdoors, it is always important to water the area lightly. Emphasis should be laid on the word 'lightly', because to use too much water, or with too heavy a force, you can easily undo all your earlier good work. Therefore, a watering can with a fine rose attached is the best piece of equipment for the job. If one is having to water a vast area, a hosepipe with the finest of jet spray nozzles attached would be easier to work.

Labels

To say that labels play a major role in the propagation of seeds outdoors is, perhaps, overstating the case, but their use is significant and without them we run the risk of misidentifying our young plants as they grow. Plastic or aluminium labels are best; the wooden type can only really be used for one season. The marker to use is practically as important as the label itself! Waterproof pencils may be purchased; and many gardeners opt for the common 'chinograph' pencil as used by draughtsmen.

Seed sowing under cover

Sowing seeds under the protection of glass or plastic is a vastly different operation from that of sowing outdoors, and a whole new set of equipment will be needed, with only the watering can and some labels in common.

Greenhouse

First and foremost, the person who wishes to propagate a wide range of tender plants, or plants that need to be started when it is too cold for them outdoors, must have some form of transparent shelter. If a large number of plants are concerned, where propagating them in the home (on window sills, in airing-cupboards, and such like) is impracticable, then a greenhouse must surely be first on the list of things to buy. It is not our purpose here to provide you with the complete guide to buying a greenhouse; after all, whole books have been written on this subject. But we can, briefly, and with plant propagation in mind, discuss the merits of the different types currently available.

The novice gardener may well believe that choosing a greenhouse is easy. It is, if one has no concern for the implications! In fact, choosing the *right* greenhouse for your requirements, is extremely important.

Cost is probably one of the main considerations, although greenhouses today represent better value than ever before; a standard 2.4 × 1.8 m (8 × 6 ft) model made from aluminium can be bought at relatively low prices from such outlets as high street supermarkets and diy superstores.

Structure

Today, aluminium alloy is the material most commonly used in the structure of greenhouses (Fig. 2.2). These types of house are good for the gardener with less time to spare, as they need no maintenance, other than cleaning the glass and oiling any moving parts (doors, vents, and so on). The use of glazing clips, as compared to old-fashioned putty, also makes aluminium greenhouses easy and quick to erect. The

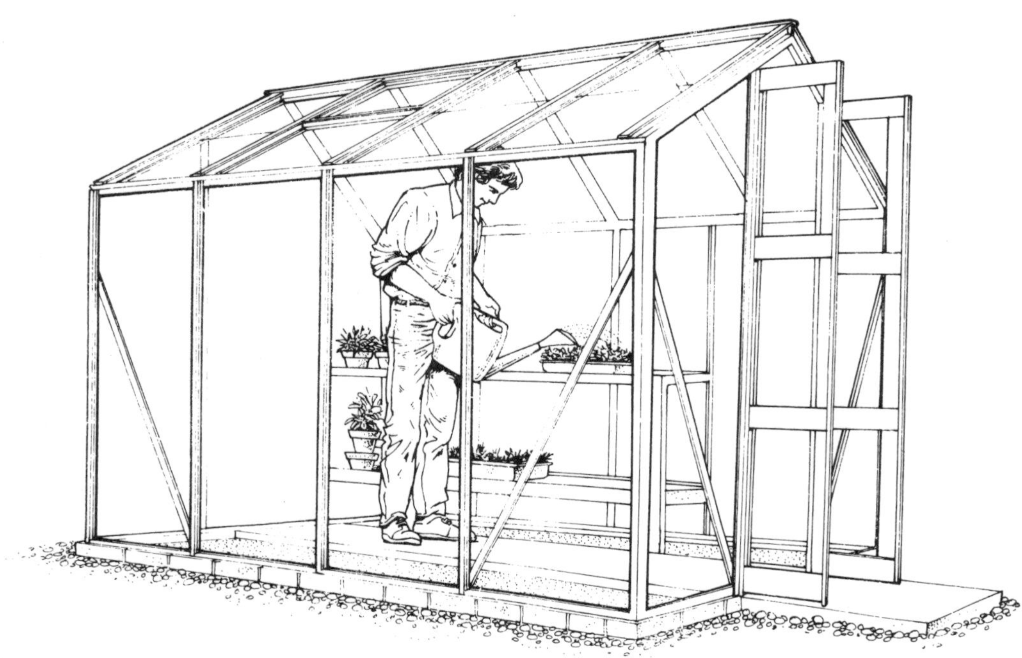

Fig. 2.2

Fig. 2.2 Aluminium greenhouses are lighter in weight than wood, and are generally cheaper.

framework is remarkably strong, lightweight, not at all bulky, and quite cheap compared to its wooden counterpart.

There are two main disadvantages of aluminium. The first is that, being a metal, it is a poor insulator and expensive heat from within is conducted outside, which can be a particular problem in spring when seedlings are being raised. This act of conduction often causes the air within to condense on the glass or plastic panelling which is, in itself, a problem. Adequate light transmission and dry air are important factors, and both are inhibited if there is heavy condensation in the greenhouse.

The second disadvantage, although rather minor, is that a certain amount of corrosion takes place to the metal when it is first exposed to the weather. This will show as a white powdery crust, almost as if the metal has contracted mildew! However, this does the metal no long-lasting harm. Some manufacturers now offer greenhouses where the aluminium frame is coated with a protective acrylic paint (usually white), which is useful if you appreciate the properties of aluminium, but for the most part dislike its appearance.

Softwood framing is the main alternative to aluminium, and this is usually the rot-resistant western red cedar from North America (Fig. 2.3). It looks very attractive in a garden setting, and is competitively priced, but one significant disadvantage is that it needs treating with a water repellant and wood preservative every few years. This chore can be made somewhat hazardous if you have one of the larger houses! With the cheaper softwood models, usually made of Baltic pine or such like, the timber is not so durable, and so may need a more thorough treatment, more often.

Glazing

What about the glazing? Should you go for glass or plastic? Again, both have advantages and disadvantages. They are fairly equal as regards to the general intensity of transmitted light. However, the sun's infra-red light can pass through plastic more quickly than glass, so the greenhouse warms up more speedily. Conversely, plastic also lets the infra-red light out quickly, so the greenhouse will cool down just as rapidly at night. The point is that rapid heat loss can only really be countered by providing the

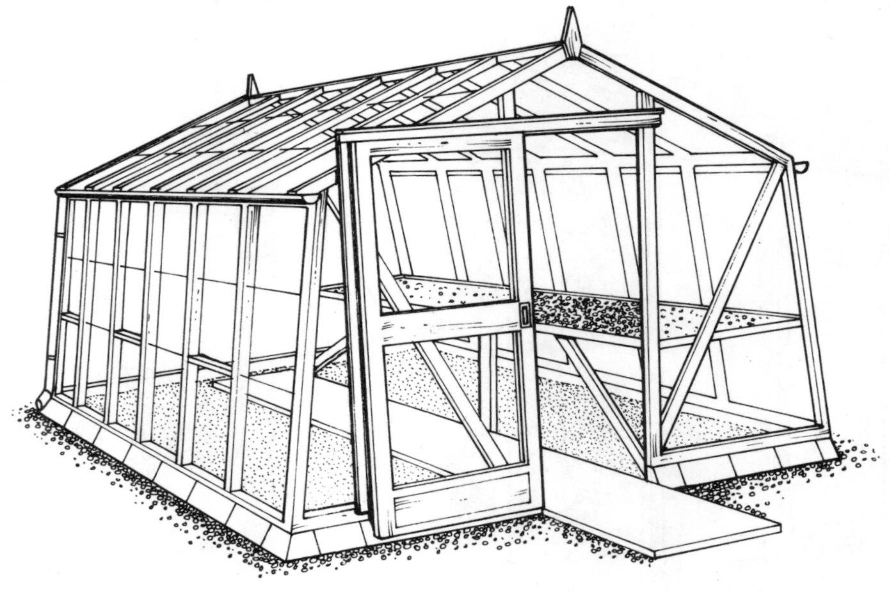

Fig. 2.3 Wooden greenhouses are usually made from rot-resistant western red cedar from North America.

Fig. 2.3

house with extra heating, and this costs. Condensation can be a further result of rapid heat loss.

If you are governed largely by the depth of your pocket, and so have to opt for plastic glazing, beware! PVC and polythene are harbingers of dirt and dust; they also quickly deteriorate in daylight until, after a few years, they have become brittle and yellow. Other more expensive plastics, such as double-skinned box-section polycarbonate sheeting, may last longer, but after six years or so even these will have to be replaced. For this reason, you may feel that to choose one of these will eventually prove a false economy.

The positive aspects of plastic when compared to glass can be summed up as follows: it's cheaper, lighter and safer. But glass has very many supporters, and with good reason. It's a better insulator, it will last indefinitely (unless it is inadvertently broken), and will afford better light transmission if it's kept clean. And here's a point worth noting: in the summer months it is relatively easy to apply a shading emulsion to glass, and practically impossible (and certainly not to be recommended) to do so to plastic.

Other factors

Once you, as an aspiring propagator, have decided on the material of the framing and the glazing what, briefly, are the next considerations? As we will see later, the sucess of your 'under cover' propagation will depend on many factors, not least of which is the amount of light your seedlings and cuttings will receive. Therefore it would be in your best interests to opt for a greenhouse with glazing down to ground level, or with sloping sides. Smart, domed greenhouses are now available, and despite their expense, these are particularly recommended for their 'all-round' accessibility.

More than likely, the size of your greenhouse will be determined by price, but if you are intending to become a serious greenhouse gardener, then buy the biggest house you can afford. You won't regret it.

Frames and cloches

Although in general gardening terms these are quite different, for the purposes of this book we will look at these items together (Fig. 2.4). A garden frame will act as a storehouse and nursery for many tender plants, and a cloche (or more likely a row of them) will bring you the results of your

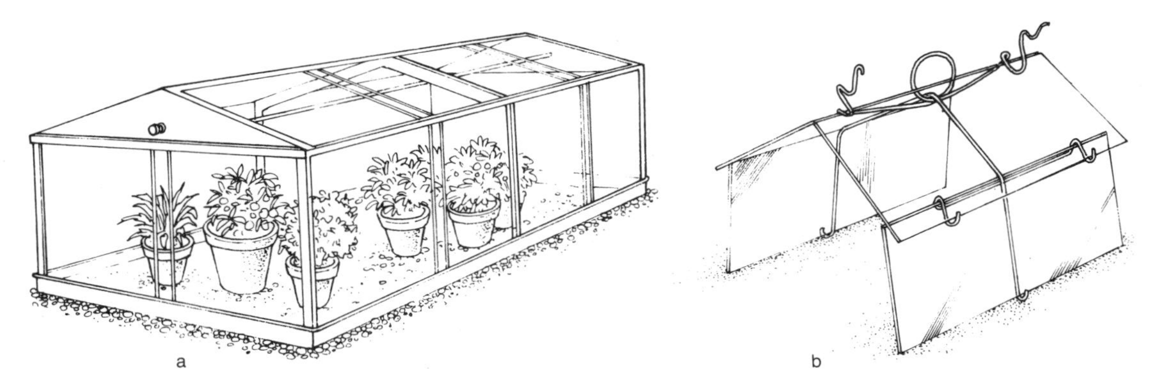

Fig. 2.4

Fig. 2.4 Garden frames (*a*) and cloches (*b*) are equally important to the plant raiser. Frames act as a storehouse for tender plants and seedlings whereas cloches, when placed over seed sown in the open, will facilitate quicker germination.

labours, such as a continuous supply of winter lettuces, or succulent strawberries, weeks before the plants would have done had they not received the luxuries of warmth and shelter.

The frame can be regarded as the half-way stage between the greenhouse and the garden. Many plants raised in the greenhouse, but which are ultimately destined for the outdoor life, must slowly be acclimatized to the often hostile environment of the garden. Wind, rain and full sun may all take their toll on tender young seedlings so it is here, in the frame, that they are hardened off.

Other uses for a garden frame, but which do not strictly come under the category of propagation include: the overwintering of plants that are frost-tender; forcing fruit and vegetables to crop out of season; and relieving the congestion of an overcrowded greenhouse (a common complaint during mid-spring).

The materials from which frames are made today, are as numerous, and as equally difficult to come to a decision over, as greenhouses themselves. Exactly the same arguments concerning wood, aluminium, glass, plastic and the price exist here. There are many designs and styles from which to choose, but the main criteria must be (*a*) good light transmission, (*b*) effective insulation, and (*c*) enough space and depth to accommodate the plants you want to be housed.

We have already seen that the Victorian bell glass constituted one of the first types of cloche, but for what purposes are cloches actually used? Essentially they were designed to protect plants in the open during the colder months of the year. They can be used to hasten the germination and subsequent development of many vegetables, and are particularly worthwhile for those which can be encouraged to mature when they would otherwise be unobtainable, or when the shop prices are prohibitively high.

Cloches can also be used in other areas of gardening. For example, there is the forcing of fruit (strawberries and melons seem particularly well suited). Cloches can be used to keep the winter rain off alpines and other such plants that, should they be subjected to prolonged wet and cold, would surely perish. In addition, cloches can be placed on their sides to shelter taller plants, such as outdoor tomatoes whose tender flowers may be protected from late frosts, or clumps of border perennials which may be encouraged to bloom earlier.

Any good cloche should possess strength and durability, it should have good light transmission and adequate ventilation, and it should be portable and easy to erect. Styles include the tent and the barn (in plastic or glass, the latter sometimes with an option for a side with netting), and the tunnel (plastic only, but with varying densities, such as rigid plastic or polythene).

Seedtrays and pots

If you are intending to undertake a degree of propagation in the greenhouse, you will need suitable containers. Trays and pots come in a good range of shapes and sizes.

Years ago, wooden trays were all that could be bought. Nowadays plastic, perhaps, is the best choice of material for both trays and pots, and it is certainly the most readily available. However, they vary in price according to their strength, and some are very flimsy and need careful handling. Others are very strong, though, and these can even be picked up with one hand when filled, without fear of them breaking – but it is always safer to carry them two-handed.

Standard trays are 35 cm × 23 cm (14 in × 9 in) and 6–8 cm (2½–3 in) deep. Seedtrays in other dimensions are available, and may prove to make better use of space in some cases. Pots come in a wide range of sizes. Plastic is preferable to clay. Pans and half-pots are shallow versions. Whereas seedtrays are used to raise moderate or large numbers of seedlings or cuttings, pots and pans are ideal for small quantities.

Peat, fibre, expanded polystyrene and black plastic bag pots, used in strips or singly, extend the range of containers. Particularly popular for propagation are compressed-peat modules. These expand when saturated with water, and they have a ready-made planting hole which is designed for taking a cutting or for direct sowing.

Wooden presser

This is a block with a short handle which is used for firming and levelling compost before seed sowing. It's a tool which can easily be made at home.

Unheated propagators

The most inexpensive way of supplying the high humidity required for germinating seeds (and rooting cuttings), is to cover your pot or seedtray with a polythene bag supported on a wire frame. At slightly higher cost you could purchase a few clear, rigid plastic domes, which are designed specifically to fit over a pot. Perhaps, though, the most economical method is to make use of several unheated propagators, or domed seedtrays. The compost is allowed to reach the right temperature, and with the high humidity it is possible to raise a wide variety of plants in a cool or unheated greenhouse from late spring onwards. However, during cold spells the added safeguard of an electrically heated propagator will pay dividends.

Electric propagators

Spring is undoubtedly one of the busiest times for the gardener, and all too often things get forgotten. With all the seeds that need sowing to provide a healthy supply of summer flowers and vegetables, it is quite common to discover that you should have sown a particular variety at least a few weeks, perhaps a month, earlier. All is not lost, however. By supplying ideal growing conditions with a degree or two of bottom heat, you can soon catch up and enjoy flowers and vegetables as good as any in the neighbourhood. The answer is an electric propagator (Fig. 2.5)!

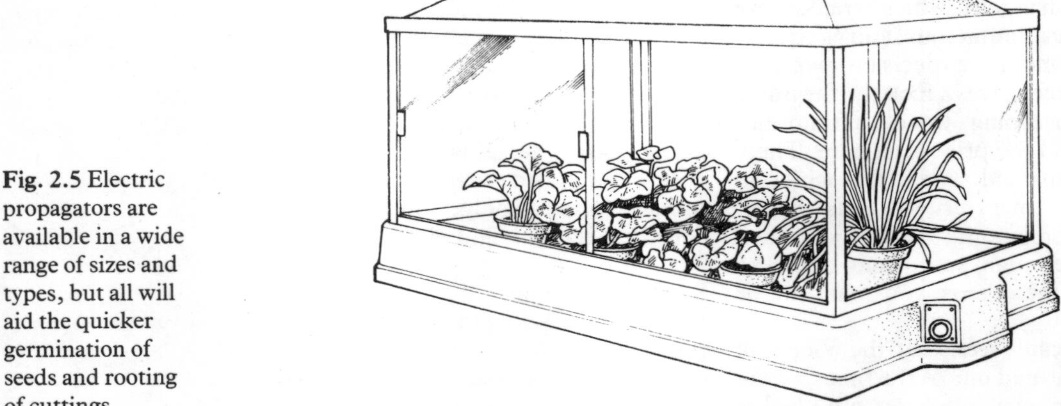

Fig. 2.5 Electric propagators are available in a wide range of sizes and types, but all will aid the quicker germination of seeds and rooting of cuttings.

There are many styles and makes from which to choose, but finding one to suit your needs – and budget – should not be a problem. The most basic are simply one step above the standard 35 × 20 cm (14 × 8 in) seedtray and plastic top. They have a 10–12 watt heating element in the base and usually some form of ventilation incorporated in the top. Into this you can put one standard seedtray, or a combination of half and quarter trays. The use of these smaller trays gives greater flexibility with the range of seedlings that you can grow; not all seeds germinate at the same rate, so if the smaller trays are used for the fastest varieties, these can be removed at the right stage for potting on, leaving room for the slower varieties to continue developing, and for further sowings.

The next type of electric propagator, and one that will prove invaluable if space is already at a premium, is the type designed for the window sill. These are usually rated at 20–25 watt, and approximately 15 cm (6 in) wide so a reasonable width of sill is needed. They have a reasonably large capacity, however, and they capitalize on the maximum amount of winter light available. It is no use encouraging seedlings to germinate, only to watch them grow tall and leggy due to a lack of light!

With the inclusion of a thermostat, the more advanced electric propagators are more versatile and therefore slightly dearer. Rated at anything up to 65 watt they will take two, three or four standard seedtrays or a number of pots. A thermostat means that you can maintain a set temperature within the propagator which will be unaffected by outside conditions. You can also select the best germinating temperature for the seed varieties you are growing. For example, some temperate plants will revel in temperatures of 15–18°C (59–64°F), while the house plants from hotter climates prefer 24–27°C (75–80°F).

As the larger and more powerful units are proportionately more expensive, some manufacturers these days sell the components separately. For example, one might buy a fairly costly 75 cm (30 in) base, and add the equally expensive frame top at a later date. Buying in pieces like this does work out slightly more expensive, but will reduce the initial outlay. Incidentally, the base of an electric propagator may be used on its own to provide bottom heat quite satisfactorily.

Soil-warming cables

If you wish to propagate your plants in a larger area than is provided by a heated propagator, perhaps in a garden frame or on a greenhouse bench (or in a greenhouse border), but you still wish to take advantage of bottom heat, then electric soil-warming cables may be laid (Fig. 2.6). Under-soil heating can work out surprisingly economical, especially when compared with air-heating from above where money tends to be wasted heating space which is not directly required.

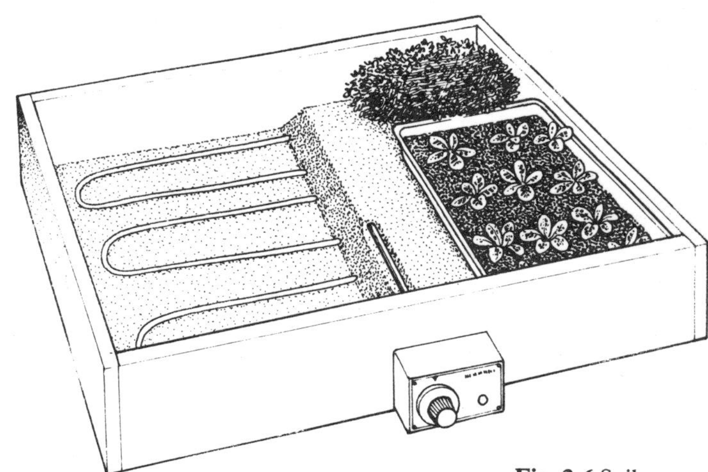

Fig. 2.6

Fig. 2.6 Soil-warming cables are available in a choice of lengths to suit the size of the area you intend to heat; most are mains operated, earthed and insulated for safety.

The cables are available in a choice of lengths to suit the size of area you intend to heat and the temperature you wish to achieve. Most cables are mains-operated, earthed and insulated for safety; but for garden frames some way from the house, a low voltage type is preferable. These comprise a small transformer and a length of galvanized wire, and although they are a little more expensive to buy, they are cheaper to run. Most systems available today include an automatic on-off switch, and a thermostat is also a good idea, to make sure the heat is controlled economically.

The wire cable should be laid on a 2.5 cm (1 in) layer of sharp sand, then covered with about 15 cm (6 in) of soil, or 23 cm (9 in) if it is the greenhouse border. Lay the cable backwards and forwards across the area to be heated, avoiding any sharp bends and without touching in case you create hot spots. Try to keep it no less than 10 cm (4 in) and no more than 20 cm (8 in) apart for good, safe coverage. Other safety points to watch out for, which may cause heat build-up and damage to the cable, are the use of ash, clinker, moss or excessive peat in the soil, or any medium containing polystyrene granules or perlite. Always have your installation checked by a qualified electrician, or your local electricity board, and remember to turn off the current when working the soil.

Once your system is installed, you will find it invaluable for establishing early salad crops and vegetables, such as lettuces, carrots, cauliflowers and radishes, to root cuttings and for growing early cucumbers, marrows and melons. Any shallow-rooted plant is suitable, provided you remember the depth of the cable! Used as a propagator in spring and autumn, then for over-wintering tender plants, it will soon earn its keep.

Artificial lighting

Plant growth depends not only on heat, but also on light. If your plants receive the former but not the latter, only weak and spindly specimens will be produced. If you can provide a suitable extra light source, you can maintain plant quality and increase activity even when natural light levels are low (Fig. 2.7). This is of most benefit during winter when the sun is lower in the sky, giving us much less in terms of light intensity and, of course, the hours of daylight are much reduced as well. In addition, many gardeners are keen to avoid massive heat loss in their winter greenhouses by putting up an insulation material against the glazing; this, too, will reduce the levels of light entering the house.

Commercial growers frequently use artificial lighting to enable them to continue producing crops throughout the winter. In recent years developments have been made in the production of smaller supplementary lighting units, suitable for the amateur gardener. This means that seedlings, cuttings and greenhouse crops, and flowering plants can be started earlier in the season, perhaps six weeks earlier in the case of tomato plants. Certainly almost all plants will grow faster and healthier with the help of some artificial lighting.

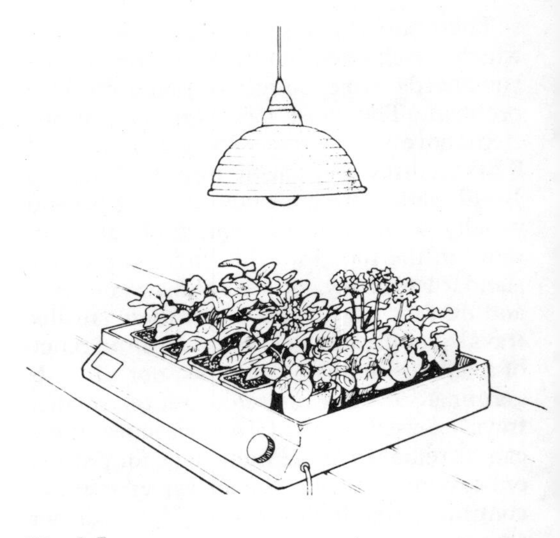

Fig. 2.7

Fig. 2.7 Artificial bulb lighting is often used by amateur and professional plant raisers, but it should only be used to supplement the natural daylight available, not substitute it.

It would be very wrong to suggest that plants should be grown with artificial lighting alone; what we must aim to do is add to the natural daylight available. To put this into context: natural light in the United Kingdom on a really cloudy day can be as low as 5000 lux (the units used to measure light). A good greenhouse fluorescent lamp, rated at 160 watts, can increase this by half as much again. A light of this strength, mounted 60 cm (2 ft) above the seeds or cuttings, should give an extra 2300 lux at the centre and 540 lux at the edge of a circle 60 cm (2 ft) across. Because of this variation, it would be advisable to rearrange your plants periodically within the lit area.

If you can afford the extra expense of a control gear, some manufacturers offer more powerful 400 watt lamps with various types of reflector. However, these require a special control mechanism, and are really intended for the commercial grower.

If you want to construct your own lighting unit, or use the space under greenhouse staging where there is little daylight anyway, it is cheaper to use warm-white fluorescent tubes, if necessary mounted in banks on wood battens. This type of lighting gives good results, but the reflectors are bulky. Remember that the lamp's efficiency is lower at the two ends, so you get better results by using, say, one 2.4 m (8 ft) tube than two 1.2 m (4 ft) tubes end-on. Standard tubes may be used where there is no problem with moisture, but some companies make corrosion-resistant fittings specially designed for greenhouses. There are no set rules on the subject of greenhouse lighting, so there is ample scope for experimenting and finding out what best suits your plants, and propagational activities.

Compost

The growing medium onto and into which the seeds will be sown must be of good quality if the seedlings are to be given the best start in life. In terms of propagation, 'compost' does not refer to garden compost from the compost heap, and it is always important when working under cover never to use garden soil, which often harbours disease and pests. Instead, always buy a specially formulated mixture; this will encourage successful germination of seed, and rooting (in the case of cuttings).

Formulated compost

Proprietary seed composts also contain the correct amount and balance of fertilizer to get the plants off to a good start. There are two basic types of formulated compost used in propagation: you can either choose from the range of John Innes seed or potting composts, which contain loam and look much like soil from the garden, or one of the soilless equivalents. Much work has been done on the developments of these latter composts over recent years. While they generally consist of peat and sand, or of pure peat alone, they are also available in mixtures comprising peat and vermiculite, or peat and perlite.

Soilless compost

Soilless composts are lighter in weight, cleaner to handle, do not require sterilization and are easier to store. They also encourage plants to grow more uniformly, due to less variability in their components when compared to the loam-based alternatives.

The main disadvantage (if one can call it that), is that soilless composts tend to dry out more quickly, so the gardener must always check on the moisture requirements. If the compost does become dry, it is sometimes difficult to moisten it again, by which time any growing seedlings or rooted cuttings may have suffered irreparably.

The same composts may be used for the later stages of propagation, that is pricking out and potting on. However, it should be noted that the soilless types become exhausted of fertilizers more speedily than those which are soil-based. So, the gardener must also apply extra plant food – if the seedlings or cuttings demand it.

Enthusiastic amateurs may wish to make their own composts, and these, if properly done, may be every bit as good as the proprietary brands. Try either of the following mixtures:

For seed sowing

The basic ingredients should comprise sphagnum moss peat and very fine lime-free sand in equal parts by volume. To each bushel* add:

30 g (1 oz) 18% superphosphate
14 g (½ oz) sulphate of ammonia
14 g (½ oz) sulphate of potash
110 g (4 oz) ground chalk or limestone

For pricking out

Three parts by volume of peat and one part of sand. To each bushel* add:

55 g (2 oz) 18% superphosphate
7 g (oz) ammonium nitrate
30 g (1 oz) sulphate of potash
85 g (3 oz) ground chalk or limestone
14 g (½ oz) urea/formaldehyde (38%N)

**A bushel is an old, and now largely unused unit of measurement. However, when discussing volumes of compost there is no modern, and suitable alternative. The best way to visualize a bushel therefore is to take a box 55 cm (22 in) × 25 cm (10 in) × 25 cm (10 in). This would hold one bushel of soil.*

Lightening a compost to improve drainage is sometimes recommended. Do this by adding small quantities of sand, but always use sharp or silver sand, never the kind favoured by builders, as this will not have been treated to eradicate any possible weed seeds or various harmful pathogens.

Soil blocks

These serve a similar purpose to small pots, but are made of compressed compost. They can be made at home easily, by using a special plastic mould which presses the compost material together. It also leaves a small depression in the top of the block, to facilitate the easier sowing of the seed, or planting of a seedling. Seed or cuttings compost may be used, but it must be moist. Some gardeners even add a small amount of washing-up liquid (say, six drops) into a watering can and then thoroughly soak the compost. This helps the compost material to bind together. Never allow the block to dry out, before or after sowing – keep it moist by spraying with water. Although the block-making tool will need to be acquired, over a season or two it will probably pay for itself, in that pots and/or trays may not be required. The only demerit for soil blocks is their tendency to dry out and subsequently collapse.

Sieves

For general gardening purposes a 1 cm (⅓ in) mesh sieve is essential for grading soils and composts. A finer one, however, such as an old kitchen sieve, should be used for lightly covering seeds with compost after they have been placed in position.

Taking cuttings

Practically all of the equipment recommended for seed sowing will be used in the taking of cuttings too. The same types of pots and compost will be acceptable but there are a few specific items of equipment.

Misting equipment

The rooting of cuttings does take time, and the success rate varies with the prevailing weather conditions. Sometimes cuttings of choice shrubs and trees show their reluctance to grow and develop in a frame or outdoors by wilting before they root. Early growers used a system of covering the cuttings with a plastic sheet; this increased relative humidity, and reduced the transpiration rate of the cuttings, so they stayed turgid. Unfortunately, plants need to transpire to keep themselves cool, so cuttings taken during the summer months can rapidly overheat when surrounded by a saturated atmosphere like this. Overheated cuttings do not root! Shading could help, but without ventilation the beneficial effect is limited, and by ventilating one is defeating the whole object of growing under plastic in the first place.

The answer? Intermittent mist systems! A fine spray of water droplets on the leaves of cuttings every few minutes prevents them from wilting (Fig. 2.8), but the evaporation of those same water droplets from the surface of the leaves during the periods between sprays cools the cuttings down. Since the rate of evaporation of water from the leaves varies with the prevailing weather conditions, an automatic control system is needed if full advantage is to be taken of a mist system.

Fig. 2.8

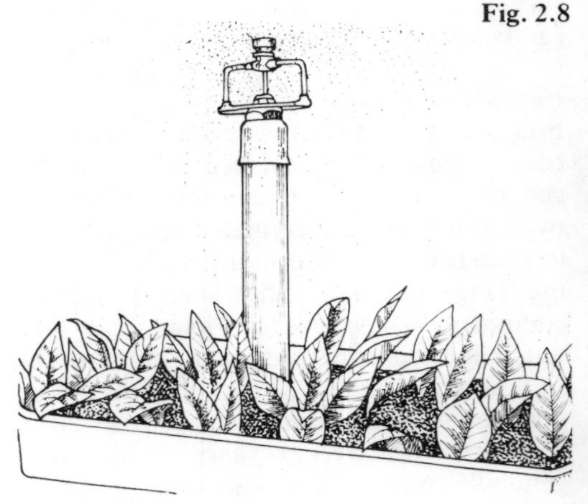

Fig. 2.8 Intermittent misting systems create higher moisture content in the air, so encouraging a better environment for the rooting of most cuttings.

The 'electronic leaf'

A simple, but ingenious device. Basically this is a sensor consisting of two electric terminals embedded in a flat piece of plastic. When the surface is wet the water connects the two terminals and so an electric current is passed. The control equipment then shuts off the solenoid valve that directs the flow of water to the spray system. When the electronic leaf dries, the electric current is broken and so the solenoid valve opens and misting starts once again.

Mist nozzles

There are a number of makes of these but all are based on the same concept, either mounted on top of risers, or slung underneath overhead pipework. The column of water emerges from a single orifice and impinges on a plate which forces it into a thin sheet spreading in every direction, at first horizontally but gradually falling so that its continuing shape is that of an umbrella.

The advantages of intermittent mist propagation are clear, but we should highlight the disadvantages too. Most importantly, the system is relatively expensive to set up, and its effectiveness depends completely on the equipment working efficiently. Therefore regular maintenance is necessary, particularly in hard water areas where problems are caused by the blocking of the jets with calcium deposit (this also may be left as a residue on the leaves of the cuttings).

Knives

Although it is true to say that cuttings may be prepared using a pair of garden secateurs, to make the fine trimmings this is perhaps a clumsy approach, and certainly a better job will be made if you use a sharp knife. If lots of cuttings are to be taken, a knife will also be easier to use. Choose one with a straight blade as it will be easier to sharpen (an oil stone will be required). Use the knife only for taking cuttings – using it for cutting into paper, string or plastic will soon blunt the edge, so making it less able to make clean, straight-edged cuts into the plant material.

If you take so few cuttings that the purchase of a special knife seems wasteful, razor blades make suitable alternatives, but be extra careful when handling these as there is no natural 'handle'.

Hormone rooting powders

These can greatly improve your success rate with cuttings, as well as speed up the process of rooting and help the plants produce stronger root systems. The powders available commercially contain synthetic versions of natural plant hormones, called auxins, which stimulate root production. The majority contain the plant hormone NAA (naphthyl acetic acid) or IBA (indole butyric acid), plus a fungicide to help protect the exposed plant tissues within the cuttings. As only very small quantities of rooting hormone are required per cutting, the powder is supplemented by the addition of powdered chalk or talcum powder. One or two modern preparations are in gel form. All of the rooting hormones available to amateurs are suitable for use on softwood, semi-ripe or hardwood cuttings.

It is worth noting that rooting hormones deteriorate with age, so it is wise to buy a fresh batch each year. Containers that appear as if they might be old stock, such as with faded labels or with a thick coating of dust on the lid should be avoided.

Grafting and budding

Although they are not difficult, these methods of plant propagation probably demand more care and practice than any other. They certainly require a good set of basic tools.

Grafting knife

As with the taking of cuttings, a straight-bladed knife with a sharp edge must be used if the job is to be done properly. Avoid knives with curved blades; although often sold as grafting knives, they are really designed for pruning purposes and are next to useless in grafting.

Budding knife

If you possess a grafting knife, investing in a budding knife is not essential, but it does make the job a little easier for novice gardeners. Usually smaller than the grafter's knife, the basic difference is in the flattened, broad, spatulate shape of the handle. Made of bone or plastic, it is used in the budding process to prise flaps of bark away from the stem. If you are proficient in the art, you will be quite happy using the blade of the grafting knife instead.

Secateurs

These are useful in supplementing the

knives, and using them requires no special skill, and relatively little effort. There are two main types: the parrot-beak or by-pass with two cutting blades, and the anvil with one. The latter type use a soft metal plate to support the stem while the single hard-metal blade cuts through it. The two-bladed secateurs cut like a pair of scissors, though only one of the blades has a cutting edge. In plant propagation, it is the art of grafting which most regularly calls for the use of secateurs: they will be needed to cut scions and the tops from rootstocks.

Tying materials

The two portions of plants which are directly involved in the grafting process must be held together securely, until such time as they are properly united by tissue. The traditional material for such a bonding is raffia, and provided it is soaked well beforehand, it is excellent for the job. Polythene tape is a modern alternative, and is easier to handle.

Small squares of rubber centred by a metal staple are known as 'budding patches', and in this process are a quick and easy way of securing the rootstock and the bud. They are designed to be used once, and to perish in just about the time it takes for the bud to start growing on its own.

Waterproofing the points of contact between scion or bud and the rootstock is essential. The professionals used to make a grafting seal by using hot wax, but much easier are the modern proprietary bitumen-based compounds usually recommended for sealing large cuts in the pruning of trees and shrubs. The buds used in budding are tender and easily damaged, and the thick black bitumen tars are not suitable for these, so instead use a dab of pure petroleum jelly. If this is smeared over the bud, it will afford perfect protection.

Other items of equipment

So far we have covered about 90% of the equipment needed by the enthusiastic propagator. Whilst the sowing of seeds outdoors and under cover, the taking of cuttings (in its numerous forms), and grafting and budding constitute by far and away the main areas of plant propagation, there are, as we shall see later, other forms and these in turn require some extra items of equipment.

Division

A trowel or a border fork is required for digging up the plants prior to dividing them. The classic way to split plants is to prise them apart using your fingers (if the plants are small, such as house plants or polyanthus) or, if it is a larger clump, to use two border forks inserted back-to-back. Few gardeners have the luxury of possessing two garden forks, but perhaps you have an old one that is beyond digging, but is otherwise serviceable. Or perhaps you feel that investing in a second fork (perhaps a smaller version) will be worthwhile. Two hand-forks, on the other hand, are quite adequate for the division of smaller plant clumps, and certainly cheaper.

Knives figure in plant division too. Some types of plant are particularly tough and will resist all attempts to persuade them gently into smaller pieces. These can then be tackled with a good sharp knife but, as stated previously, not with one that is used for taking cuttings, as this will soon be blunted.

Layering

This form of propagation happens most freely in nature, so relatively few items of special equipment are required. However, a knife is needed to slit the stems of branches, and a trowel for digging the hole which will receive the layer. In addition, wire bent to the shape of a hairpin is used for pegging down springy branches. Thinner wire than the type used for coathangers is best, as it is strong enough to hold the branch in place, while being thin enough to bend easily.

Air layering

As we shall see later, the simple process of air layering requires the use of a small wad of moss. The type generally recommended for this is sphagnum moss, which is readily available from garden centres and shops. It is, actually, the dehydrated and often shredded residue or living portions of acid bog plants in the genus *sphagnum*. It is light in weight, relatively sterile and has a very high water-holding capacity, being able to absorb 10 to 20 times its weight of water.

CHAPTER 3

Hygiene

The term 'garden hygiene' is taken to refer to the clearing away of rubbish and debris outdoors, where insect pests and slugs and snails are known to breed. But the greenhouse, too, is a haven for all manner of plant pests and diseases, and in plant propagation, being able to sow seeds, take cuttings or expose inner plant tissue in order to increase stocks requires strict controls in terms of hygiene and initial nurturing.

Fortunately, young plants from cuttings or seed seem to thrive within a rather generous range of conditions. If your charges show plenty of good green foliage, compact growth (short distances between leaves and/or stems) and slow but steady development, then there are no problems. Occasionally, though, an excess or deficiency in one or other of the young plant's life requirements will cause concern, and a plant in trouble will invariably let you know. If the problem is diagnosed and treated at the earliest opportunity, there is a good chance of saving the plants. But of course prevention, through hygiene, is far better than cure.

Let's look at the various ways we can provide the cleanest possible conditions for our seeds and cuttings.

Garden hygiene

Not only are seeds and seedlings at risk from the dangers of the garden, but so are buds, grafts, layers and hardwood cuttings that have been taken and placed outdoors in a sheltered spot. This six-point plan will guide you towards a cleaner, healthier garden in which to propagate plants.

1. Rubbish

One of the commonest places for the hibernation of insect pests is in the rubbish of hedge bottoms and in corners of the garden that have been left untended. In an ideal garden there should be no dirty areas. Unfortunately, rubbish can build up too quickly; old pots, boxes, compost bags, weeds and dead wood will make ideal breeding grounds for pests.

Try to think of garden refuse as 'compost' and not 'rubbish'. Rubbish should be disposed of in the dustbin, while the compost should be formed into a nicely contained heap for rotting down naturally.

2. Dead wood

Most gardens have at least one tree or shrub and it is safe to assume that each year some wood will be cut off and disposed of, or it will fall off as dead, damaged or diseased material. A good gardener will not allow dead or damaged wood to remain on plants; it should be removed immediately it is noticed.

The same applies to other plants. Always remove diseased leaves and, if the plant is damaged beyond repair, take up the whole thing and see that it is destroyed. Weak plants are far more likely to be attacked by enemies than sturdy, strong ones so these, too, should be discarded.

3. Adequate drainage

It is impossible for garden plants to grow properly if their roots are constantly standing in water. With a heavy clay soil it is quite possible for a plant's 'feet' to be in water all winter, and then baked dry in a rock-hard soil in summer. Good drainage, therefore, is essential. If you have problems with your garden becoming waterlogged, then rectify this as soon as you can by laying land drains or by lightening the soil.

4. Air to breathe

In too many cases plants are not given sufficient room for development, and when they are crowded together they are more susceptible to disease. Vegetables, ornamental plants, and fruit trees and bushes all need light and air, and where a gardener takes the trouble to space his plants out well, or to thin them early, he generally does not have to spend so much money on pesticides.

5. Feeding and conditioning

If plants are properly fed, then there is a

very good chance that they will grow strong and robust. If possible, the soil should be treated annually with well-rotted organic matter (which also aids drainage) and, in addition, it may be necessary to supplement the three normal plant foods found in the soil: nitrogen, phosphorous and potash. An excess or deficiency of one or more of these can cause plants to go soft, making them liable to both insect attacks and diseases.

6. Choosing healthy plants

Always obtain new plants from reputable suppliers, and examine them carefully straight away to ensure that they are free from obvious signs of trouble. It is not a bad idea, if space permits, to keep new arrivals in a remote part of the garden – a sort of quarantine area – for a few months. This will allow symptoms of latent trouble to become apparent, so affording you the choice of corrective treatment before transplanting, or immediate disposal. Pay particular attention when buying house plants, bulbs, corms and tubers, as these can easily carry pests and diseases. With tree, bush and cane fruits, ensure that they are healthy, preferably by purchasing certified stocks.

By adopting these six practices, your garden will be as 'clean' as possible, and any outdoor plant propagation will have the best opportunity for success.

Fig. 3.1

Fig. 3.1 Greenhouse hygiene: ensure the glass panels and glazing bars are cleaned of dirt and sheltering insects, at least once a year.

Greenhouse hygiene

Basically the same six practices for hygiene can be adopted under glass, but obviously with modifications for the limited size and space. The subject of greenhouse hygiene, however, can really be summed up in terms of the spring or autumn 'clean-up'.

Autumn is, arguably, the important time if you are to make only the one clean-up in the year. The greenhouse is a nice, cosy place for pests and diseases to overwinter. Don't let them! Choose a fairly warm, preferably bright day to do your cleaning. First, to give you more room take out all the bits and pieces such as watering cans, dibbers, pots and trays. Old compost and growing bags should be emptied (the contents, if any, could be put on the compost heap), while the bags themselves can be thrown away. Then make a start on the plants.

If the weather is suitable, these can be put outside, as a temporary measure. However, if it is cold, windy or misty it is not a good idea to risk your most tender plants; these should be brought into your home for the duration, or keep them in the greenhouse and move them from side to side as you work through it. Keep a wheelbarrow or rubbish bag as near to you as possible, as this will be extremely handy. As you pick up each plant to move it, check it for any dead, dying, infested or diseased parts. Prune it back if required, and remove any weed seedlings from the compost.

When the greenhouse is empty, use a stiff brush to go all over the interior structure and staging. Brush out all of the dirt and cobwebs, getting right into the corners, where pests like to hide. Rake and brush up the debris from the floor.

The next operation is to wash down the whole structure with a horticultural disinfectant solution (Fig. 3.1). Try not to splash too much water about, as this will create a damp atmosphere which may not do the plants any good once they are reinstated. Wooden framed greenhouses are worse than aluminium for harbouring pests, so take extra care cleaning these.

The glazing, whether of glass or plastic, will need cleaning both inside and out. There are a number of proprietary products that deal very efficiently with algae and general grime, and are easy to use. The outside of the highest panes can be reached with a soft broom, a 'squeezy' floor mop or with a brush attachment as used for hoses.

If you have a greenhouse border, this should be dug over, removing any weeds. Both under glass and in the open garden, weeds should be destroyed as soon as possible, for not only do they act as hosts to certain pests and diseases, but they also check the growth of crop plants. It may be a good idea to sterilize the soil in the greenhouse border, and although there are a number of ways this can be done, where the private gardener is concerned using a horticultural disinfectant is probably the most convenient. Always follow the maker's instructions. Do not be over-zealous with the chemicals, and if recommended *always* wear protective clothing.

If your greenhouse is wood-framed, then this is the ideal time to apply a preservative treatment to the interior. There are many brands available, but for the greenhouse it is essential you invest in one that is safe in the vicinity of plants – many are not! The exterior would be best tackled on another day, particularly when the weather is warm and dry.

Once you have thoroughly cleansed the inside, use an electric fan heater (if you have one), to help dry the atmosphere. Otherwise, leave the door and vents wide open – provided it's a sunny day! While the inside of the greenhouse is airing, take the opportunity to wash down all of the pots and trays, and let them drain well (Fig. 3.2). Put aside any cracked or broken ones; these may perhaps be useful as crocks in clay pots at potting time in spring.

At this stage the plants and other bits and pieces should be brought back. There is still one final thing to do, and that is to use a greenhouse smoke pesticide, or fumigant. This type of application is available in the form of pellets or smoke canisters which, once ignited, discharge a smoke impregnated with active ingredients. Be very careful, however, as the fumes are poisonous and must not be inhaled. Keep the greenhouse sealed, with vents and doors closed, and do not allow anybody to enter it for at least 12 hours after the fumigant was set into action (Fig. 3.3).

If all of this is accomplished once a year, any problems you have with overwintering pests and diseases will be minimized. A twice-yearly clean-up means you should hardly ever have problems with these at all!

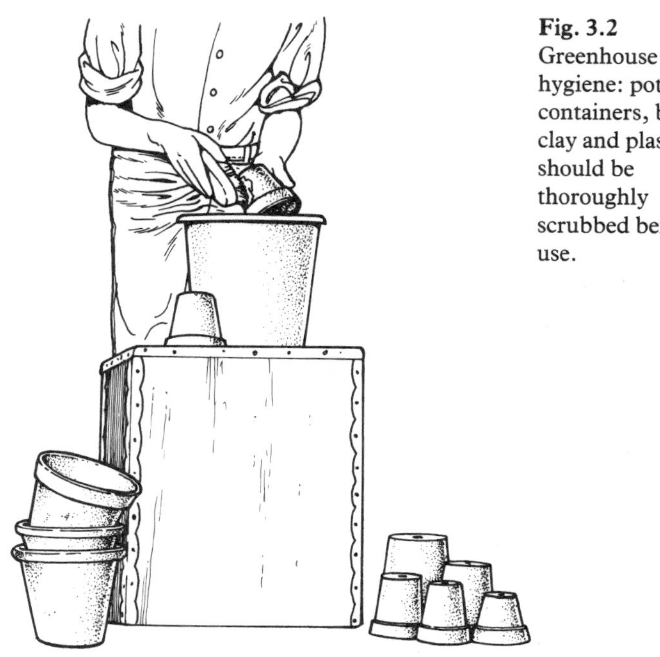

Fig. 3.2

Fig. 3.2 Greenhouse hygiene: pots and containers, both clay and plastic, should be thoroughly scrubbed before use.

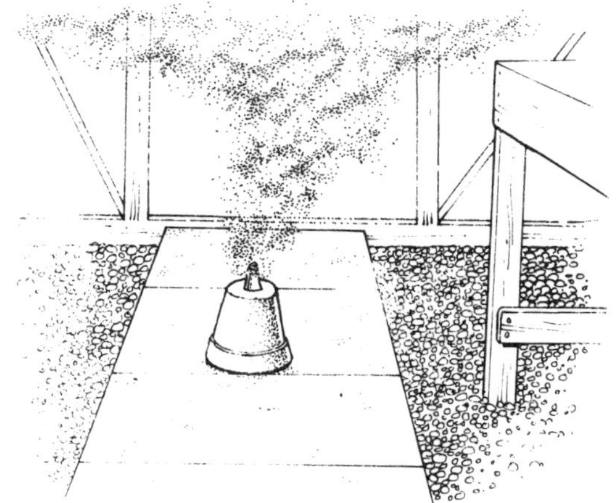

Fig. 3.3

Fig. 3.3 Greenhouse hygiene: a smoke pesticide, or fumigant, should be used to eradicate the remaining pests after the house has been cleaned; but always follow the safety instructions.

Seeds and seedlings

Many problems affecting seeds in their dormant state (and subsequently the germinated seedlings) may be prevented by a dressing or 'treatment'. In some cases the seed suppliers will have carried this out by coating the seeds with a special insecticide or fungicide formulation. The idea is to protect the seedlings from certain pests and diseases, while at the same time help to get the young plants established as quickly as possible, with a minimum number of losses.

A number of seed dressings may be applied by the private gardener. Try coating brassica seeds with the insecticide gamma-HCH to protect them from flea beetles; peas and beans with gamma-HCH/captan against the bean seed fly; and onions with benomyl against the neck rot fungus.

Viruses, such as lettuce mosaic which affects tomatoes, are problems which can be transmitted by seed, and unfortunately seed dressings in this instance are completely useless. Indeed, there are no viricidal chemicals comparable with fungicides and insecticides, so the only 'control' is a preventative one – using seed that is contamination-free. It should be noted, however, that seed claimed to be 'virus-tested' or 'mosaic-tested' may still carry the virus, but in lesser quantities (unlike when the former term is applied to fruit, such as raspberries, where it really does refer to the plants being of the highest quality).

Seedlings may be attacked by a variety of living insects and animal pests. Let's consider the symptoms and the controls:

Leaves and growing points eaten away

Slugs and snails

There are several species responsible for the destruction of plant seedlings, and the most effective way of combatting them is to eliminate their breeding grounds by adopting strict garden hygiene (Fig. 3.4). Once they are in existence, however, the most effective way of killing them is by laying down poisonous bait containing metaldehyde or methiocarb. If you cover the bait with a slate or block of wood, raised off the soil, the bait or pellets will be less accessible to pets. Some manufacturers claim that their baits are safe for pets, but current British law states that these claims cannot be made without reference to the active ingredient and a warning of the dangers to other life. Alternative, and arguably less effective, controls are traps made from upturned orange or grapefruit skins, or margarine tubs filled with beer and sunk into the ground around susceptible crops. Old-time gardeners used to hand-pick the culprits and drop them into buckets of salt water, but this is perhaps the most inhumane method of them all!

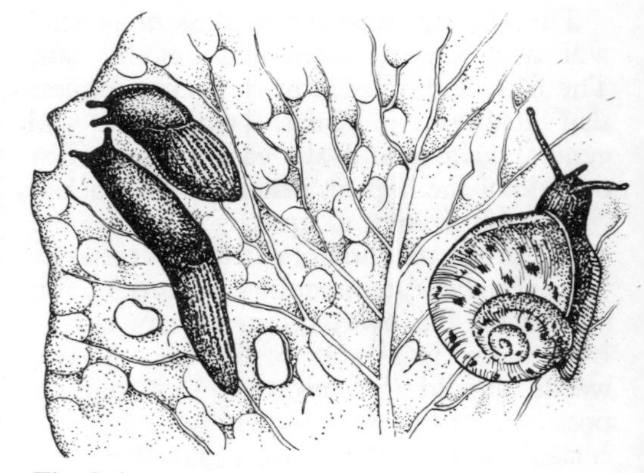

Fig. 3.4

Fig. 3.4 Slugs and snails will soon devour a row of tender seedlings.

Woodlice

Again, garden hygiene is essential to reduce the breeding opportunities. These insects do not move too far from their breeding sites, so a concerted campaign against them in winter or early spring should reduce numbers for the rest of the year. Boiling water poured onto concentrations of woodlice should kill many of them, or you can puff gamma-HCH or carbaryl dust over them. Baits containing boiled potato, turnip, bran or sugar, as well as gamma-HCH or carbaryl, can then be left in strategic places for the individual roaming woodlice to feed off.

Cockroaches

These are most troublesome in heated greenhouses, where there is plenty of room and warmth for them to breed and overwinter. Greenhouse hygiene is therefore essential. Baiting, similar to that used in the control of woodlice, can be adopted. Gamma-HCH or malathion are effective

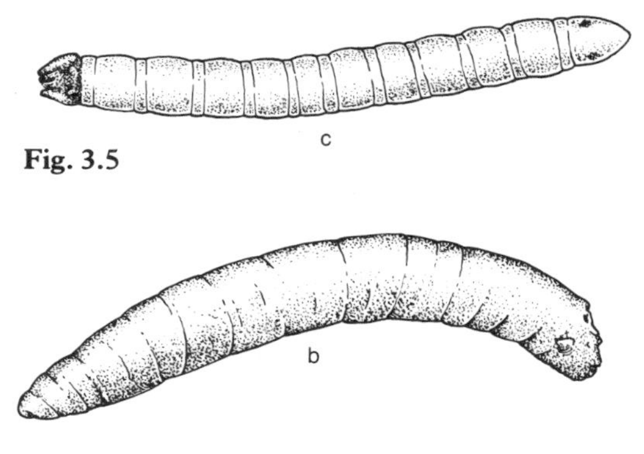

Fig. 3.5

Fig. 3.5 Cutworms (*a*), leather jackets (*b*) and wireworms (*c*) all hide under the soil surface, and between them are responsible for eating through plant stems and roots.

chemicals, and these can be dusted over the areas the cockroaches are known to traverse at night, such as paths, brickwork, ledges, ducts, and so on. Some bait mixed with insecticide in the bottom of jam jars, which are then set in the ground, will entrap and kill the pests; or you can lure them into the jars with ordinary bait, and kill them later using an alternative method like boiling water.

Stems eaten into at soil level
Cutworms
These are the light brown caterpillars of various moths, and are so called because of their habit of eating (or cutting) right through the stems of young outdoor plants and seedlings at ground level (Fig. 3.5*a*). Prevention is better than cure: turn over the soil before or during winter to expose the grubs to birds and other predators; keep down the weed population, for these encourage egg-laying; and protect susceptible plants in their early stages by applying a suitable insecticide (such as bromophos) to the soil around the stems. Individual grubs may be hand-picked off and destroyed.

Leatherjackets
Grey-brown caterpillars of crane-flies (or daddy-long-legs), these feed on the roots of most plants, and are particularly damaging to lawn and meadow grass, and any available seedlings (Fig. 3.5*b*). The hygiene notes described for cutworms also apply here, and you could also fork a little gamma-HCH insecticide into the soil around young plants.

Wireworms
Click beetles are so called because of their ability to flick themselves into the air, so making the characteristic 'click' noise (Fig. 3.5*c*). Their rather more silent larvae are known to us as wireworms, and are often the cause of eaten-away potato tubers, and the damaged stems and roots of such food plants as strawberries, carrots, lettuces, onions and tomatoes, and ornamentals like chrysanthemums, dahlias and gladioli. They are also partial to any available succulent seedling stems. To make matters worse, they are almost as troublesome inside the greenhouse as they are outdoors. Good garden hygiene renders infestation less probable. Gamma-HCH seed dressings work well, or work an appropriate insecticide (gamma-HCH, diazinon or bromophos) into the soil around the roots of susceptible plants.

Seedlings with small, round pits and holes
Flea beetles
There are a number of commonly seen flea beetles which attack many garden plants, but they particularly go for members of the brassica family. Damage to seedlings may be so severe that they eventually die. A seed dressing with gamma-HCH usually works, or apply gamma-HCH or derris to the plant's leaves and surrounding soil.

Seedlings with leaves minutely pitted
Springtails
These are tiny animals with jointed feet, related to insects. Different species will attack outdoor and greenhouse plants, but both do the same sort of damage: very small holes, similar to those caused by flea beetles but smaller, eaten into young leaves. In extreme cases the leaves are reduced to a skeleton of veins. They are most troublesome in wet, acid conditions, so there is less likelihood of severe damage if the areas (or pots) most susceptible are given good drainage, are not overwatered, and perhaps given

an application of lime. If the pest is already doing its damage, seedlings and soil may be dusted or sprayed with gamma-HCH, diazinon or malathion.

Seedlings scratched out of the soil
Cats
Our feline companions, either wild or domestic, are attracted to freshly cultivated soil which they use as toilet areas. As they prefer dry soil, to a degree they can be deterred by keeping susceptible areas well watered. There are a number of proprietary repellants on the market, and they all have a deterrent effect, to a greater or lesser extent.

Seedlings pulled from the soil, pecked and mutilated
Birds
Nearly all common garden birds will take it as an invitation if they see a nicely unprotected row of seedlings, specially if it is early in the season and other food is scarce. In the United Kingdom all birds are protected by law, so the gardener must rely on preventative measures. Man has devised many different methods of preventing or reducing bird damage over the years, but this has often been matched by the birds' ability to avoid or outwit them. The two most effective measures are easily summed up as being either chemical (proprietary repellants containing ingredients like alum or anthraquinone), and nets and cages. The latter is better for the environment (and the birds!), is longer-lasting and more permanent, but is very much more expensive.

Fig. 3.6 Pea and bean weevils (*a*) feed on leaf margins of many plants, while millepedes (*b*) often gnaw at young plants, so leaving entry points for diseases and rots.

Other seedling pests
Other creatures which cause damage to seedlings and young plants include aphids, weevils (particularly the pea and bean weevil and the turnip gall weevil), millepedes and root fly. Let's look at each.

Aphids
Greenfly and blackfly will not attack just seedlings – they go for plants of any age! They feed from the plant's sap, and if there are enough of them will cause severe debilitation. They also spread viruses. Non-chemical methods of control are ineffective once the populations are high. Using one or more of the many suitable types of insecticide will work, but they must be applied frequently, and before damaging populations have built up. Look for proprietary chemicals containing diazinon, fenitrothion, gamma-HCH or pirimicarb. The latter is especially useful as it kills aphids rapidly, but does no harm to beneficial insects (the predators and parasites).

Pea and bean weevils
These cause damage to broad bean and pea seedlings, as well as to lupins, laburnum, carnations, and others (Fig. 3.6*a*). The weevils feed on the leaf margins, which eventually take on a scalloped appearance. Treat the seed with a dressing of gamma-HCH powder or, if necessary, dust the plants' leaves with derris or gamma-HCH.

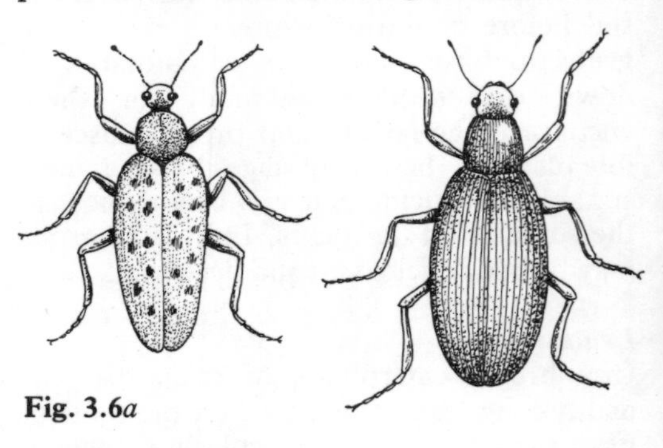

Fig. 3.6*a*

Turnip gall weevil
Young brassica plants are affected by this pest. It causes the development of rounded galls on the roots (not to be confused with clubroot disease). Prevent attack by treating the seed with a combined fungicide/insecticide seed-dressing before sowing. Discard any seedling transplants that show signs of attack.

Millepedes
Although these generally live on dead plants, they sometimes eat seeds and young seedlings, particularly enjoying peas, beans and beetroot in early spring (Fig. 3.6*b*). The eating of the plant at or below soil level is not, in itself, enough to cause death. But the rot that quickly ensues is! Dust along outside seed drills with gamma-HCH, but be sparing as too much use of this chemical in soil to be used for vegetables may taint the

eventual crops. In the greenhouse, crops and potted plants may be watered with a diluted solution of gamma-HCH.

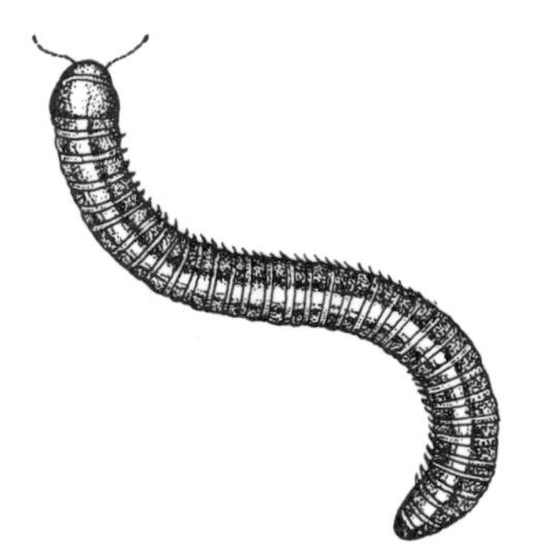

Fig. 3.6*b*

Root fly
There are three insect pests which come under this heading: cabbage root fly, onion fly and carrot fly. The first is a serious pest of cultivated brassica crops in many parts of the world. White larvae feed on the roots, and severe attacks to young plants in spring can cause death. Maggots of the onion fly feed on the stems and bulbs of onions, leeks and shallots, causing wilting and death in young plants, and rotting in older plants. Carrot fly larvae feed on the roots of parsnips, celery, parsley and, of course, carrots. Light infestation can be ignored, but severe attacks may render the roots unusable.

Treatment varies. Cabbage root fly is difficult to control. Good hygiene will reduce numbers of overwintering pupae. Calomel dust, which is recommended for controlling clubroot disease, may give partial control if it is used to dust the planting holes, as will granules of bromophos or

Damping off
Where fungus diseases are concerned, there is really only one that plays a major role in the life of seedlings: damping off. One of the most troublesome of garden diseases, it will affect seedlings of many plant types, but particularly vegetables such as tomatoes, lettuces, peas, beans and brassicas, and annual bedding plants such as petunias, stocks, tagetes and antirrhinums. We are not actually talking about one specific fungus disease, but a number of similar types with the same symptoms and results.

Seedlings and young plants may die out in patches across the pot, propagator or seedbed. Some of the larger seedlings which contract the disease may develop lesions on their stems at about soil level. Roots of some seedlings may rot away completely. Fortunately the disease is not as widespread as it once was, due no doubt to the increased use of sterilized or peat-based composts which greatly reduce the risk of infection.

As with so many other problems associated with plant propagation, the control of the damping off disease is essentially synonymous with good garden hygiene. The affected plants (and the medium in which they were growing) should be thrown away. Sadly, even apparently healthy seedlings, if they were grown in the same container, or extremely close to the diseased seedlings, should be disposed of, for they too may carry and further spread the disease.

In addition to hygiene, try to prevent the disease from getting a hold on your seedlings by creating the conditions which it finds unfavourable. In greenhouses and unheated frames ensure you do not over-water and cause waterlogging or high humidity. This will put the seedlings under a degree of stress and so render them more liable to attack. Indeed, water itself can be a source of infection, so it is best to use water straight from the mains rather than any saved in a butt.

Chemicals may be used as a preventative treatment, with varying degrees of success. Annual flower seedlings may be sprayed with thiram fungicide, two or three times at weekly or 10-day intervals. For extra insurance, apply a single spray of benomyl to pricked out seedlings; this will help prevent infection of the grey mould, or botrytis fungus.

diazinon. On a small scale it may be possible to protect young brassica plants by cutting discs out of cardboard, plastic or felt, and fitting them tightly around the bases of the stems to stop the female insects from moving up and down the plants and laying their eggs.

Onion fly can be combatted by using a seed-dressing of gamma-HCH before sowing, combined with a programme of turning the soil during winter to expose the resting pupae to birds.

Carrot fly can be similarly controlled, but it is also advisable with this pest to dust along the drills every two weeks or so with gamma-HCH. Also, avoid thinning out the seedlings as this encourages attack. If the operation is necessary, remove the thinning immediately, firm the soil around the remaining plants and water the area thoroughly, to disperse the odour of crushed roots (which is an irresistible attraction to the fly!).

Cuttings and other vegetative propagation

Plantlets, runners, cuttings and bulblets are all ways to propagate plants, and they are all carried out risking attack by a host of new pests, diseases and viruses. Many such problems occur only after the propagation technique has been carried out. On the other hand, several pests and diseases are carried into the young plants because they were already present in the material used for the propagation.

Pests

Aphids, as we have already mentioned, will attack all plants in all stages of growth. Treatment is described on page 30.

Two of the many types of eelworm are, arguably, the worst pests which are likely to be transmitted during vegetative propagation. Stem and bulb eelworm, as well as chrysanthemum eelworm, are microscopic worms no more than about 2 mm (1/16 in) long, which feed on and inhabit the plant tissue.

Stem and bulb eelworm is by far and away the most troublesome, having been recorded in over 400 different plant species throughout the world. Among these it will infest the following garden plants: strawberries (runners and plantlets may carry the pest); onions and all related vegetables (bulbs); rhubarb (crowns); narcissus, tulips, hyacinths, snowdrops and many other flowering bulbs; perennial phlox, campanulas, hydrangeas, irises (leaves and stems), and so on.

Different plants respond with different symptoms of attack. The bulbs start to decay when badly affected; herbaceous plants develop twisted and distorted stems and leaves.

Chrysanthemum eelworm is also known as leaf and bud eelworm. Although it uses the chrysanthemum as a host plant, it has been known to use over 150 different plant species, including garden favourites like dahlias, delphiniums, peonies, buddleias, perennial asters, saintpaulias (African violets), blackcurrants and strawberries. Leaves are the main parts to be affected; the lower ones first, spreading upwards. The foliage turns brown, hangs limply, and eventually dies. Cuttings are stunted, and usually perish.

The control of both eelworms is quite specific. To start with, ensure that you only propagate from healthy material; this is not difficult to deduce! *Never* propagate from material knowing that it may contain this kind of pest. Although treating plants with warm water to control eelworm is commonplace with commercial growers it is, perhaps, too critical for the average home gardener. It is an excellent form of control, however. Dormant narcissus bulbs, for example, are immersed for three hours in water maintained at 40°C (115°F). Chrysanthemum stools should be immersed for five minutes, and strawberry runners for ten; the temperature must be low enough not to damage the plant tissues, but high enough to kill the eelworm, so thermostatic control is essential. After this hot treatment, the plants should be plunged into cold water, and then replanted or propagated into sterilized compost.

Not all plants are suitable for this warm water treatment. For example, perennial phlox, which is so valuable as a border plant, is particularly prone to eelworm presence, so the method of propagation used is that of 'root' cuttings, of which more later.

Unfortunately the chemicals used by commercial growers to combat eelworms are too toxic for use by amateurs, so they are not available.

Diseases

There are a number of fungus and bacterial diseases which can be transmitted by propagating infected plant material. Possibly the greatest area of occurrence is with bulbs, tubers and corms. For example, tulips are subject to, amongst others, grey bulb rot and tulip fire; narcissus bulbs are subject to white mould, basal rot, storage rot and smoulder disease; gladiolus corms are liable to infection by dry rot, hard rot and neck rot, and potato tubers are subject to blight. Correct identification and diagnosis of these problems is of paramount importance if you are to get the best from your plants. If you are unsure which disease is prevalent, seek the help of an expert: gardening magazines or the Royal Horticultural Society will advise you on the best course of action.

There are a number of routine measures which may be taken to prevent or treat most of the above diseases. Bulbs, tubers and corms must always be handled gently – particularly at lifting time and before they are stored – to minimize injury which will precipitate infection. Any damaged bulbs should be destroyed, and only sound, healthy specimens kept. Storage conditions are important if we are to reduce the chances of storage rot: they should be dry, cool and well-ventilated.

Dipping bulbs, prior to planting, in a mixture of thiophanate-methyl and benomyl for 15–30 minutes will help protect them from rots during growth. Ensure that any growing bulbous plants showing symptoms of disease are dug up, together with the soil surrounding them, and destroyed. If your beds or borders *do* contain diseased plants, clear them immediately and avoid planting fresh stocks for at least three years (five in the case of tulip grey bulb rot).

Potato blight

This is a serious disease, particularly in wet seasons; it was a series of severe blight epidemics which led to the Irish potato famine of the 1840s. Tomatoes may also contract it. The chances of your plants getting blight will be minimized if you avoid the potato varieties that are most prone: 'Arran Comet', 'King Edward' and 'Ulster Chieftain'. A preventative chemical-spraying programme may be worthwhile in damp years: spray with maneb or zineb fungicide in early to mid-summer, and repeat at fortnightly intervals.

Galls

These are unusual growths of plant tissue, caused as a direct result of attack by groups of specific mites, midges, wasps or eelworms. There are also, however, bacterial disease galls, and at least two types may be transmitted during propagation. The most common is crown gall, which may affect nearly all woody fruits, as well as herbaceous plants (chrysanthemums, hollyhocks, lupins, etc) and vegetables (beetroot, marrow, tomatoes, etc). The characteristic galls, of various sizes, develop at or near soil level. It is not regarded as a serious disease, but severe cases will cause plants certain stress. Obviously, affected materials should not be used for propagation.

The other type is known as leafy gall, and will use the following plants as hosts: carnations, chrysanthemums, dahlias, geraniums, sweet peas and strawberries. A dense mass of distorted shoots are produced, at or around ground level, instead of (or as well as) normal growth. Affected specimens should be destroyed, and susceptible plants should be avoided on the contaminated soil for several years. Hands and tools should be washed thoroughly after handling diseased plants.

Viruses

Although some virus diseases are known to spread from plant to plant by the action of sap-sucking insects (such as aphids, mites, thrips, weevils, etc.), and soil-borne eelworm and fungi, others are transmitted by vegetative propagation of already diseased material. A virus will invariably occur throughout the whole plant, and will not necessarily be confined to the parts of the plant showing the symptoms. Therefore the transmission of the virus by vegetative propagation is commonplace, but not easily detected.

Where insects are the 'vectors', the severity of the virus infection is, to a degree, dependent on the climate; in cool, wet seasons the insect population is slow to reproduce, so slowing down the spread.

In young plants or rooted cuttings, the leaves provide the tell-tale symptoms of virus infection: mottling and blotching in shades of pale green, yellow and white; a general yellowing, either across the whole leaf or in angular patches between the veins (known as 'mosaic'); ring spots, or perhaps there is a crinkling, curling or narrowing, or a complete malformation of the leaf.

Because there are no known viricidal chemicals, the onus is on the gardener to minimize the risk of infection mechanically. Therefore, the six-point plan that follows should become routine:

1. Only introduce to the garden or greenhouse virus-free stock. There is a continually extending range of such plants available.
2. Do not propagate from plant materials known to be infected with virus, or that which has been in contact with virus-infected plants.
3. Where possible, take action to prevent or control the insect vectors *before* they are able to spread the virus
4. Any plants diagnosed as carrying virus should be immediately 'rogued' (removed and destroyed by burning).
5. Do not plant on soil that has grown virus-infected crops: use fresh stock away from the contaminated site.
6. Initiate a programme of weed control, and of quickly removing old plants after they have cropped or flowered.

CHAPTER 4

Seed Collecting

Before we embark upon the fascinating study of raising plants from seed, it is important for us to know at least a little of the biology of the subject. First, we must understand why plants produce seeds. Of course, the answer is that a plant is a member of a species, and it exists only to preserve the species. The leaves, stems or branches and roots exist to provide food and energy, and anchorage. But that is not all. They also act as a framework, upon which the plant's reproductive organs are displayed in the best possible way to atract the pollinators. The organs, or flowers, have evolved over the centuries to attract the best and most efficient insect and bird pollinators.

Of course, some plants may also be pollinated by the action of the wind, and these rarely find it necessary for such attractively-built reproductive organs.

What is a seed?

Once fertilization has taken place, the seeds will form. Each individual seed may be considered as a living organism, perhaps more correctly thought of as an embryo plant. Whereas in the animal world an embryo is generally cared for and nurtured by its mother, in the plant kingdom it is practically ejected from its parent, sometimes with great force, and is left to survive as best it can in the harsh world.

In Nature, seeds may have to endure all kinds of conditions before they settle and germinate. Some may travel many miles (even thousands), by wind or sea, often taking months or years before they find a resting place. Others may endure fire, or freezing conditions, or perhaps they will be consumed by a bird or animal and pass through its intestines. Some become buried in the ground, and do not consider germinating until decades or more have passed.

These seeds are not always launched into the world without some parental blessing. Usually a seed is equipped with a sort of life-support system: a supply of enough stored food to carry it through its dormancy, protected by a hard, dry seed coat. It has sufficient enzymes within to convert this stored food into energy and a form the tissues can use. Within the cells of the seed, all of the genetic information is held, for directing the growth patterns when germination occurs. More evidently, many seeds are given little 'extras' in the form of wings or paddles, so they may be propelled through air or water to their resting place (Fig. 4.1).

But why is the offspring dispersed so unkindly in this way? The answer is because it is undesirable for a seed, or seeds, to germinate too close to the stationary parent plant. After all, plants do not issue single seeds, but often thousands (or millions in the case of some orchid capsules). For seeds to germinate within the immediate vicinity of the parent would mean both generations fighting for the same space, food and moisture supplies. The parent wishes for the offspring to grow without hindrance, and once germination takes place there is no chance of changing location. Consequently, plants have a built-in desire to disperse the seeds and so prevent the inconvenience of overcrowding.

Dormancy

It may seem unlikely, but the continuation and preservation of the species is the sole reason why seeds require a period of dormancy before they germinate. The ability of seeds to remain in a dormant state, or in varying degrees of dormancy, has contributed to the survival of seed-bearing plants as we know them.

For example, a plant that needs around 100 days of warm weather to mature and set seed, will indeed be doomed to failure if the second generation seed sprouts straight away (which is likely to be at the end of the summer, around the time of the first frost).

The reasons why dormancy occurs have

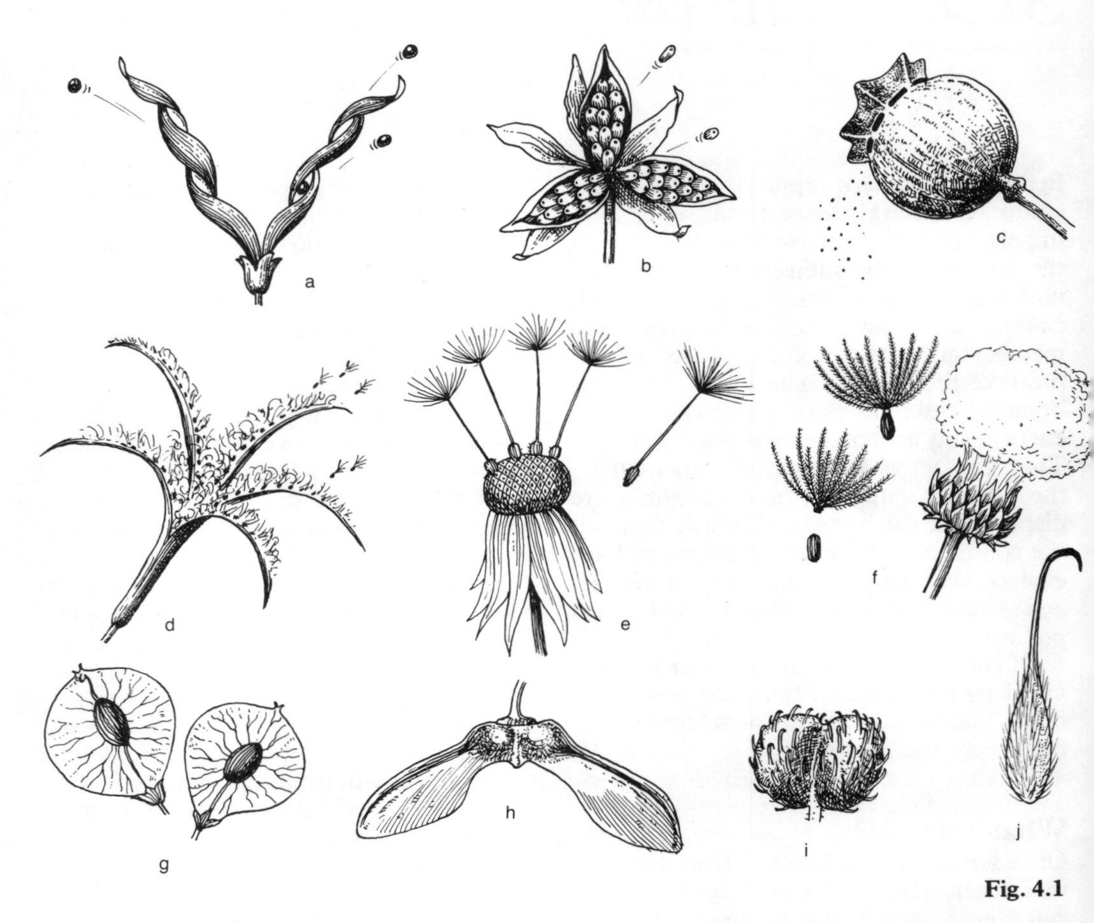

Fig. 4.1
A selection of the many types of seeds and seed cases that are found in gardens: (*a*) sweet pea, (*b*) violet, (*c*), poppy, (*d*), willow herb, (*e*), dandelion, (*f*) thistle, (*g*), elm, (*h*), sycamore, (*i*) cleavers and (*j*) common avens.

Fig. 4.1

become clear to us, but there is still much to learn about the 'way' some seeds effect their dormant periods. Perhaps some have immature embryos, or the seed coats are impermeable to water or gases, or the coats are too unyielding to permit embryo growth.

We know that heat, light and moisture each has its own part to play. As gardeners, we can often use this to our advantage and break the dormancy, by chilling or subjecting the seeds to light. In the case of the hard seed coat, we can file, chip or abrade it, although if done carelessly this can damage the seed beyond repair.

If you save your own seeds, you'll find that beans, mustard and many other vegetable seeds never go dormant. A few of the root crops such as carrots and parsnips, for example, need around a month or so of after-ripening following the harvest, before they will germinate. Of course, seed that you have purchased from a garden shop, or from a mail-order seed-house, will have had time to undergo any necessary after-ripening in the months between harvesting and sowing.

Methods for collecting seed

As has already been stated, seeds depend on stored nourishment to carry them through their dormancy period, which generally tends to extend over the coldest time of the year. Seed that is harvested too early may be deficient in embryonic development, and is likely to deteriorate in storage. If it does survive, the resultant germination may be uneven, or the seedlings produced inferior in some way.

However, you must be sure to collect the seed before it succumbs to the elements outdoors and rots. The time of harvesting is crucial, particularly for those plants which release their seed the moment it is ripe. Again, preservation of the species is all that matters, because some plants that ripen and release their seeds rapidly, do so one stalk at a time over a period of weeks – inconvenient for us, but a wonderful survival mechanism for the plant.

Incidentally, if you are intending to collect seed from brassicas, onions or lettuce – all of which set seed in this way – tie small paper bags, with air holes punched in them, over the fading flowerheads to catch the seeds as they ripen and fall, in case you are not able to make daily inspection.

Of course, the fruiting vegetables – tomatoes, cucumbers, marrows, peppers, aubergines, and so on – each produce their seeds within a soft shell of edible flesh. If you wish to collect the seed from these, leave the fruits to over-ripen on the plants (a figure of speech, as most fruits we consider to be marketable are not completely ripe). As an indication, cucumbers will pass the green stage and turn yellow, tomatoes should be deep red and soft, and the red skin of peppers should be wrinkling. Do not attempt to collect seed from rotting fruits, as the act of decomposition raises the temperature within the fruit, and the seed may have become damaged.

Collecting seed from trees is, for some, a fascinating hobby, but can be extremely dangerous. Foresters may climb into the canopies of trees to collect the bounty, but the average gardener is not an expert tree-climber, so should make do with what he can get from the lower branches or from windfalls. Some seeds are easy to spot, either because of their size (as with acorns and horse chestnuts), or their colour (some magnolias, for example, produce bright orange seeds).

Choose a dry, sunny day for your collecting. The seeds should be free from moisture in the way of rain or dew. Frost is not the deadly danger you might have expected, as long as the seeds are dry. If seeds are left outside for a period of time, and subjected to moisture, frost and thaw, then there is every possibility that they will rot. Put the seeds into clearly labelled paper bags, packets or envelopes (Fig. 4.2). This will save getting the batches of collected seed mixed up, with the ensuing confusion!

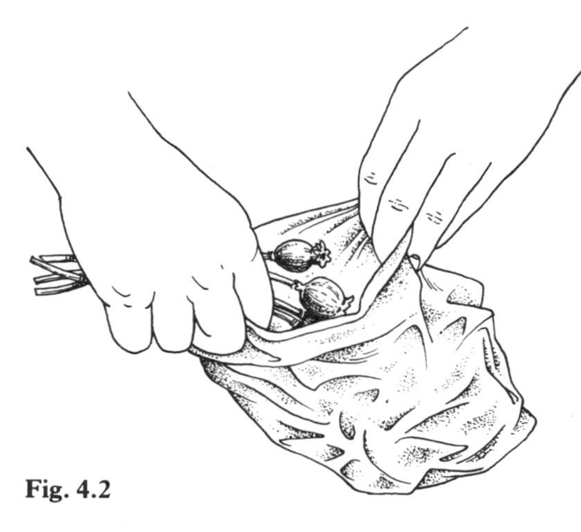

Fig. 4.2

Fig. 4.2 When collecting seeds, put the faded flowerheads from the plants in question, into clearly labelled bags; this will avoid mixing up the species or varieties.

Seed collecting is an activity most gardeners can do throughout the year, starting with the early flowering bulbs such as *Tulipa tarda* and *Aquilegia alpina* (alpine columbine), which set seed in early summer. *Euphorbia myrsinites* and *Pulsatilla vulgaris* (pasque flower) are ready in mid-summer. Most of the mainframe perennials release their seeds in late summer and autumn – *Alchemilla mollis* (lady's mantle) is one such – but if winter overtakes them before they have ripened, the stems may be cut from the plants, put into water and taken indoors to finish off the process.

There are always exceptions to the rule, however. Agapanthus, or African lilies, are latecomers and need to be picked green at the end of summer to ripen indoors. Then there are those plants which take a long time to ripen their seed. *Cyclamen hederifolium*, for example, will flower from late summer to late autumn, but will only release its seeds after almost a year has passed. It is easy to forget to collect them! An added problem is that cyclamen seeds are coated in a sugary film that ants find irresistible. They will heave and pull the seeds down into their nests, and the seeds will be lost to you forever. You must, therefore, recognize immediately when the seeds are ready, and collect them. Cyclamen have a curious looking seed pod. The stems bearing the rotund capsules coil tightly, tucking the pods close in to the mother plant. When the seed is

Fig. 4.3 The stems bearing the rotund seed capsules on hardy cyclamen coil tightly, tucking the pods close in to the mother plant.

ready to be dispersed, the coil unwinds, and the capsule rolls away (Fig. 4.3). It then begins to split apart and the seeds are released.

Later in the year a variety of berries, fruits and cones may be collected. The latter can be obstinate in holding on to their seeds. Place the cone in a warm airing cupboard, or by warm pipes, to encourage the scales to open and release the seeds.

Drying and storing

Once the seeds have been collected they should go into storage, but not before they have been completely dried. Moisture in heaped-up seed will cause a warming which may harm the delicate embryos, and moisture will also speed up the seed's metabolic rate, causing it to use up its stored nourishment too quickly. All seeds should, therefore, ideally undergo a drying period of around a week, no matter how dry or shrivelled they already look. Simply spread the seed on newspapers in a dry, well-ventilated room, greenhouse or outhouse (Fig. 4.4). It is a good idea to change the newspaper once or twice during the drying period if the seeds are moist at the start.

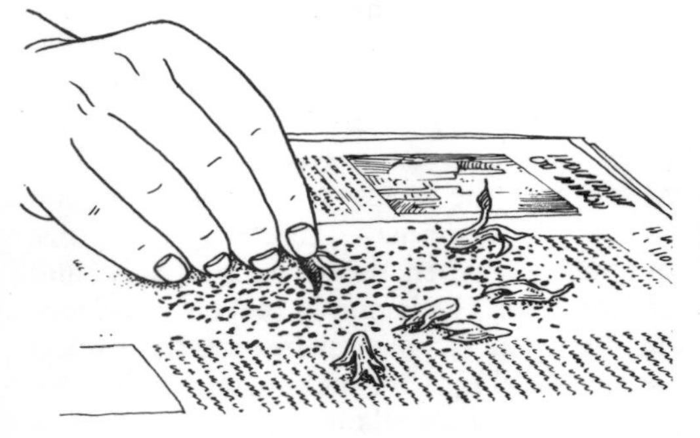

Fig. 4.4 Collected seeds should be dried before they are stored. Spread them out on newspaper, in a dry, well-ventilated room.

Exposure to sunlight is not unwelcome. Indeed, if you live in a damp climate, the seed may actually be absorbing an excess of moisture from the air (it is natural for it to absorb *some*). Sunlight, or artificial heat from a light bulb, will help dry the atmosphere, as well as the seed. But do not expose the seed to extreme temperatures or naked flame. This will cause them to dry out too rapidly, and they are likely to crack, shrink or develop a hard seed coat.

Do not try to store seed that is mixed up with undesirable bits of vegetation or animal life – leaves, stems, seed-pods, capsules, insects, and so on. These items will rot and possibly infect the seed. If the insects are alive, they may eat it! Cleaning the seed, therefore, is a vital part of the propagating process but regrettably, if great quantities of seed are concerned, this can become a real chore.

Screening is the term for passing seed through holes in a sieve or mesh. This will separate the coarse, heaviest bits of debris, such as pebbles, sticks or leaves. On the other hand, winnowing is the term for pouring the seed from one container to another while blowing off the lightweight chaff. A stout pair of lungs may suffice, alternatively use an electric fan.

We already know that a seed is a living thing. Thus when it is in storage it must not be allowed to die. Its metabolism should be kept operating at the lowest possible level, to keep the seed on 'hold'. To do this, as many of the following damaging influences as possible should be kept out.

Heat

If the atmosphere is too warm there will be enzyme activity within the seed, and it will not keep. Under laboratory conditions, scientists have discovered that for every 9°F the storage temperature can be lowered, between 0–44.5°C (32–112°F), the seed's period of viability will double.

Seeds may be successfully stored in a domestic freezer; they will survive at −18°C (0°F), but they must be dry.

Moisture

If a seed is very dry, it can endure extremes of heat (for short periods) and cold. Needless to say, it is essential not only to have the seed well dried *before* storage, but also to keep it dry *in* storage! It is very difficult for home gardeners to control humidity levels, within percentages, so we should follow a rule of thumb. Store dried seed in an air-tight (which should also be damp-proof) container, and store seed that is not thoroughly dry in an open container.

Pests and diseases

In closed, air-tight containers the seeds may be vulnerable to bacteria and fungi, especially if the seeds were not cleaned well before incarceration. Fortunately, both of these potential problems are inhibited by cold, dry conditions.

Animal and insect pests are really only troublesome with open-stored seed. The majority of insect predators need around 4°–10°C (39°–50°F), and a reasonable degree of humidity to maintain their activity. Birds and rodents can make serious inroads on a bulk of seed stock, particularly beans, peas and corn. If this is likely to be a problem traps may be set. With small-scale stored seed, prevent the loss by using tightly closed metal or glass containers.

Viability

Most seed collected from garden plants and vegetables will remain viable (capable of germinating) for several years after harvest, provided they were collected, cleaned, dried and stored properly. It is, though, impossible to be absolutely specific about the keeping qualities of seeds, as any one of these operations can adversely affect them. The vegetable seed information chart below will give you some idea of what to expect. Occasionally some seeds will satisfactorily germinate after even longer periods than indicated on the chart, but don't count on it!

Seed may still be used after the average time recorded for it, but the expected rate of germination is considerably lower. So as not to waste too much time, effort or space, sow the seed thickly, or 'dilute' it with new seed of the same species.

Vegetable seed viability

If stored in a cool, dry and dark place, vegetable seeds will, on average, last for a number of years:

SEED	*YEARS*
Asparagus	3
Aubergine	5
Beans	3
Beetroot	4
Broccoli	5
Brussels sprouts	5
Cabbage	4–5
Carrot	3–4
Cauliflower	4–5
Celeriac	5
Celery	5–6
Chicory	5
Chinese cabbage	5
Cucumber	5–6
Fennel	4
Kale	5
Kohl rabi	5
Leek	3
Lettuce	4–5
Marrow	5–6
Melon	5
Onion	1–2
Parsley	2–3
Parsnip	1–2
Pea	3
Pepper	4
Pumpkin	4
Radish	5
Salsify	2
Scorzonera	2
Seakale	1–2
Spinach	5
Sweet corn	1–2
Tomato	4
Turnip	5
Watercress	5
Water melon	5

Stratification

Many seeds from trees, shrubs and other berrying plants (such as roses) will refuse to germinate until they have undergone a period of exposure to low temperatures in moist conditions. This operation, or stratification, acts to soften the outer covering of the seed.

It is not a difficult process, and extremely worthwhile. To start, break open the outer shell of the berry or hip, to expose the inner seeds. Place these opened-up fruits in a small seedtray covered with polythene, or a modern transparent cling-wrap, and leave it in the refrigerator for about two months.

There are, however, two alternatives. First, cover the bottom of a tray or pot with sharp sand and scatter the seed on top. Add other layers of sand and seed until you reach about 2.5 cm (1 in) from the rim of the container. The other method is to mix the crushed berries thoroughly with about twice their bulk of sand.

In either case the container should be left outdoors in an exposed position where the winter weather will cause alternate freezing and thawing. Set the pots or trays in sand or soft soil to keep them sturdy and level, and cover the tops with wire netting to keep mice away (Fig. 4.5). Ensure that the pots do not rest under the drips of trees or broken gutters, nor in collected rainwater. During late winter or early spring remove the pots or trays containing the stratified seed from the sand bed, or the refrigerator. Remove the seed and sand mixture from the containers, and sow in the normal way – it is not necessary to separate the seed from the sand.

Fig. 4.5

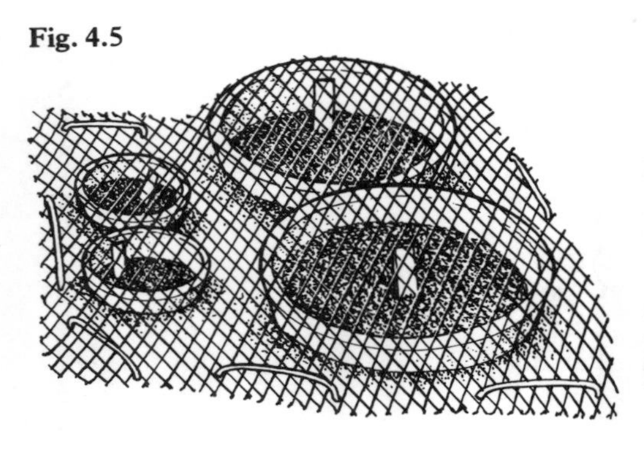

Fig. 4.5 Many tree and shrub seeds must undergo 'stratification' before sowing. If outdoors, cover the pots with wire netting to keep mice away.

F_1 hybrids

Many of our garden flowers and vegetables are hybrid plants, bred by commercial growers for superiority in colour, scent, yield, taste or resistance to disease or the elements. They are brought about by the crossing of one species of plant with another in the same genus, or between distinct varieties within the species. These carefully nurtured offspring are usually the sort known as F_1 hybrids. Should we save seed from such plants?

The answer to that is only if we have the time and the space to experiment. The problem is that seed set by these hybrids, when germinated, produces plants that have a tendency to revert to the highly inbred, imperfect qualities of the recent ancestors. The seedlings will contain a haphazard mixture of genetic forms from the original crosses, many of which would be regarded as completely useless.

CHAPTER 5

Seed Sowing Outdoors

It goes without saying that all gardeners have the facility for sowing outdoors, and if they have a greenhouse, or an abundance of light window sills, there is a bonus in that they will also be able to sow seed under cover. Indoors and under glass, one has the luxury of not having to rely particularly heavily on the weather, but outdoors the gardener must place his total faith in the elements. So often, patience is the order of the day.

For example, there is absolutely no merit in attempting to sow seeds of vegetables or flowers if the soil is cold and wet: wait until the weather is mild, and the ground medium-dry, warm and more workable. By sowing before the soil is ready, you risk the chance of seeds rotting away, and there is certainly no gain in terms of earlier flowering and/or cropping. On the other hand, germination rates are likely to be much higher, and the resultant seedlings that much sturdier, if one waits for the right soil and weather conditions.

Sometimes the seed is sown in a nursery bed and transplanted to a different site for flowering or cropping; in other circumstances the seed is sown where the plants are to remain and mature.

Most hardy trees, shrubs and plants commonly found in gardens may be propagated successfully by seed sown outdoors. Each type or group of plants will require slightly different techniques and conditions. In this chapter we will consider them all, but we must start with the common denominator: getting the soil in the right condition for sowing.

Soil preparation

Long before sowing can take place, the soil will need to be cultivated so that it has enough nutrition, oxygen and moisture for the seeds to germinate and develop. Warmth, the other requirement of seed germination will come later. Digging or tilling the soil is necessary if you are to provide these factors, but not if it is saturated by rain or melted snow: soil is easily compacted by foot traffic crossing it or, worse, wheelbarrows or machinery. Also, if large spadefuls of wet soil are turned over, they will often dry into impervious clumps which will be difficult to break up later.

How, then, can you tell when your soil is dry enough for digging? The best test of soil condition is to gather up a handful and squeeze it! Using both hands, form it into a ball. If it shatters or crumbles when you drop it from about 1 m (3¼ ft), or if you can press into it with your fingers, it is ready. However, if the ball keeps its shape, or merely breaks in two, as opposed to completely crumbling, then it still contains too much moisture.

Autumn digging

The ground should really be dug over in autumn. By leaving it turned over, the actions of frost and thaw during winter will help to break up and mellow the surface, so making it far more workable at sowing time (Fig. 5.1). Whilst digging, remove any perennial weeds. If the whole weed, or any part of the root, remains in the soil, it is likely you will be troubled by it throughout the year.

Fig. 5.1

Fig. 5.1 Autumn is the best time for digging the soil, especially if the ground is heavy. Winter frosts will help to break down the large clumps into a more workable 'tilth'.

It is not essential to add large amounts of bulky organic matter to the soil at digging time. Seeds do not require a heavily enriched soil – this would make the seedlings undesirably soft and lush. Therefore, it is only necessary to add organic matter (such as well-rotted garden compost or farmyard manure) to poor, under-nourished soils. If it is felt that the addition of some material *is* necessary, make sure it is well-rotted and not fresh. Fresh matter will decompose in the soil, and in so doing will rob it of nitrogen.

Peat is a natural substance favoured by many gardeners. It is an organic material which has received a considerable amount of publicity recently. Anti-peat extraction lobbyists have done everything in their power to bring to the attention of the public notice that by the end of this century the remaining natural peat bogs will have disappeared. Conservation of such areas are important to future generations, so it is left to the individual gardener to decide whether or not large quantities of peat should be applied to his soil. Regardless of the moral implications, peat is appropriate for adding to some soils in advance of seed sowing. It will lighten a heavy soil, and improve water-holding in sandy free-draining soils. Unlike manure and garden compost, however, it does not have stored within it large amounts of plant food.

Spring digging

The majority of outdoor sowings are made during spring, when the ground and air are both warming up. But it is wrong to sow straight onto soil that has been left rough for four or more months. A 'tilth' will need to be created. This refers to the condition of the soil surface after the processes of digging, weathering, raking and (sometimes) rolling. So far the soil has been left, rough dug, over winter. Now you must go over the area, systematically, with a large wooden or metal-tined rake. Level the soil, and remove any fresh weeds. It may also be desirable at this stage to remove the larger stones that have come to the surface.

Fertilizers

You will probably remember from school science lessons that the three main foods required by plants are: nitrogen (N), potash (K20) and phosphate (P205). While none of them are needed for the process of germination itself, they are all required during the development of seedlings, particularly phosphate as it promotes root growth. Roots, of course, provide stability and anchorage, and only through the roots will the seedlings be able to take in food and water.

Nitrogen promotes rapid growth so is relatively important but where the development of seedlings are concerned, fertilizers containing it should be used judiciously. Slow-acting organic forms of nitrogen, like hoof and horn meal, are better for seedlings than the quicker nitrate of soda, or sulphate of ammonia. Potash, meanwhile, is least important. It promotes flowering and fruiting, so has no particular significance with seedlings. It is still, however, an important plant food, and as such adequate supplies of it should be available to the young plants.

A good base fertilizer for an outdoor seed-bed can be made at home from the following: 4 parts superphosphate, 1 part sulphate of potash, 4 parts hoof and horn meal (all parts by weight).

Lime plays an important role in seed-bed preparation. It not only tempers the acidity of the soil, but it also helps to minimize the toxic effects of some plant-inhibiting soil elements. The quantity to apply varies with the existing acidity of the soil, which may be determined by using a soil testing kit, and also which plants are to be sown. The kit will indicate the acidity or alkalinity of a given area of soil, and will suggest the right course of action to attain the desired reading.

Alpines

Ornamental plants native to the Earth's mountainous or high-altitude areas are termed 'alpines', and many have been introduced to gardens on lower grounds as rock garden plants, because of their charm and delicacy. The usual methods of propagation are by seed, cuttings or division. However, some alpines, such as certain species of androsace are impossible to increase by any method other than seeds.

The sowing of alpine seeds is relatively

Fig. 5.2

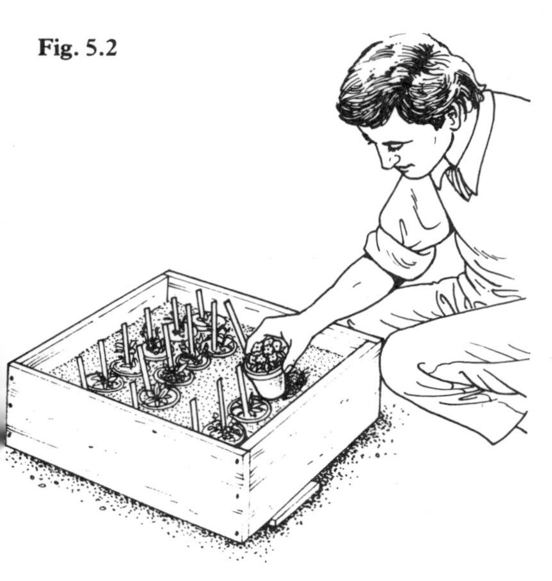

Fig. 5.2 Alpine plants are best started off in small pots. Many will later be transplanted to the rock garden, but some are at their most attractive growing as small containerized plants.

easy, and fairly cheap. You may save seeds from your own plants, or buy from a recognized seedhouse. A further way of obtaining seeds is by joining a group such as the Alpine Garden Society. Each year it sends to its members lists of appropriate plants, the seed of which may then be requested and distributed to them.

The soil preparation techniques so far discussed in this chapter actually do not apply to alpines in any major way. Many species, such as *Leontopodium alpinum* (edelweiss) or *Erinus alpinus* (fairy foxglove), may be sown in specially prepared seed sowing compost under cover in spring, when they will germinate quite quickly with the aid of gentle heat. But they can also be sown outdoors.

Others, meanwhile, prefer to be sown into containers of similar compost, outdoors during autumn or winter, to be followed by a period of cold weather, before they germinate in warmer conditions during spring. This cold weather treatment is a form of stratification, as discussed in the last chapter. Small amounts of seed may be sown in 9 cm (3½ in) pots (Fig. 5.2), while pans or half-pots should be used for larger amounts of seed. Plastic containers are perfectly acceptable, but the alpine growing fraternity tends to prefer the more traditional clay pots or pans. This is, perhaps, because the plastic containers generally remain wetter than the clay, and it is overdamp compost that often proves fatal to alpines.

Soil-based composts, like the John Innes seed compost, are suitable for alpines, but the overriding factor is that whichever compost you choose, it must be free-draining. If you are in any doubt about the drainage properties of your compost, add a little extra coarse sand before use to give a really open, airy mix. Alpines, generally, will not do well in a peat-based compost as this tends to hold water and remain too wet. Exceptions to this rule are the *Ericaceous* alpines – species of heather or rhododendron – which require an acidic, peaty, lime-free compost and eventually soil, in which to thrive.

Techniques for sowing seeds in containers will be discussed more formally in the next chapter. Needless to say, with alpines the seed should be sown thinly to avoid overcrowding. Do not cover very small seeds like those of, for example, meconopsis, but larger seeds can indeed be covered, either with a thin layer of sieved compost (in thickness equal to about twice the diameter of the seed), or with a layer of coarse grit or chippings. Opting for one or other of the latter two will help prevent the compost from drying out, while at the same time preventing the unsightly growths of moss and/or liverwort smothering the surface (particularly in the case of those alpine species which are slow to germinate, and so remain in their container for, perhaps, a year or so).

Of course, a way to avoid the over-use of pots and pans is to sow seeds direct into pockets of soil on the rock garden. Species of linaria (toadflax), for example, are easily propagated in this way. However, true alpine species are particular as to their growing conditions, and to be sown straight into perhaps hard, stony and inhospitable soil, will not suit all.

Aftercare

Treat the container-grown seeds as if they were growing naturally in the mountains. Place the pot or pan, after it has been watered and allowed to drain, in an open but sheltered part of the garden – perhaps north facing – to ensure that the seeds will get cold. This will then allow germination to take place, unhindered, when the weather warms up in spring.

Generally speaking, alpines raised from seeds will take up to a year to germinate. If pricking out is considered necessary – for it is not always – then this may be done several months after germination. Perhaps a year should pass after pricking out before the young plants are of a suitable size for planting on the rock garden. The best time of year for final planting is spring, although some specialists prefer early autumn.

Hardy annuals

Annual plants grow to maturity, flower and release their seeds within one year. After this, they die. Annuals are divided into two groups, half-hardy and hardy. The former type are not frost hardy and are usually sown under cover in spring, for planting outdoors once the danger of frost is over. These will be discussed more fully in the next chapter. Hardy annuals, however, will stand some frost, and can be sown direct in the open, where they are to flower.

Annuals are colourful plants for summer display, and are often grown in their own beds or borders to provide colour from late spring until mid-autumn. Although their displays come relatively cheaply, hardy annuals have rather fallen from favour in recent years. This may be as a result of the traditional methods of sowing, which involve a degree of hard work and preparation, and the possiblility of disappointing results.

Seed catalogues are issued each autumn, with time for the gardener to consider which varieties he would like to choose and order. Good catalogues list and illustrate perhaps hundreds of hardy annuals. Some will be hybrids in a range of mixed colours, while others, such as *Atriplex hortensis rubra*, or *Euphorbia marginata* (snow-on-the-mountain), will be true species, from which you will be able to collect and save seeds each year instead of purchasing new.

Choosing the right spot in the garden for annuals is important. Many originated in hot lands, such as Mexico, or the Mediterranean border countries, so they should really be sited in areas receiving full sun. If they are in the shade, development and growth of the plants will be weak, and flowering will be poor. The soil need not be too rich; indeed some annuals, like the ubiquitous tropaeolum (nasturtium), will thrive on very poor soil.

Hardy annuals may be sown outdoors in spring once the soil is in a workable condition. These include *Amaranthus caudatus* (love-lies-bleeding), lupinus (annual lupins) and *Reseda odorata*, (mignonette). Some of the hardiest types will endure the complete winter outdoors, and these include iberis (annual candytuft), *Clarkia elegans* and nigella (love-in-a-mist). All things considered, if your soil is heavy and lies wet over the winter, autumn sowing is not really to be recommended, as the seeds will most likely rot away. The best soil for a pre-winter sowing is a light, sandy free-draining one.

If you want to achieve a bed or border display, it is wise to plan the planting scheme beforehand to avoid unsuitable colour clashes or plant height combinations. The area to be sown can be marked out with a stick or cane just prior to the operation, according to the plan. Many old books recommend that lines of sand are laid along the marks, to help you visualize the finished bed.

The actual sowing operation itself is not difficult. Sowing the seed in straight rows is much better than sowing 'broadcast' (Fig. 5.3.), which makes subsequent thinning of seedlings and weeding extremely tedious and practically impossible. The drills, into which the seed will fall, can be taken out with the corner of a draw hoe, and should be spaced about 15 cm (6 in) apart (Fig. 5.4). Be sure not to make the drills too deep – remember that the seeds should be covered with soil only around twice the diameter of the seed! As the seedlings grow and mature, the straight lines gradually disappear; the plants merge together and form a block of colour.

Sow the seeds as thinly as possible to minimize the amount of seedling thinning later (Fig. 5.5). The seeds may be mixed with sand so you can gauge the rate of sowing; ideally, one seed should come to rest within the drill every 1 cm (½ in). Different gardeners adopt different sowing techniques, and whichever you use is a matter of personal preference. You can sow straight from the packet, or pour a quantity of seed into the palm of your hand and allow individuals to fall between your thumb and forefinger.

After sowing each group, draw soil back over the seeds, using a rake (Fig. 5.6). It is not necessary to firm the soil at this stage: it can create a hard surface to the soil, especially after watering, and this may inhibit germination. Talking of watering, this should be the next operation, particularly if the soil is dry. Make sure you use a watering

Fig. 5.3

Fig. 5.3 Sowing seeds 'broadcast': make a sweep with your hand and let the seed scatter randomly.

Fig. 5.4

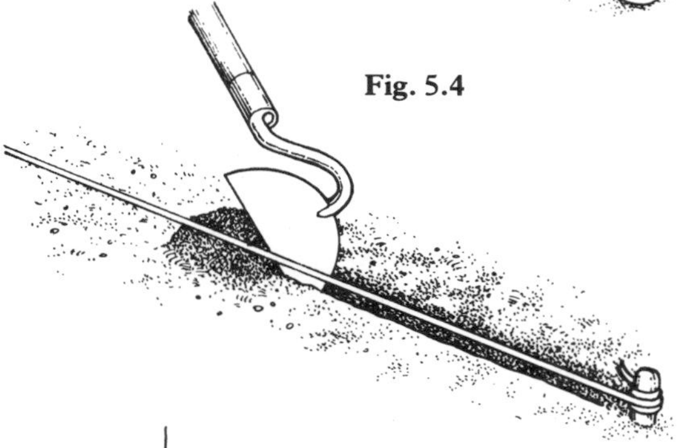

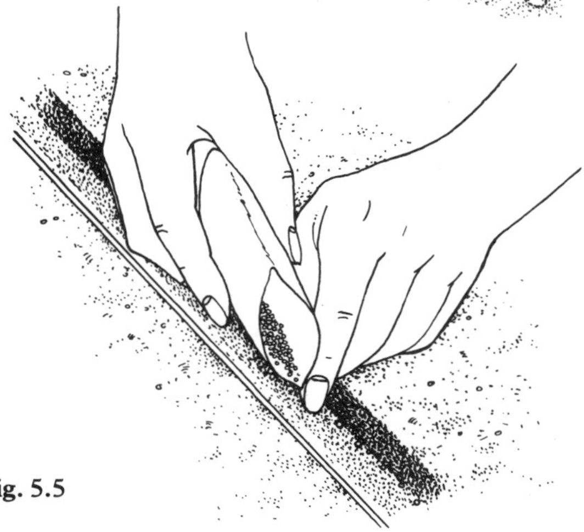

Fig. 5.4 Sowing seeds in a drill: make a shallow drill by using the corner of a draw hoe against a taut line.

Fig. 5.5

Fig. 5.5 Sow the seeds as thinly as possible, to minimize subsequent thinning.

Fig. 5.6

Fig 5.6 Using a rake, draw back the soil over the seeds.

Fig. 5.7 As soon as the seedlings are large enough to handle they should be thinned, but only if you feel it is necessary.

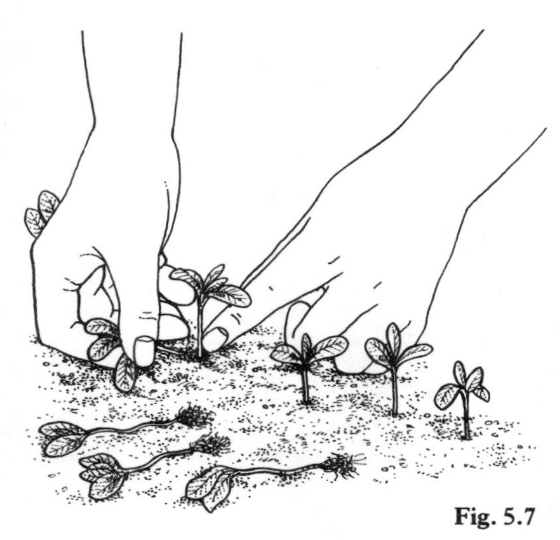

Fig. 5.7

can with a rose attachment, or a garden sprinkler, so as to avoid washing away the seeds. From now on apply water whenever it is required during the growing and flowering months.

Aftercare

As soon as the young seedlings are large enough to be handled they should be thinned, but only if it is deemed necessary (Fig. 5.7). For most varieties of hardy annuals at this stage of growth the ideal distance between seedlings is 15 cm (6 in). Simply pull out the unwanted seedlings with your fingers, or use an onion hoe if there is a large amount of thinning to be done. It is likely that soil around the seedlings will be disturbed during the thinning, so settle it back by watering.

The taller annuals (you will know the ultimate height of varieties as this information is provided on the seed packet) may need to be supported. Fail to do this and the mature plants will flop to the ground, or snap completely, and the whole appearance of the bed will be ruined. One of the best, and most natural forms of support, is to use twiggy pea sticks. Place them in position early, so that the young plants can grow through them and completely hide them.

All annuals, whether hardy or half-hardy, will be past their best by autumn. It is at this stage that the plants should be pulled up and consigned to the compost heap – but only after you have robbed them of any seed you wish to save!

Hardy biennials

Whereas annuals complete their life cycle within one year, biennials require two growing seasons to do the same. Most of the spring bedding plants, such as myosotis (forget-me-not), *Dianthus barbatus* (sweet william) and digitalis (foxglove), are true biennials. In addition, a number of short-lived perennial plants are best treated as biennials as they have a tendency to decline in vigour, or their appearance is less than desirable if they are kept growing for longer than two seasons. These include cheiranthus (wallflower), bellis (double daisies) and *Primula vulgaris excelsior* (polyanthus).

Biennials, and their perennial comrades, are not difficult to grow from seed, but the timing is important: sow too late and the plants will be killed by a severe autumn frost, or they will never reach the right size for flowering. Hardy biennials are, therefore, best sown out of doors in late spring, although they may be started as late as mid-summer, with reasonable resultant success.

The basic difference between hardy annuals and biennials, as far as the initial growing is concerned, is that the former flower in the place they were sown, whereas the biennials are sown and later transplanted.

The soil should be prepared in much the same way as described for hardy annuals. Choose a piece of spare, but well-cultivated ground. Sow the seed in drills drawn 30 cm (15 in) apart, and if the soil is dry, it is a good idea to water it well prior to sowing.

Aftercare

When the young seedlings are large enough to be handled they should be pricked out in rows or nursery beds, 15–23 cm (6–9 in) apart (Fig. 5.8). Here they will grow on and develop into larger plants, until mid-autumn, when they will be ready for transplanting to their flowering positions in the bed or border.

The nursery bed should be in an open, sunny position, although it is fair to say at this stage that polyanthus prefer their formative months to be spent out of full sun: dappled shade being ideal!

Fig. 5.8

Fig. 5.8 Seedlings of biennial plants should be transplanted to a nursery bed.

Fig. 5.9

Fig. 5.9 Make sure the transplants are watered in to their new quarters.

The technique for moving the seedlings to the nursery bed is quite specific. Lift the young plants by using a small fork. Avoid damaging the roots or leaves as far as it is possible. The plants will be lifted more easily if the soil is nicely moist, not sodden, so choose your time well. Do not allow the roots of the plants to dry out during the transplanting process; it is surprising how quickly this could happen on a warm, sunny day. Keep a bowl or bucket of water handy to keep the roots moist.

Use a hand trowel to make the holes for the new plants, and upon placing each one in its hole make sure that its roots are allowed to rest easily in the soil, not cramped and bent backwards. Some gardeners, when pricking out wallflowers, break off the long, central main tap-roots to create a more fibrous root system and so help to ensure a more successful transplanting in the autumn.

The next stage is to use your fingers for firming the young biennial plants into their new homes. Only gentle pressure is required; do not become over-zealous and compact the soil around them. Keep the plants moist at all times (Fig. 5.9) to ensure they grow steadily, and be sure to remove the competitive weeds regularly by hand or hoe.

Bulbs, corms and tubers

Before embarking on the propagation by seed of these plants we should understand, at least in basic terms, what it is about a bulb that makes it such a valuable plant structure. We should also address the question in a strict botanical sense, of what the differences are between bulbs, corms, tubers, swollen rhizomes, bulbils, and the other structures that gardeners tend to group together as 'bulbous' plants.

To begin, all of these structures are swollen tissues that serve as food stores for plants whose above-ground parts die down for a period during the year, usually winter. Thus the plants have sufficient fuel to enable prompt and speedy growth when it is time, usually in the spring.

Bulbs are actually telescoped stems accompanied by a mass of swollen inner leaves that contain the stored food. There is an outer covering of thin, brown, dry scale leaves. One can see the whole as a stem if one cuts through a bulb lengthwise: the growing tip is right in the centre of the structure. Examples include narcissus, tulipa (tulips), galanthus (snowdrops) and hyacinthus (hyacinths).

A corm is similar to a bulb, but comprises the swollen basal part only of the stem. Acidanthera, gladiolus and freesia are examples of plants grown from corms. A tuber, on the other hand, may be either a swollen stem (as in a cyclamen or potato), or a swollen root (as in dahlias).

Rhizomes are also underground stems, but usually seen running horizontally just below the soil surface: the most familiar examples in the ornamental garden are those produced by bearded irises.

We have seen that there are a few botanical differences between these plants, but for the purposes of this chapter we will asssume that they are treated, in terms of seed propagation, in the same way.

The usual methods of reproducing them are by such means as offsets, bulbils and scale propagation, all of which are covered later in this book. However, raising them by seed is not difficult; indeed, many flowering bulbs set their seed and reproduce naturally in our gardens. As with annuals and biennials, if exact replicas of the parent plants are desired, only true species should be raised from seed, not the hybrid varieties. The main demerit with bulb propagation by seed is that you will be required to exercise patience, for very few types will build up a sufficiently large bulb, and adequate food reserve, to enable them to flower in less than three or four years from seed. Some of the lily species take appreciably longer!

The best germination of bulb seeds will be attained if they are sown as soon as they are ripe and have been collected from the plants – during summer or autumn. Obviously this will not be possible in the case of seed purchased from a seedsman. However, excellent results are still possible with bought seed, and in this case the best time for sowing is in late winter or early spring, similar to the time recommended for some alpines.

The procedure to follow is the same as for alpines, with the majority of seeds being raised most successfully in containers. Seedlings of certain species of bulb are tiny and difficult to handle, so it is important at sowing time not to sow too thickly and cause overcrowding in the container. For precise sowing instructions, please refer back to the section on alpines.

Aftercare

Place the pots of sown seeds in a cold, yet sheltered part of the garden, where they will receive the cold conditions which are required prior to germination. Many types will appear to do nothing for a year or more, and then they will slowly begin to show their original seed leaves. Equally, once germination takes place, it is not unusual for bulb seedlings to be left, untouched, for another year or so, in order for them to grow to a suitable size for transplanting.

Except for when the foliage dies down for its annual period of dormancy, make sure the seedlings are kept moist (but not waterlogged).

When they have reached the stage for planting out – during the dormant period in late summer or autumn for most subjects – they should be gently knocked out of their containers and separated. It is recommended that these young bulbs are planted rather more shallowly than the larger fully mature shop-bought bulbs. You now have two options. First, you may plant the young bulbs in nursery rows and leave them here for another year or so, until they reach flowering size. They will then be re-planted into their final flowering positions. Or second, plant them straight out into the garden and be patient. Whichever method you choose, it is purely a matter of personal preference.

Hardy perennials

These are also known as herbaceous or border plants, and are essentially non-woody. They are different to annuals and biennials in that they die down to the ground each autumn, overwinter at or just under the soil surface, and throw out new shoots again in spring. This is a generalization, however, for some perennials, such as helleborus (hellebores), bergenia (elephant's ear) and some of the ornamental grasses, are evergreen.

The majority of perennials, whether hardy or tender, are propagated by means of division or cuttings. However, many are unsuitable, as they have single crowns and are therefore impossible to divide, or they are particularly 'fleshy' and soft, so do not provide suitable stem material for cuttings. With these, the easy and cheap answer is to propagate by seed.

The hardy species may be sown outdoors during late spring and early summer. Choose a site that is open, sunny and well-drained. The soil should have been prepared (as explained in the first part of this chapter), and a fine tilth created. As with hardy annuals, it is generally better to sow the seed in shallow drills taken out with the corner of a draw hoe. This will facilitate easier weeding and thinning during seedling growth. Sow the seed as thinly as possible, to minimize thinning, and cover it with a very thin layer of soil.

Always keep the seedbed damp. A drying soil could do irreparable harm to a germinating seed. Moist soil will also help achieve an even germination across the bed.

Aftercare

When the seedlings are large enough to be handled they should be gently lifted with a small fork and replanted into a nursery bed for them to continue developing. Again, this should be an open, sunny and well-drained site. A day or two before transplanting takes place, a handful or two of general fertilizer could be spread evenly over the surface, and lightly raked in. Try to cause as little disturbance to the delicate seedling roots as possible during the process, as damage at this stage can cause the plant to die or grow awkwardly throughout its life. For the same reasons do not allow the roots to dry out.

Line out the seedlings in rows 30 cm (12 in) apart with a 15–20 cm (6–8 in) spacing between the plants. Follow the same procedure as described for hardy biennials.

By the second autumn after sowing, the young plants will be ready for transplanting to their flowering positions where they will, hopefully, reward you by blooming the following summer.

Vegetables

Although not 'ornamental' plants, vegetables for use in the kitchen can generally be propagated by seed, in a similar fashion to the hardy annuals already described. However, they differ in their cultivation and growth habits to the decorative garden plants so far covered, so let's look at vegetable propagation from seed a little more closely.

Most vegetable seeds will not germinate until the temperature of the soil reaches 5°C (41°F). Some require much warmer conditions, around 10°C (50°F), and these include celery, sweet corn, French and runner beans. Cucumbers and marrows like it even warmer, at 13°C (55°F). For this reason many vegetables can only really be successfully started under some form of cover – greenhouse, frame or cloche.

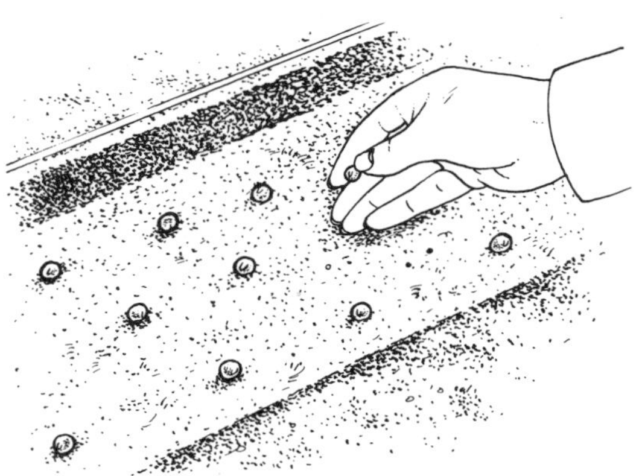

Fig. 5.10

Fig. 5.10 Large seeds like broad beans, runner beans and peas (as seen here) should be 'space sown'.

With the largest seeds, such as broad beans, marrow and sweet corn, it is best to 'space sow' (Fig.5.10). In other words, seed is sown in the drill, or in holes dug with a trowel, at the places where they are to grow. It is usual to sow two or three seeds at each 'station', thinning all but the sturdiest seedling after germination.

The sowing technique for the vast majority of seeds is exactly the same as the drill methods already described for hardy annuals, biennials and perennials. But in recent years, experiments in fluid sowing and chitting have made better the rates of germination, and this has greatly assisted both the gardener and farmer.

Chitted seed

This, in strict terms, should come under the heading of the next chapter, ie. seed sowing under cover. For chitting is an operation applied to seeds that are difficult to raise outdoors – they may also be costly to buy, so losses should be minimized. Included here are such vegetable seeds as summer lettuce, parsley (notoriously difficult to germinate outdoors), carrots and parsnips (avoids unnecessary thinning), salad onions, some outdoor tomatoes, cucumbers and melons.

Let the seed fall, evenly, onto moist blotting or tissue paper in a shallow dish or plas-

Fig. 5.11 Chit seeds by allowing them to germinate on moist blotting paper before they are 'sown'.

Fig. 5.11

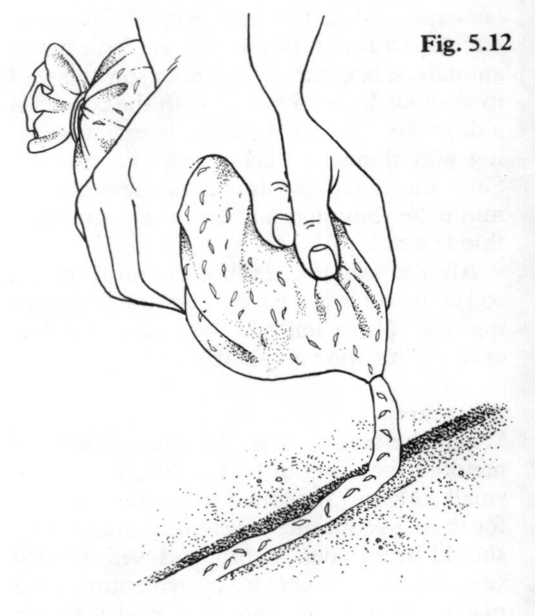

Fig. 5.12

Fig. 5.12 Chitted seed should be mixed with a fungicide-free wallpaper paste, and put into a piping bag (or washing-up bottle). This is termed 'fluid sowing'.

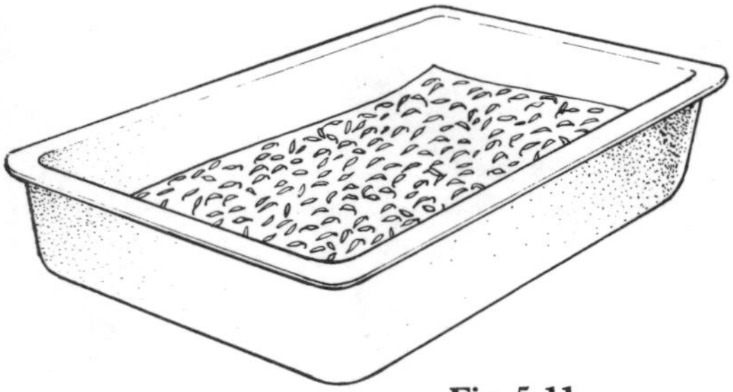

tic ice-cream container (Fig. 5.11). Put the lid on, or some other suitable covering, and place them in a warm cupboard, or somewhere where the temperature reaches 16–18°C (61–64°F). Soon the tiny radicles (roots) and plumules (shoots) will form. At this stage, and before the roots are 1 cm (½ in) in length, 'sow' them individually in rows outdoors, but make sure the soil is damp and fine, not 'lumpy'. At all costs be extremely careful, as the tiny chits are very fragile.

Some merchants are now selling these pre-germinated or chitted seeds. Packs of six seedlings are sent, by post, and upon delivery they should be pricked out or planted. Such young plants stand a really good chance of succeeding if they are attended to immediately. They are invariably more costly per plant than packeted seed, but the advantage is guaranteed survival which, depending on the gardener's performance, may represent a saving in the long-term.

Fluid sowing

This is a modern method which, if used, follows on from the chitting process. As soon as the seeds have germinated they are carefully mixed with a jelly-like material, usually a fungicide-free wallpaper paste. The glue should be mixed up to half its recommended strength. Stir the chitted seed slowly, but thoroughly, to ensure an even distribution throughout the paste; aim for around 60 or so seeds in a half pint. The sticky mixture is then poured into a container, such as a polythene bag or a well-cleaned plastic washing-up liquid bottle. In the case of the former, cut a small hole in one corner, and then sow the mixture into outdoor drills, as if icing a cake (Fig.5.12). Cover with soil in the usual way. Noticeable germination and subsequent development of chitted seeds sown in fluid is very rapid, and the young plants get off to an excellent start. But always take particular care not to allow the soil to become dry in the first few weeks.

The advantages of fluid sowing vary with different types of vegetable. Quick-maturing tomatoes, such as 'Red Alert' for example, can be sown directly outdoors in mid- to late spring (in cold areas cloches should be used to keep the worst of the weather off). Higher yields of bulb onions can be expected, and earlier salad onions. Summer fluid sowings of lettuce give more reliable results, while carrots need not be thinned, and parsnips are earlier to mature.

Seed problems

Many gardeners blame the seed when vegetables fail to grow from an ordinary outdoor sowing. Of course, it is quite possible to purchase a batch of poor seed, but the real reason is more likely to be one of the following four points:

Sowing too deeply

Carelessness in the sowing process can result in the seeds being left too deep in the soil, so that they rot before they germinate. Additionally, the seeds along the row may be covered unevenly when the soil is drawn back over them.

Sowing too early
After the cold and damp of winter, the soil needs to warm up and dry out. If the seed is sown prematurely, the conditions will cause rotting.

Poor soil preparation
If the ground is rock hard, and was not dug, say, the previous autumn – and if it has not been levelled and raked to form a tilth – then it is practically impossible for small seeds to urge their first roots into it.

Poor garden hygiene
Perhaps you have taken insufficient precautions against pests and diseases, for example, using dirty pots and old compost? See Chapter 3.

The specific requirements governing indoor and outdoor seed propagation of vegetables are covered in Chapter 21.

Woody plants

The term 'woody plants' really refers to trees, shrubs, conifers and climbers. These are all relatively expensive to purchase from a nursery or garden centre as container-grown plants – particularly conifers, because they contain resin which makes propagation by cuttings quite difficult, even though this is regarded as being simpler than seed sowing! The cost of such plants often drives the gardener to consider propagation at home, and sowing seeds is often attempted. This is particularly the case if you want a number of plants, say, for a hedge.

In the case of trees, seed is often the only practical method of propagation, and for minimal cost you can produce specimens in two or three years that are likely to be every bit as good in quality and size as those being sold in nurseries.

As with the other plant groups discussed in this chapter, the seed either may be collected from garden plants, followed by a period of cleaning and storing, or packeted seed may be purchased from garden shops or by post from seedsmen. Many types of seed will require cold treatments prior to germination, so refer to the section in Chapter 4 on stratification.

On the subject of seed collection, many readers may wish to collect conifer cones. These should be placed in boxes or other similar containers. You never quite know when the cone will release its seeds, and you don't want them scattered over a wide area! Some cones will open quite satisfactorily, but others will appear more reluctant to open. In forestry commerce, these are artificially heated to a safe temperature in a kiln. This will encourage the cone to open quickly. Amateurs will not be in a position to do this, so the best advice is to keep the cones in a warm place, such as the airing cupboard, over heating pipes or a radiator – indeed, anywhere the heat can circulate around them. Even so, the process may take several weeks.

At opening time, separate the seeds from the pods and capsules; sometimes these are reluctant to release the seeds without a struggle. If this is the case, gently crush the pods between your fingers, but do not damage the seed inside. Of course, the seed will now be mixed with the 'chaff', fragments of leaf, pod, capsule, and any other item of unwanted debris. Separate the chaff from the seed by using a sieve (or sieves), until only cleaned seeds remain. Older gardeners may well advocate the 'blowing' method, where the mixture is spread onto a sheet of newspaper, and the chaff is gently blown off. This is perhaps a quicker method, but may not be quite so efficient!

Seeds of shrubs, climbers, trees and conifers may be sown in a heated greenhouse or propagating case (see Chapter 6), or outdoors in seedbeds. We will deal with the latter here.

Prepare the bed in the normal way, finishing with the application of a general fertilizer raked into the surface to create a fine tilth. If the soil is particularly heavy, and drainage poor, one (or more) raised beds will enable you to introduce some fresh workable topsoil into the garden scheme. Even if you use ordinary garden soil, you will find it easier to work, and drainage will be improved significantly. Tree and shrub seeds can be sown during early to mid-spring, as soon as the soil is in a workable condition. Again, take out drills with the draw-hoe, ideally 10 cm (4 in) apart.

Unlike vegetables and annual seeds, these will be in the ground for many months before they show in a significant way. Heavy soils are likely to consolidate during this time, so providing an open, well-drained growing medium is very important. Once the seed has been sown thinly in the drills do not draw the normal soil back over them, but cover them with a layer of gravel or pea shingle. The benefits are threefold: the soil will not dry out so easily; it will be prevented from 'capping' in heavy rain; and the seeds will be protected from the ravages of birds.

As with most other outdoor seed propagation so far discussed, ensure you initiate a programme of watering in dry weather, and regular weeding. In very hot, prolonged sunny weather, it is not such a bad idea to erect some form of shading over the seedbed to prevent the seedlings from getting scorched.

Aftercare

Let the seedlings remain untouched until the autumn or following spring. At this stage use a garden fork to carefully lift and transplant them. Your choice now is to transplant them to a nursery bed where they can further grow and develop into more mature specimens, or you can plant them straight into the garden, in their permanent positions.

Sowing seed of woody plants is not always straightforward, and you may find that some seeds have not germinated. Do not dig up the seedbed; leave it intact. You may find some of the seed will germinate much later than the others, and by the second spring you will have a further crop of seedlings.

Lawns

Most books on increasing plants seem to ignore the act of sowing a lawn, yet this is as much a form of plant propagation as any we have so far covered. Creating a new lawn is a thrilling first step in the layout and design of a garden. There are a number of different types of grass seed mixture, designed for different uses. Some people want fine quality lawns, so they should opt for a blend of fine-leaf fescue and bent grasses which produce a slow-growing, thick sward, but which will not take hard wear. Others wish for their lawn to be used as play areas for children. In this case, choose a hard-wearing mixture of dwarf ryegrasses, fescues and bents. Finally, orchards and shaded lawn areas require a seed mixture containing rough-stalked meadow grasses.

The best time to sow grass seed is early autumn, when the ground is still warm and the chance of the soil drying out is less of a problem. Alternatively, go for mid-spring, but if the summer is dry, regular watering may become a problem.

The first thing to remember is that grass, like any other plant, needs a constant supply of food and moisture, and a reasonable depth of soil. Thorough preparation of the soil is therefore a must! Follow the same guidelines as before, but this time you should remove from the soil not only perennial weeds, but also large stones, bits of concrete, brickbats, buried pieces of wood and any other debris left by builders.

The digging need not be tackled months in advance as with ornamental and vegetable seeds; instead it can be done about 10 days prior to sowing. Combine the digging with levelling and grading. The lawn may gently undulate and rise to create interest, but it must present a smooth surface to the mower, or the blades will scalp the grass.

Choose a dry day to rake the soil level, roll it flat, apply a fertilizer and rake it in. The fertilizer should be rich in phosphates, to promote root growth.

The sowing operation itself is not difficult, but is completely different to all that has gone before. Again choose a calm, dry day. Have at the ready enough seed to allow an application rate of 28–45 g (1–1½ oz) per square metre (yard). Sow half the amount of seed in an up and down direction across the plot, then repeat the sowing with the rest of the seed, sowing from side to side.

Alternatively, you can mark out the area into metre or yard squares, using canes or string (Fig. 5.13). Sow the correct amount

Fig. 5.13 Before sowing grass seed, mark out the area into metre or yard squares, using canes or string.

Fig. 5.14 Sow the correct amount of seed to each square, until the whole area is seeded.

Fig 5.15 Remove the canes and lightly rake the ground over, covering the majority of seed.

of seed to each square, until the whole area is seeded (Fig. 5.14).

The next step is to rake in the seeds lightly, just covering them (Fig. 5.15), and to protect the seed from the birds, cover the area with light netting or brushwood. Now carefully water the whole area, either with a rosed watering can (if the area is small), or with a fine spray from a hose. However, in either case try to avoid treading on the freshly sown seed.

Aftercare

Give the grass a light rolling once it is 5–8 cm (2–3 in) high. When it is 10–15 cm (4–6 in) high, give it a first cut by lightly trimming off the top couple of centimetres or so, to encourage the grass to spread and thicken. Although the young grass may look healthy and durable, a new lawn should be nurtured and not subjected to heavy foot traffic for the first year. If seedling weeds appear in the first few months, mowing will help to kill the annual types. As a result of recent legislation in the UK concerning pesticides, there is now no suitable weed-killer for use on new lawns. More than ever, good soil preparation and the complete eradication of perennial weeds before sowing is of paramount importance.

CHAPTER 6

Seed Sowing Under Cover

We have so far looked at the wide-ranging subject of seed propagation outdoors. There is, in addition, a whole new area of opportunity if you are able to provide a degree of shelter, warmth and protection to your seeds against the elements. This can be achieved by using such devices as frames, cloches and tunnels and, of course, the greenhouse and your home. In Chapter 2 we examined the many specific types and models available; we can now discuss them in greater detail.

Seed of hardy plants may be sown outdoors, or under cover of any of the devices mentioned above – the gardener has a choice of location for the seed sowing. Tender plants, however, are almost always sown indoors, and with some artificial heating (and possibly lighting) to speed things along. The sort of plants included here are house and greenhouse plants, cacti and succulents and orchids, tender perennials and shrubs. Summer bedding plants, or 'half-hardy' annuals, should also be raised under cover, for planting outside once they are large enough, and the weather is warm enough.

Vegetables, too, may be sown under glass to provide crops extra early in the season, and possibly for a higher yield. The seedlings should be planted outside once the weather has warmed up or, alternatively, they may be of varieties that are particularly recommended for greenhouse growing. Marrows, cucumbers, tomatoes, lettuces, cabbages, cauliflowers, onions and leeks are just some of the vegetables that can be treated this way.

Many true greenhouse plants, such as melons, cinerarias and gloxinias, for example, must always be propagated under glass.

There is a very fine distinction between raising seeds under *cover* and under *glass*. Many seeds may be germinated in a warm room in the house. However as soon as the first seed leaves are visible, the container should be moved to a well-lit window. Even here it is not always sufficiently light for the developing seedlings, and through the actions of phototropism (the tendency of plants to grow toward a light source), you may need to turn the container 180° every day.

The overriding advantage of seed propagation under glass is that the perils of the weather and some pests are considerably lessened. Of course, there are a number of glasshouse pests to watch for, but if strict greenhouse hygiene is observed, as outlined in Chapter 3, the effect of these should be minimized. The gardener can also exercise far greater control over the seeds' environment which, in turn, results in more rapid germination and speedier development of the seedlings. Thus, even with completely hardy plants, there are certain advantages of sowing under cover. Indeed, the seed of some hardy shrubs, heathers and rhododendrons, for example, are so fine and small that sowing them outdoors is most certainly doomed to failure.

Let's continue to look at groups of plants, to see how they should be treated.

Tender annuals and perennials

These are easily raised from seeds in a heated greenhouse. The main sowing period is early to mid-spring, although some subjects which require a longer growing time (like pelargoniums) can be sown during late winter.

Standard-sized plastic seedtrays are, perhaps, the most suitable of containers. They allow the seed to be evenly spread across the compost, and they are easily positioned on a window sill or greenhouse bench without wasting space. They are also shallow when compared to standard flowerpots, which is acceptable as seeds do not require a great depth of compost in their early stages. Pans or half-pots are available, however, and these are ideal for small quantities of seed.

The decision concerning the best type of compost to use for seed sowing is made entirely on the preference of the gardener.

There are dozens of proprietary brands available – loam-based and peat-based – and, of course, you can always make your own. It is worth noting, however, that with seed propagation the soilless or peat-based kinds often produce better results. See Chapter 2 for further details.

Whichever compost and container you opt for, the sowing technique itself is the same, and relatively straightforward. The following sequence should be adopted:

1. Fill the container
Heap the compost well up in the centre of the container, then push it into the corners and spread the heap out until the whole tray or pot is filled level to its rim. Strike off any excess compost by running a piece of wood along the edges.

2. Firm the compost
Use a wooden presser (Fig. 6.1) to firm the compost lightly so that you end up with an even, level surface just below the rim. Make sure that there are no bumps or hollows, especially in the corners.

Fig. 6.1

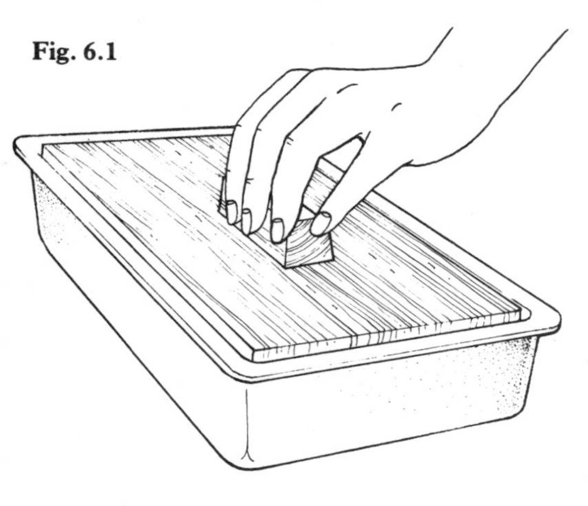

3. Water the container
The compost should now be watered, using a fine rose on the watering can. The rose should be fitted so that the holes face upwards; the water then fountains up and out, giving a much gentler spray. If the rose does not fit the spout properly, it either dribbles irritatingly, or falls off disastrously! Either problem can ruin your level compost surface. Allow around 30 minutes for the containers to drain.

Instead of using plain water, you could use a solution of a preparation formulated to prevent damping-off diseases (see Chapter 3). It is worth guarding against these right from the start. A garden shop will help you find a suitable proprietary product.

Fig. 6.2

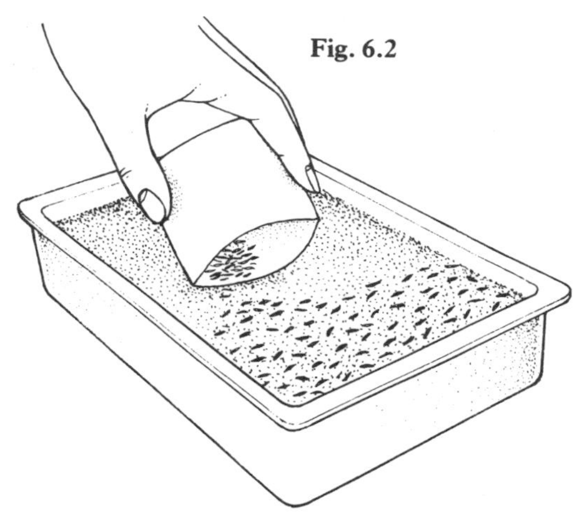

4. Sow the seed
If you are sowing packeted seed, tap the packet so that the seed falls to the bottom before you tear off the top. Pinch the lower half of the packet so that the open end is gaping, then shake it gently so that the seeds fall out (Fig 6.2). They should thinly cover the whole surface of the compost. An alternative is to shake all the seeds into the palm of one hand and then tap the edge of it with your other hand to dislodge the seeds and allow them to fall onto the compost.

It is most important to sow the seed thinly. Always remember that each seed, with luck, will develop into a healthy seedling, and that to do this it needs space around it.

Seed sizes vary considerably. Some, such as begonias and lobelia, are so small that they are literally as fine as dust. Obviously this means that sowing them 'thinly' is practically impossible. So, the best thing to do is mix them with a 'carrier', like fine dry silver sand. Put a layer of the sand in the bottom of a clean jar, and tip the seeds into it

Fig. 6.1 Sowing seed in trays: the compost can be levelled by using a wooden presser.

Fig. 6.2 Sow the seed thinly over the surface of the compost.

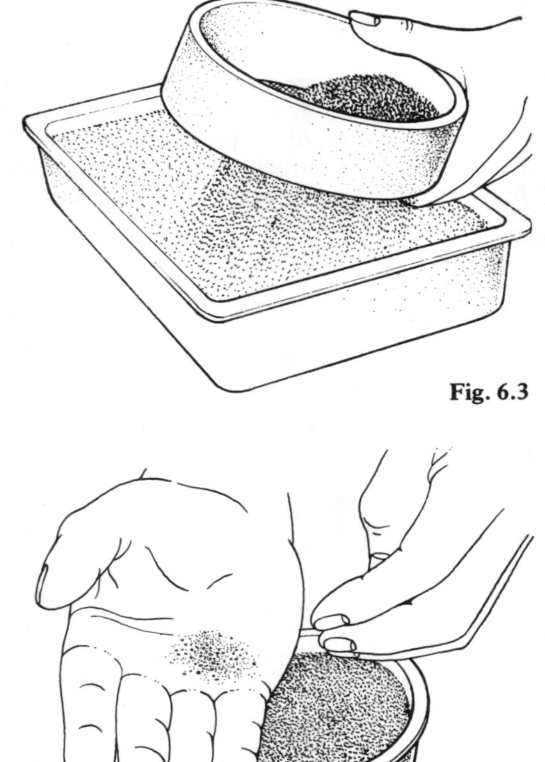

Fig 6.3 Lightly cover the seed with fine, sieved compost.

Fig. 6.3

Fig. 6.4 Small quantities of seed may be sown into a 7 cm (3 in) pot.

Fig. 6.4

(choosing a draught-free place for this!). Stir the mixture with a stick to distribute the seed as evenly as possible. Now spread the mixture, again evenly, over the tray of compost.

Large seeds are much easier to deal with. They have the advantage that they can be spaced by hand in the trays, so they will not require pricking out until much later. Some large seeds are best sown straight into pots, as they can subseqently be planted without too much disturbance to the roots.

5. Cover the seed

Most seeds of standard size must now be covered with a fine layer of compost to prevent them from drying out (Fig. 6.3) and to anchor them in place. Use an old kitchen sieve to obtain an even covering. Do not 'bury' the seeds too deeply; it is safest to stop sieving as soon as the last seed disappears from view.

The finest seeds – those that are mixed with a carrier – should not be covered with sieved compost (Fig. 6.4). The larger, individually pot-sown seeds can actually be pushed into the compost, and for these you can just use your fingers to draw compost over them.

Finally, a sheet of clean glass should be placed over the sown container to keep the humidity and temperature high: for extra warmth, a layer of newspaper goes over the top of the glass. Place the covered trays on the greenhouse bench, room window sill or frame floor. Different annuals and perennials have varying temperature requirements, but 18–21°C (64–70°F) is a good average.

House and greenhouse plants

The main period for sowing these tender plants is from late winter to early spring, although certain subjects are raised at other times of year to ensure they flower in the right season. In addition, many may be sown either before or after their recommended sowing time, to provide a succession of flowers.

In order to germinate, many house plants need a minimum temperature of 18°C (64°F), but some need as much as 27°C (80°F) or even 29°C (85°F). The gardener should not be too quickly discouraged if nothing seems to be happening. Large seeds, in particular, sometimes take over two months to germinate.

Exactly the same sowing procedure should be followed for the majority of these pot plants. Specific groups of plants do, however, require slightly different conditions. Cacti and succulents, for example, may be sown at any time of year if heat is available; otherwise sowing should take place between early spring and midsummer. On the whole, it is best to aim for a germinating temperature of around 21°C (70°F).

Loam-based composts are much better than peat-based ones where succulents are concerned. The John Innes seed compost is ideal. Too much lime can be harmful to most of these plants, so a slightly acid compost is preferred.

Some succulents, like opuntias and cerei grow rapidly and may reach a height of 15 cm (6 in) in a year. For the most part, five or six years will elapse before these plants are mature, although many will flower when younger.

Orchids are something of an oddity. Growing them from seed was once regarded as practically impossible – an orchid pod contains up to 4 million seeds, finer than dust! In Nature they may be carried by wind for thousands of miles, until a few settle and are lucky enough to find a root fungus that provides ideal germinating conditions. In the 1920s, it was discovered that orchid seeds could be germinated in an artificial compound called Knudson's Formula C, a sterile growing medium containing trace elements.

On the whole, orchids are not easy to grow from seed, and there are other, more promising methods of propagation, such as division of the pseudobulbs. It is, perhaps, best to seek specialist advice from an orchid book, or expert.

House plants, generally, are slightly more receptive to light. This is not a vital consideration prior to germination, but the use of fluorescent lighting after germination is useful. It provides bright, all-round light, without drying out the compost (as would sunlight). Tiny seedlings, such as those from the gesneriad family (gloxinias, African violets, and so on), benefit particularly. The ideal situation is to keep seedlings 25–30 cm (10–12 in) below the fluorescent lights for 15 hours per day.

If no artificial lighting is provided, your seeds and seedlings may still come to no harm. But with it, you will be providing them with a much better environment for developing, and so the resultant plants will be superior. Inadequate lighting will cause seedlings to grow spindly, with a danger of them toppling over and rotting, while light from only one side will have a similar effect or, at least, make for unbalanced growth.

Vegetables

There are really only six types of vegetable that are harvested in the greenhouse: aubergine, cucumber, melon, sweet pepper, tomato and winter lettuce. Some are not always economically viable on a small scale, and it is only really enthusiastic vegetable gardeners that would attempt more than one of these. However, there are rather more types of vegetable that may be raised from seed in the greenhouse, and transplanted outside when large enough. These include Brussels sprouts, early cabbage, cauliflower, leeks, parsley, sweet corn, French, dwarf, broad and runner beans, and lettuce.

The principles for sowing these are exactly the same as for half-hardy annuals. The only diffference is that at the time of planting outdoors, there is an even greater desire not to cause disturbance to the roots, and therefore it is wise to sow vegetable seeds, (*a*) as thinly as possible, and (*b*) wherever possible, individually in their own little container. Larger seeds, such as sweet corn, cucumber and marrow, as well as various vegetable seeds that are sold as 'pelleted' are best sown individually in soil or peat blocks, or small peat pots. However, due to the demands of greenhouse space being made on the gardener at the time of year when most vegetables (and other plants) are being sown, it is usually impossible to grow all seed in their own pots.

Do remember that it is generally more economic to sow only the few seeds of each kind that are sufficient for the average garden. Resist the temptation to over-sow. Twenty or thirty lettuce, tomato and aubergine and similar plants may be sown and germinated in a 10–13 cm (4–5 in) pot.

Cold frames are also useful for raising seeds, particularly early vegetables like leeks, onions and cauliflowers. Whether you sow seeds in trays of compost, and then place them in the frame, or whether you sow straight into the soil underneath the frame, is a matter of personal choice. In the case of the latter, however, the soil will need to be thoroughly cultivated as described in Chapter 5. In frames that are not artificially heated in any way, it is not wise to sow too early, as the cold weather is likely to cause much damage, or even prevent germination. A later sowing in such frames usually means better and quicker germination and more rapid development of the seedlings.

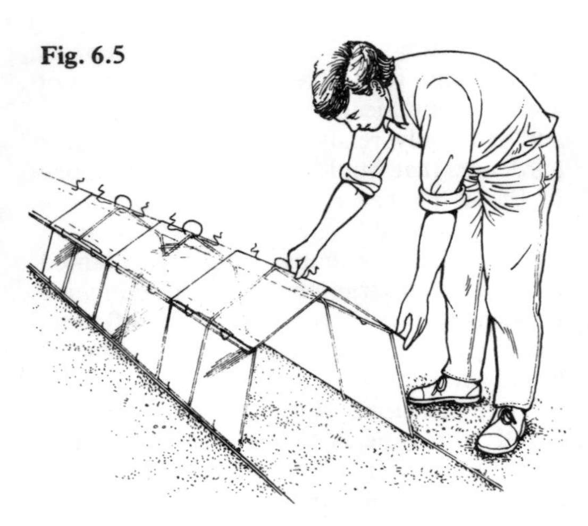

Fig. 6.5
Vegetables can be harvested earlier than normal if the seed is sown outdoors in drills, and then covered with a cloche to create a mini-greenhouse.

Seed sown under frames or cloches may be drilled in rows (Fig. 6.5), as if they were sown straight outside, or some may be scattered thinly, along the lines of a method known as 'broadcast'. This latter technique is suited to lettuce, carrots and radish.

The tables that follow show at a glance the yearly calendar for food crops that may be raised under cover. Remember that the reasons for starting vegetables under cover are to produce crops earlier than the outside-raised varieties. Therefore, the table will reflect earlier than normal harvesting times.

TABLE 1: YEARLY CALENDAR FOR FOOD CROPS GROWN INSIDE A GREENHOUSE

Crop	*Sow in greenhouse*	*Harvest period*
Aubergine	Mid- to late winter	Mid-summer to mid-autumn
Bean, French	Early spring	Mid- to late summer
Bean, runner	Early spring	Mid-summer to mid-autumn
Brussels sprouts	Late winter	Early to late autumn
Celery	Late winter to early spring	Late summer to late autumn
Cucumber	Late winter to early spring	Early summer to mid-autumn
Lettuce	In succession from early autumn to early spring, depending on the variety	Mid-spring to late summer
Marrow	Early spring	Early summer to mid-autumn
Melon	Early spring	Early summer to mid-autumn
Sweet corn	Early spring	Mid-summer to early autumn
Sweet pepper	Late winter	Mid-summer to mid-autumn
Tomato	In succession from early winter to early spring, depending on the variety	Early summer to mid-autumn

TABLE 2: YEARLY CALENDAR FOR FOOD CROPS GROWN UNDER CLOCHE

Crop	*Sow under cloche*	*Harvest period*
Bean, broad	Late autumn to late winter	Early to late summer
Bean, French	In succession from early to late spring, depending on the variety	Early to late summer
Bean, runner	In succession from early to late spring, depending on the variety	Early summer to mid-autumn
Brussels sprouts	Early spring	Early to late autumn
Carrot	Mid-winter to early spring	Late spring to mid-summer
Cauliflower	Early to mid-autumn or mid- to late winter, depending on the variety	Early to late summer
Lettuce	As for greenhouse sowing	
Peas	Mid- to late winter	Late spring to mid-summer
Radish	Mid- to late winter	Early to late spring
Turnip	Early to mid-spring	Late spring to early summer

Pricking out

We have so far seen how seed should be sown into pots, pans and trays. With the right conditions they will germinate well and grow into sturdy seedlings. But what then? Let's take each stage separately.

Firstly, remove the covering of paper and glass as soon as the first seedlings touch it, or when the majority of the seeds have germinated. The first 'leaves' to be seen are usually a pair (in the case of dicotyledonous plants; or singly in monocotyledonous), and they are somewhat coarser than the usual leaves for the species. The next pair of leaves (called 'true' leaves) have the shapes of the young plant.

Once the seedlings have reached the stage where they can be handled easily, they are ready for transplanting or, in propagation terms, pricking out or pricking off. Experiments have shown that the younger a seedling is, the less it will suffer from the move, for the pricking out operation does cause a slight check to growth. Thus, the fewer moves a seedling has, the better. The main purpose of pricking out is to give the seedlings room to grow; by leaving them scattered randomly where the seeds fell, many will not survive, or will grow spindly.

Standard plastic seedtrays (or wooden if you can still find them) are suitable containers for this purpose. Developing seedlings appreciate a slightly greater depth of compost, so if you can use slightly deeper trays, say 5 cm (2 in) deep, so much the better. With half-hardy annuals, tender perennials and any occasions where a large quantity of seed are to be pricked out, it is not unreasonable to set out 40 to 54 seedlings per tray. This can be achieved with

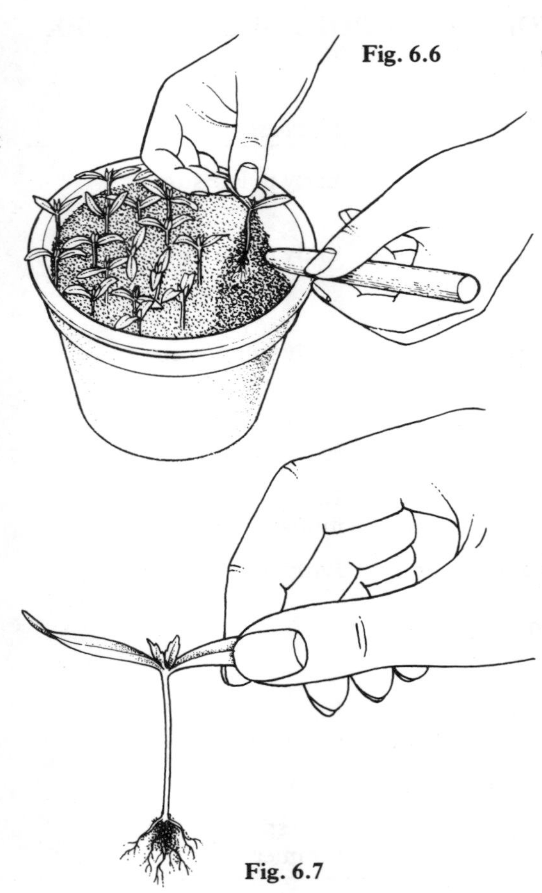

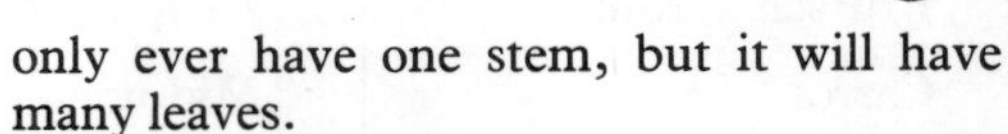

Fig. 6.6 When pricking out, use a pencil or specially-made dibber to loosen the compost around the seedling's roots.

Fig. 6.7 Never hold a seedling by its stem, as this is very soft and easily damaged. Always hold it by one of the leaves.

Fig. 6.8 Very lightly firm the seedling into its new position, using your fingers and the pencil or dibber.

five rows of eight seedlings, to six rows of nine seedlings. The rows should be kept straight, and the spacing even. If you are growing flowering pot plants, or small quantities of seed, the seedlings could be transplanted to individual small pots.

Fill the tray with compost in much the same way as you did for the seed sowing, but use a potting compost as opposed to one designed purely for sowing. Now dib holes in the compost at the places where the seedlings will be planted. The holes should be deep enough to allow the roots of the seedlings to dangle straight down. *Handle the seedlings very carefully*. Loosen the compost underneath them with a pencil or a specially made dibber, and pull the young plants apart by grasping their seed leaves (Fig. 6.6). Never hold them by the stems as these are very soft and easily damaged (Fig. 6.7). Always remember: a seedling has and will only ever have one stem, but it will have many leaves.

Place each seedling in a new hole, so that its roots are in contact with the compost, and are not bent backwards – the root should not be restricted in any way. Very lightly, firm it in with your fingers (Fig. 6.8). The pair of seed leaves should be almost resting on the compost surface.

After pricking out, water the trays with a can and fine rose. Again, it is a good idea to water with a solution of fungicide to prevent damping-off diseases, which may still attack the seedlings. Shade the young plants from direct sun for a few days; you can lay a single sheet of newspaper or tissue paper on top of them. This will help them to recover quickly from the move. Remember, the weather is possibly getting warmer at this stage, so check the moisture requirements of your trays regularly.

The seedlings are now grown on, either sited on the greenhouse bench, house window sill or in a garden frame. Some plants, such as half-hardy annuals, may be planted outside directly from the trays after being slowly acclimatized to the outside temperatures. This 'hardening' is made easier by using a garden frame, which serves as an intermediate stage between the heated greenhouse and open garden.

After pricking out greenhouse and house plant seedlings, the next stage is usually to transfer them from the trays singly into pots. If they had been pricked straight out into small pots, they will eventually need repotting into larger ones.

CHAPTER 7

Raising Plants from Spores

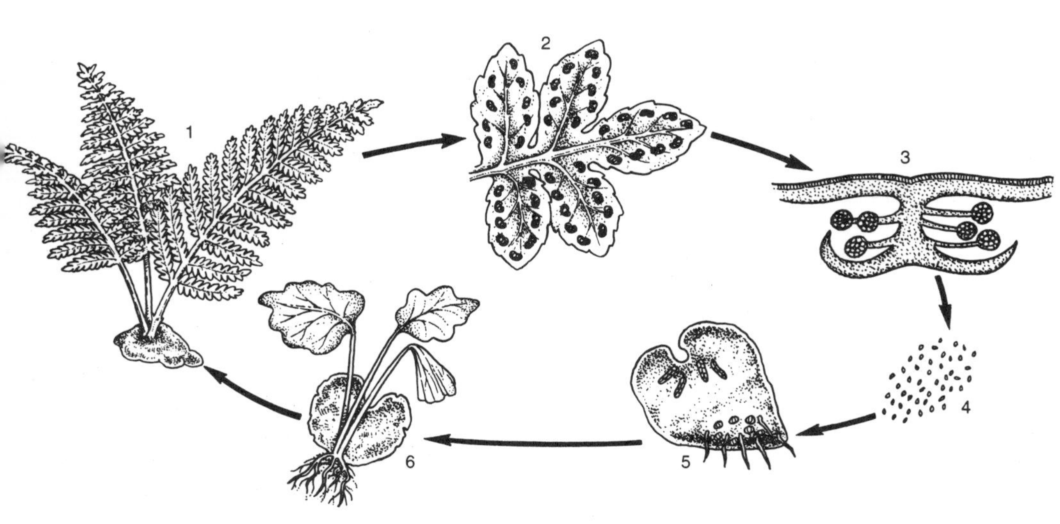

Fig. 7.1

Fig. 7.1 The life cycle of a spore-sown fern. The mature frond (*1*), carries sporangia on its underside (*2*). A close-up of a spore case (*3*), which is capable of emitting tiny individual spores (*4*). On germination, each spore develops into a prothallus (*5*), from which a young fern plant will eventually develop (*6*).

It would be wrong, in a book about propagation, to ignore the subject of plant spores and how they relate to plant reproduction. Basically, spores are the reproductive units of such plants as lichens, mosses, fungi and ferns, and they perform a similar function to that of seeds in other higher plants. 'Higher', in this context, refers to plants that are higher up the evolutionary scale, for it is now known that the earliest forms of vegetation on earth were, in fact, spore-bearing plants.

Ferns

Ferns are the most likely of all spore plants to be cultivated by amateur gardeners, and for those enthusiastic enough to propagate them, there is much satisfaction to be gained. Curiously, ferns are 'neutral' plants in that they do not have sexual organs, but instead develop minute reproductive bodies – the spores – which, unlike seeds, do not immediately produce a new plant once germinated (Fig. 7.1). This is explained more fully later.

Fern spores are very small. Indeed, they are virtually invisible to the naked eye, and they are produced in vast quantities by the parent plants. Each year a single fern plant may produce many millions of spores which are carried far and wide by the wind. As would be expected, only a tiny proportion succeed in germinating in the wild; were this not the case it is safe to assume the vast majority of the earth's land surface would be fern-covered! An enormous percentage of the wild spores land in hostile places, where germination is impossible. And of those that do manage to germinate, most are destroyed as a result of attacks by insects, fungi or draughts.

However, due to the vast number of spores dispersed, many ferns are extremely successful at natural reproduction. Having evolved over some three hundred million years, today there are thought to be some 10 000 different species spread throughout the world. Most of these are natural to semi-tropical countries, which means they make most acceptable house plants in the colder climates.

Spores are invariably borne on the undersides of the fern's leaves (usually called 'fronds') in specially developed spore cases known as 'sporangia'. Clusters of sporangia are called 'sori'. To the novice it must seem as though a whole new language and vocabulary should be learnt, but of course

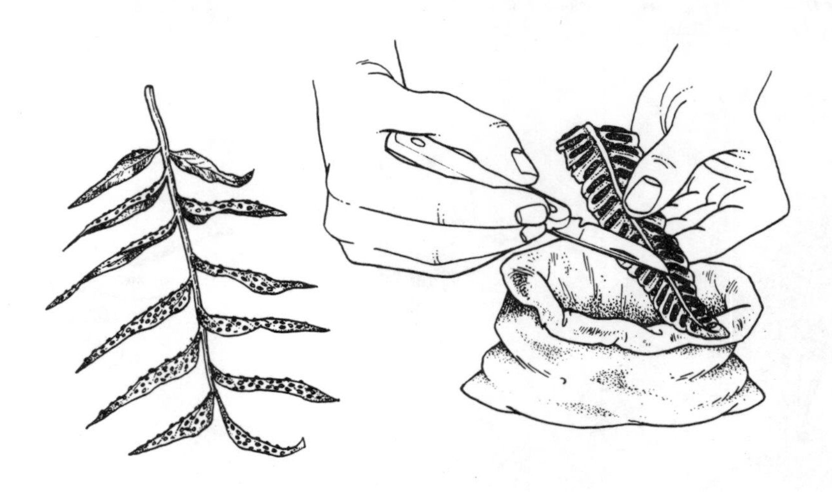

Fig. 7.2 The sporangia may be clearly seen on the underside of many fern fronds. Carefully place the chosen frond into a paper bag.

this knowledge is imparted here merely to explain what happens in nature. The actual technique of propagation by spores is not difficult.

Collecting fern spores

Most spores are easily collected and saved by the amateur, using the same principles as in seed collecting. When the sporangia begin to turn brown, usually from mid-summer onwards, they are 'ripe'. If there is any doubt, it is a good idea to make quite sure by examining those near the base of the frond with a hand lens. If they are showing signs of opening, then cut the frond off the plant and dry it by placing it in a paper bag (Fig. 7.2). Ragged-looking sporangia are likely to have already opened and therefore dispersed the spores. Fasten the bag at the top to prevent the spores from escaping. Do remember that as the spores are so light, they can easily be dispersed, both in the bag, and while they are still on the plant!

Method for sowing

Best results are achieved by sowing freshly ripened spores: Bear in mind that the spores from the osmunda genus are viable for only a few days, and some less common genera for only a matter of hours. The spores of most ferns, however, remain viable for much longer, sometimes years, but it is better to err on the side of caution and sow as quickly as possible. As with normal seed sowing, ordinary seed trays, pans and pots are suitable containers. However, they should be scrupulously clean, and preferably sterilized. In addition, the water used for damping the compost and for supplying the growing plants should be boiled, or instead use pure rainwater. These precautions are necessary if you are to avoid the growth of algae, mosses, liverworts and fungi which thrive in the same conditions as the germinated spores.

The best compost is one consisting of equal parts by volume of sphagnum moss peat, leaf-mould, sterilized loam and sand. Alternatively, go for John Innes seed compost, and add half as much (by bulk) moss peat.

Cover the base of the container with a layer of rough peat and fill up with the seed

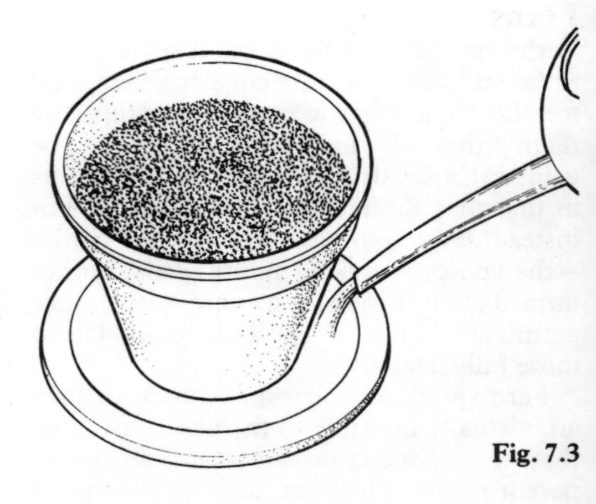

Fig. 7.3 A pot containing freshly sown spores should be watered from underneath.

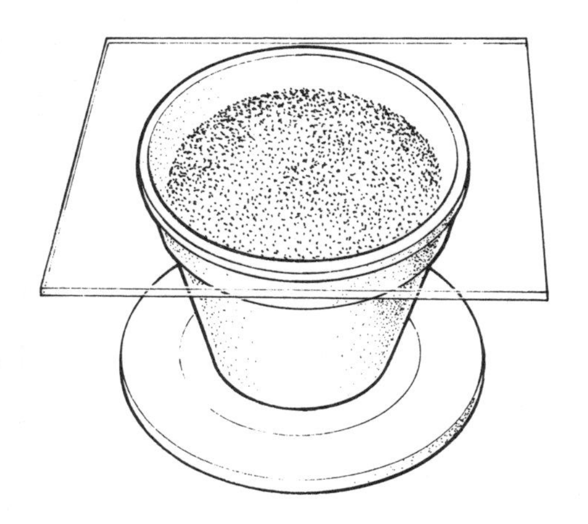

Fig. 7.4

compost to within 1 cm (½ in) from the rim. Now soak the whole container with water; some gardeners use a watered solution of permanganate of potash which helps to prevent disease. Allow it to drain for about 30 minutes.

Now everything is ready for the actual sowing of the spores. Choose a place that is away from draughts, yet in good light. Sow them thinly and evenly over the compost, and gently press them down into the surface with the base of a sterilized pot. After this, stand the container in a couple of centimetres of water for a few minutes, to make sure the compost is thoroughly moist (Fig. 7.3). Then cover it with a clean pane of glass, making sure to remove and wipe it of condensation each day (Fig. 7.4).

Germination

Leave the sown spores in a place receiving moderate light, but always ensure protection from direct sunlight. Tender, or greenhouse ferns, require a temperature of 21°C (70°F) to germinate. As a rule, this will be after six weeks or so, and it will appear as though a thin green slime is covering the compost. This is made from large numbers of sporlings, each being a small independent, scale-like plant called a 'prothallus'. Male and female organs are produced on the underside of this prothallus, and from the union of male and female cells a new fern develops.

The first young fern fronds may appear four weeks or so after the prothallus has matured. If the prothalli come up too thickly to form into well balanced little plants, they will have to be pricked out into new containers, in little bunches 1 cm (½ in) or so apart, and kept in the same conditions until they mature. But until the prothalli mature, they have no roots to speak of and it is sometimes difficult to get them to stand up! This is why watering should only ever be done from below.

When the tiny fern plants are about 2.5 cm (1 in) in height they should be pricked out and spaced at around 5 cm (2 in) apart. Keep them in warm, fairly humid conditions – a heated propagating case would be ideal. When they are large enough to fill their allotted spaces, they are at a suitable stage for potting up or being planted out (in the case of hardy ferns). The replanting operation is a delicate one, and should be done with great care.

Potting up

Once the pot-grown baby ferns have filled their containers with roots, they should be moved to a larger pot. Do not be tempted to leave them in their first pot too long, for a fern starved of food and oxygen in its early stages rarely turns into a good plant. Curiously, this does not apply to mature ferns, when over-potting can sometimes cause suffering. Before and after potting, at any stage in the fern's life, make sure it is well watered.

Fig. 7.4 Create a moist environment for the spores to germinate, by placing a pane of glass over the pot.

CHAPTER 8

Stem Cuttings

Propagating by cuttings, that is, by separating portions of a plant from the parent and getting them to root and so produce new plants, is just as common a practice as sowing seeds and, in its own way, just as successful. Whereas seed propagation is 'sexual', so raising plants from cuttings is termed 'asexual' or 'vegetative'.

Most cuttings are stem cuttings but, as we shall see later, roots and leaves may be used in their own ways to increase plants. Stems are the most popular of plant parts to be used as propagation material, probably because so many plants lend themselves to this particular method. You may choose from alpines, climbers, conifers, some fruits, greenhouse and house plants, perennials, shrubs and trees. Stem cuttings are quick, easy and inexpensive to take.

Selecting the parent

One of the greatest advantages of this form of propagation is that the rooted cutting will be identical in all respects to the original, or parent plant. This is especially important if you are raising hybrid plants and cultivars, as these cannot be increased true to type by sowing seed. Largely because of this, take great care when selecting the parent. An unhealthy, malformed specimen is unlikely to produce a superb example of the species from a cutting. Only the healthiest, most sturdy plants that are available should be selected.

Preparation

Practically all stem cuttings are prepared in the same way before they are 'planted' and left to form roots in their own time. They should all be severed from the parent plant with a razorblade or sharp knife – not your fingernails! It is crucial for the blade to be sharp, as a blunt knife or your nails will more than likely bruise the plant's inner tissue, which will subsequently rot and cause the cutting to fail.

Discoveries and realizations are being made all the time, and today plants that were once considered impossible or difficult to propagate by cuttings are being raised most successfully by this method, provided they are given the right environment and conditions for rooting.

There are basically three types of stem cutting, and a number of slight variants, all of which are covered in this chapter. The three of immediate concern are: softwood, semi-ripe and hardwood cuttings. Which type you use will depend on the plant in question, the time of year, and the propagation material available to you. It is worth remembering that cuttings from young plants, particularly with trees and shrubs, for some reason tend to produce roots more easily than if they were taken from older specimens. For specific plant details, refer to Part II of this book.

Softwood cuttings

These are usually taken in late spring and early summer from young shoots before they start to become ripe, or woody and hard. Cuttings from tender greenhouse plants may, however, be taken much earlier, in fact from mid-winter onwards, provided suitable shoots are available. Beyond the early part of summer, the stems are starting to ripen and although cuttings may still be taken, they will be treated slightly different, and termed 'semi-ripe' cuttings.

Softwood cuttings are soft and fleshy, and easily lose moisture, so they need to be kept in a constantly close, damp atmosphere. Generally, too, they need at least a little bottom heat to encourage development of the roots. A propagating frame with the temperature controlled at 16–18°C (61–64°F) is ideal for most. Whilst soft cuttings are commonly used commercially for certain shrubs, where they can be rooted in a large, professional mist unit, they are of limited use to the home gardener, except in the case of easy-to-root plants such as philadelphus, fuchsias and hebes.

In addition to these, gardeners may take softwood cuttings of alpines, some hardy and tender perennials, climbers, and greenhouse and house plants. The best material for use as cuttings is generally from the soft, new sideshoots. Collect this material early in the morning, that is before the sun becomes too hot and the plants start to lose water from their leaves. You need the stems to be full of water and sturdy, not limp and drooping (Fig. 8.1). Although the finished cutting is going to be around 7.5 cm (3 in) in length, when gathering the material from the parent plants make sure the pieces of stem you collect are around double this length. Immediately put them into a clean polythene bag, and either seal or tie it. By doing both of these things, you will help to prevent the cutting from wilting.

If you are unable to trim and plant the cuttings straight away, leave the plastic bag in a cool shady place, but for no more than a couple of hours. With the exception of species, such as the cactus group and zonal pelargoniums, which can be left to dry a little before planting, the ideal of course is to prepare the cuttings straight after they have been removed from the parent. Again, choose a cool, shady place under cover, or indoors, to make the final adjustments to the cutting. If you undertake the delicate trimming operation in full sun, or a drying wind, you may be causing untold damage to the plant tissue, and rooting will fail.

To prepare the cuttings, take your sharp knife or razorblade, and cut each selected shoot below a node, or leaf joint (Fig. 8.2). The tips of the shoots should be left intact. As already stated, the majority of cuttings will be in the region of 7.5 cm (3 in) long, but with such small plants as alpines, the cuttings could well be a mere 2.5–4 cm (1–1½ in) in length. In either case, the lower third, or even half, of the cuttings should be stripped of foliage by cutting them or pulling them off. Be careful when doing this, making sure you do not take a strip of bark or stem skin with the leaf. Most cuttings should be left with two to four leaves. If you do not remove the leaves, when the cuttings are inserted in the compost, the leaves will become submerged and will rot.

Cuttings of a great many plants root more quickly and develop a stronger root system if they are treated with a hormone rooting powder prior to insertion. With the most difficult plants to root from cuttings, such as camellias, hollies and rhododendrons, using a powder can be the deciding factor between success and failure.

The lower 6 mm (¼ in) of the cutting should be dipped into the powder – which should be chosen specially for softwood cuttings. It is not necessary, as some gardeners think, to dip the bases of the cuttings in water before dipping them into the powder; there is always sufficient moisture from the

Fig. 8.1

Fig. 8.1 Always choose a strong, healthy shoot when selecting material for softwood cuttings.

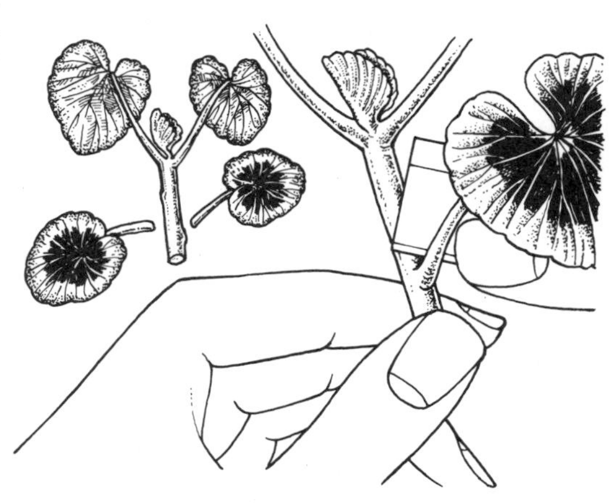

Fig. 8.2

Fig. 8.2 Each cutting should be trimmed immediately below a node, and the lower leaves removed.

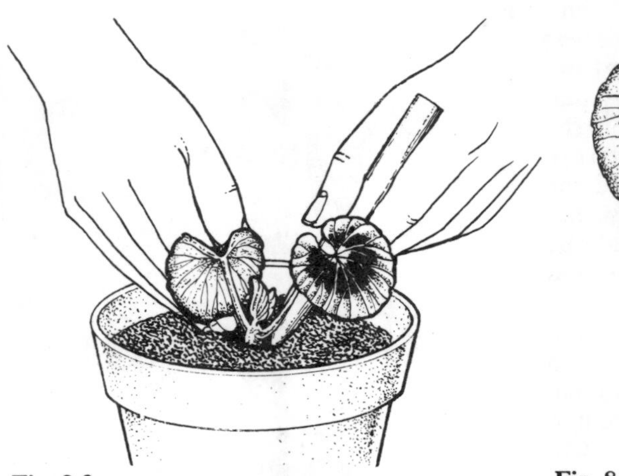

Fig. 8.3 When 'planting' the cutting, make sure that its base touches the bottom of the hole, and that it is firmed in well.

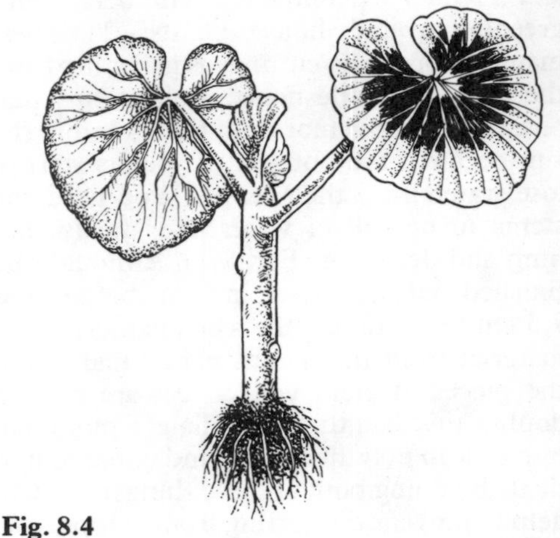

Fig. 8.4 A well-balanced and healthy rooted cutting.

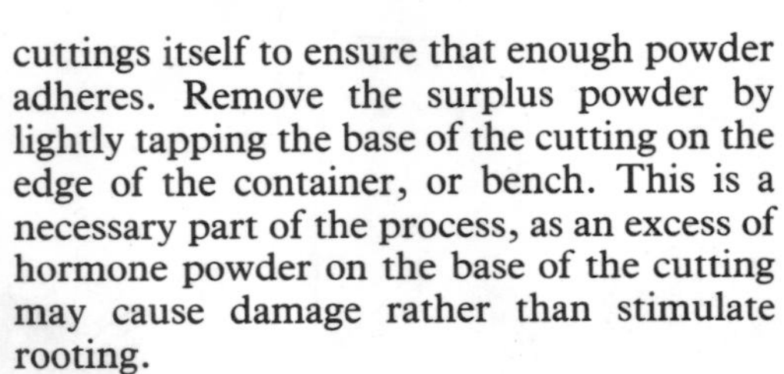

cuttings itself to ensure that enough powder adheres. Remove the surplus powder by lightly tapping the base of the cutting on the edge of the container, or bench. This is a necessary part of the process, as an excess of hormone powder on the base of the cutting may cause damage rather than stimulate rooting.

An open, well-drained compost of equal parts moist sphagnum peat and coarse horticultural sand, or grit, suits softwood cuttings best. This may be mixed at home, or there are a number of excellent proprietary makes of cutting compost available. Fill the container, seed tray or pot, with the compost to within 1 cm (½ in) of the rim. Make sure the rooting medium is thoroughly moist before inserting the cuttings, but do not make the compost too wet or they may rot.

If you are rooting cuttings in pots, you should be able to get up to six in a 10 cm (4 in) pot, and around 28 in a standard size seedtray. Cuttings are inserted in holes made with a wooden dibber. Ensure that the base of each cutting touches the bottom of the hole, and that the lower leaves are just above the surface of the compost (Fig. 8.3). Use the dibber again, this time to firm the compost around the stem of the cutting. Now water the container with a can and fine rose, allow it to drain and put it into the place it will stay while the cuttings form roots.

Unless mist propagation is used (see Chapter 2), practically all softwood cuttings must be given the protection of a glass or plastic covering to create a warm, moist atmosphere. An ideal air temperature is 18°C (64°F). Rooting is also promoted by providing botttom heat, from soil warming cables or an electrically heated propagator (see Chapter 2). In such conditions, softwood cuttings have been known to root in only two weeks.

When the tips of the cuttings start to grow rapidly, this is a fairly good indication that rooting has taken place. Inspecting the bottom of the container, looking for the white roots, is another way of finding out, but do not be tempted to pull out the cuttings to examine their bases as you could damage any little rooting that has occurred.

Once a good root system has developed (Fig. 8.4), the cuttings should be gently removed from their containers and potted individually into 9 cm (3½ in) pots, using a good loam- or peat-based potting compost. Firm them into their new homes, and water them in. Return them to the same warm place until they have become established.

If the cuttings are of hardy, outdoor plants, you will need to acclimatize them slowly to the outdoor temperatures, which may take several weeks to do properly.

Basal cuttings

We have seen that softwood cuttings are taken from young sideshoots. A basal cutting is a young shoot cut or carefully pulled from the base of a parent plant. With some perennials, of course, sideshoots are difficult to locate, or they simply do not exist. Therefore go to the centre, or crown of the plant, to take young shoots.

Remove them as close as possible to the crown and preferably with a 'heel' (a small tail of older wood). This method applies to such half-hardy perennials as dahlias and chrysanthemums. Here the dormant plants are started into growth with some form of heating during mid- to late winter, and the cuttings are taken as soon as they are large enough to be handled. Generally, such cuttings need little preparation other than the removal of a few lower leaves before planting. In all other respects, they should be treated in exactly the same way as normal softwood cuttings.

Semi-ripe cuttings

These are also referred to as 'half-ripe' or 'semi-hardwood' cuttings, and are usually rooted in the period mid-summer to mid-autumn. The terms basically mean that the cuttings are prepared from wood of the current season's growth, that is when it is becoming ripe and woody at the base, while the tops are still soft and unripened. This type of wood is usually found on the sideshoots or on top of the main stem of the parent plant.

The same types of plants may be propagated from semi-ripe cuttings as with softwood. Some conifers may actually be tackled during mid- to late winter, out of the main semi-ripe season.

When collecting the cuttings' material, apply the same principles as described for the softwoods. The only extra factor to bear in mind with semi-ripe cuttings is that they are taken during the warmest months of the year, when insect pests and diseases are very much in evidence. *On no account should you take cuttings of material that is infected with either.* If you *must* use this material, then spray it with a suitable pesticide at least a few weeks before you propagate from it.

The preparation of a semi-ripe cutting is, again, just as if it were a softwood. However, the lengths of the cuttings will vary, according to the plant. For the majority of shrubs, conifers and so on, they may be slightly longer than the normal softwoods, at 10–15 cm (4–6 in). On the other hand, heaths and heathers are smaller plants with smaller stems, so semi-ripe cuttings of these should be around 3–5 cm (1¼–2 in) in length. It is advantageous for most semi-ripe cuttings to be taken with a heel, while some root more readily if they are 'wounded'. This operation involves the cutting away of a thin sliver of bark from one side of the stem at the cutting's base. The stem can also be scored with a sharp knife to produce the same effect.

The tips of most cuttings should be left intact, but some gardeners prefer to pinch out the soft growing ends of such plants as heathers, laurels and berberis. Also, pinch off any flower buds that are seen, because if these are allowed to stay on the cutting, energy will be put into the production of flowers, which would more beneficially be put into root formation.

Plants with really thin stems, such as species of berberis, may be propagated from a special type of cutting known as a 'mallet' cutting. A thin shoot is taken with a portion – about 6 mm (¼ in) – of the older wood attached at the base. Remove the lower leaves as for a softwood cutting, and in the case of berberis, remove the spines as well, as this makes handling and inserting them much less painful!

Conifer cuttings are easy to take and are not difficult to root. Remember, though, that the base of each cutting should be well ripened and ideally have at least 1 cm (½ in) of brown wood (Figs. 8.5 and 8.6).

Prepare a tray of rooting compost in the normal way, but cover this with a layer of fine silver sand. This will improve the drainage immediately round each cutting. Dip the cuttings in hormone rooting powder (Fig. 8.7), insert them, firm them, and water them in.

A garden frame with a base of sand is a good place to root semi-ripe cuttings (Fig. 8.8). Paint the frame lights with greenhouse shading paint, or cover it with shade netting, to keep the strongest of the sun's rays off. The frame should be kept closed for the first few days to create high humidity, but ventilate if the temperature climbs to levels above 24°C (75°F).

Fig. 8.5 With semi-ripe cuttings, choose a side shoot consisting of current season's growth. Tear it away from the main stem with a small heel of rind.

Fig. 8.6 Trim the base of the cutting, neatening the heel, and removing the lower leaves.

Fig. 8.7 Dip the base of the cutting into hormone rooting powder, and tap off the excess.

Fig. 8.8 A garden frame with a base of sand is a good place to root semi-ripe cuttings.

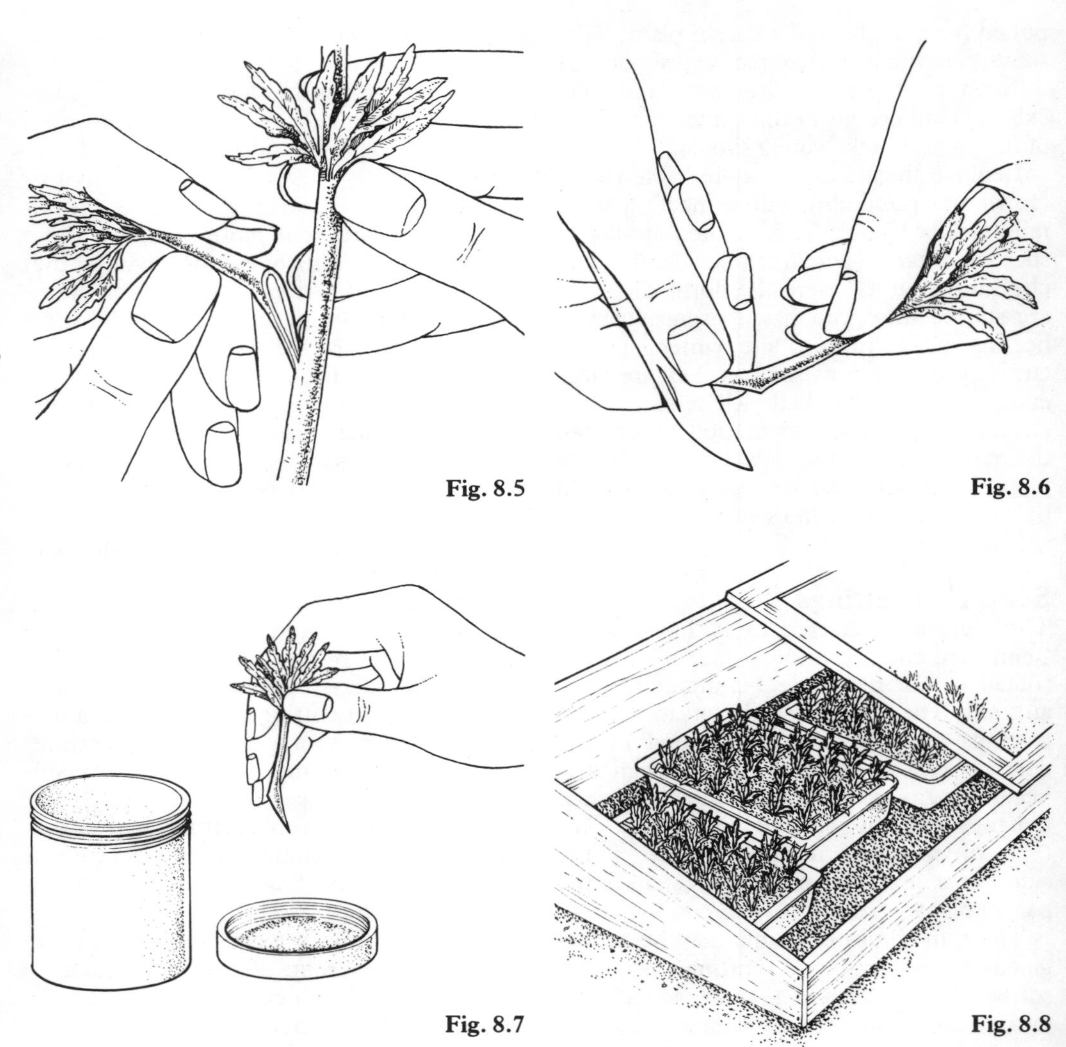

Fig. 8.5 **Fig. 8.6** **Fig. 8.7** **Fig. 8.8**

If you do not possess a garden frame, then the greenhouse may be used, but try to imitate the conditions already described. If you do not have a greenhouse, then a kitchen window sill would be suitable. The use of clear polythene bags will help to create the humidity. Some semi-ripe cuttings, especially from greenhouse plants, need to be rooted in heat (see Part II).

During the weeks after insertion, regular checks should be made on the progress of the cuttings, but again do not be tempted to remove them from the compost just to see if they have formed roots. Remove any fallen leaves, but unless the cuttings are very advanced, they can be left in their trays over winter. The time taken for the various types of cuttings to root is, of course, affected by the conditions under which they are kept after insertion, and depends also on the plant species. On average, semi-ripe cuttings root from 5 to 25 weeks after insertion. Growth will, therefore, start in earnest in spring, and the rooted cuttings will soon show signs of life.

Ensure good ventilation as the weather becomes warmer, and be prepared to harden the cuttings if applicable so that they can be planted out in 'nursery rows' during the spring. Choose a sheltered spot outdoors for these nursery rows, for it is here that the rooted cuttings will spend the next six months, at least, while they build up their root systems and develop a good framework of branches. Do be warned, though, you cannot afford to forget these plants. They should be weeded, watered in hot weather, and checked over regularly for dead leaves, pest and disease presence, and so on.

Hardwood cuttings

Having discussed softwood and semi-ripe cuttings, the type known as 'hardwood' are, as might be expected, made from fully ripened wood. A surprising number of hardy shrubs and soft fruits can be propagated by simply putting in some hardwood cuttings in the autumn, yet it is a method of plant propagation that is rarely used by the amateur. You do not need a greenhouse or any specialist equipment, at least for the vast majority of subjects. Not even pots, trays or potting compost! All you will need is a sharp pair of secateurs, some coarse gritty sand, and perhaps a cold frame or cloches. Many subjects may be rooted in the open ground with no real protection.

The technique for taking the cutting is quite simple. A hardwood cutting is made from a length of healthy, ripe, one-year-old wood, that is, a length of stem that grew in the current year, with bark that is fully coloured and firm enough not to 'give' when squeezed between your finger and thumb. The length of each cutting should be around 20–25 cm (8–10 in), but this is just a guide; it may be twice this length if the distance between the buds is great.

Take your cuttings' material – prunings are most acceptable – and with each piece of wood make a straight cut below a bud or leaf joint at the base (Fig. 8.9). Now make a sloping cut just above a bud at the top. Apart from anything else, this will help you to remember which end is which. It makes handling much more comfortable if the spines are removed from prickly plants, such as gooseberries. Discard the unwanted tips of the severed shoots.

With gooseberries, and red and white currants, and indeed any ornamental plant that needs to have a short trunk, or leg, remove all but the top three or four buds by rubbing off the others or nicking them out with a knife. This will help prevent the subsequent plant from suckering. In addition, remove the lower leaves on evergreen cuttings that would otherwise be buried. Do not forget to dip the bases of the cuttings in a hormone rooting powder especially formulated for hardwoods. Then tap off the surplus.

Fig. 8.9

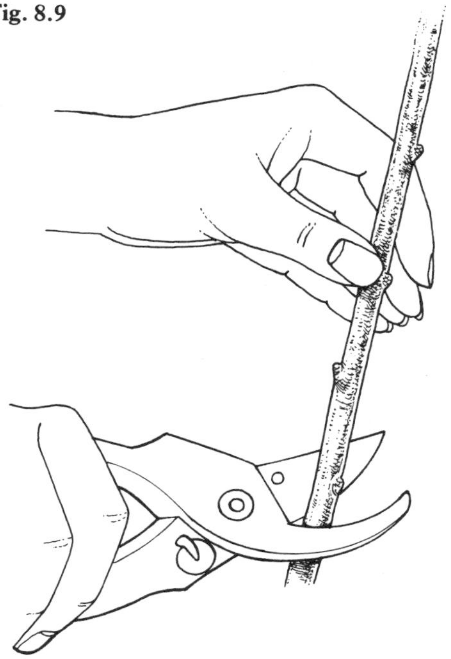

Fig. 8.9 With hardwood cuttings it is important to select healthy, ripe, one-year-old wood.

Now the cuttings will need to be inserted in their rooting site. Choose a well-drained, sheltered spot in the open garden, preferably south-facing to take full advantage of the sun. The warmer the site, the quicker the ground warms up and the sooner rooting will start. Put down a guide line, then open a slit trench beside it by thrusting a spade vertically into the ground and pushing the handle forward (Fig. 8.10). If your soil is heavy, it is a good idea to sprinkle an inch layer of sharp sand along the base of the trench, to aid drainage.

Each cutting should be spaced at around 15 cm (6 in), and stood in the trench, pushed in so that their bases rest firmly on the trench bottom. Ideally this should be about a half to two-thirds of their length. Now tread the soil firmly around them (Fig. 8.11), and then follow over the surface of the soil with a fork or hoe, loosening it to a depth of about 2.5 cm (1 in) or so. The

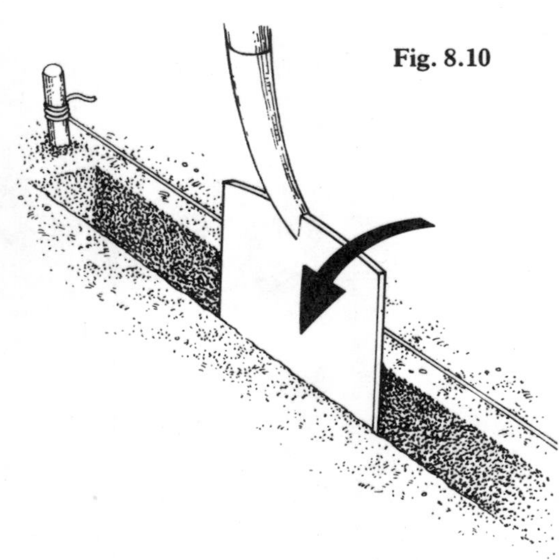

Fig. 8.10 Make a slit trench by inserting the spade vertically and pushing it forwards slightly.

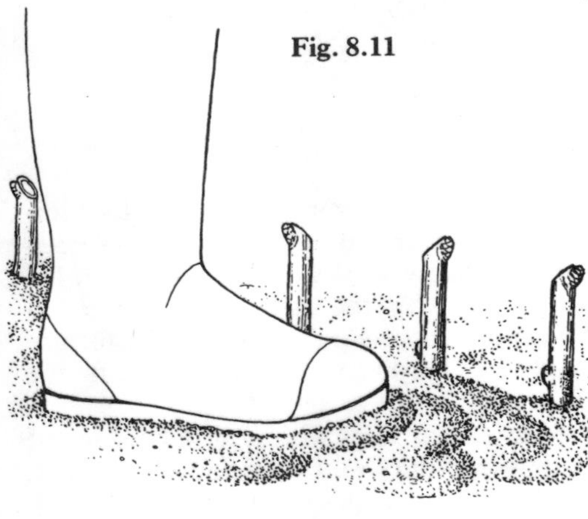

Fig. 8.11 Place the cuttings in the trench, so that a maximum of two-thirds their length is buried.

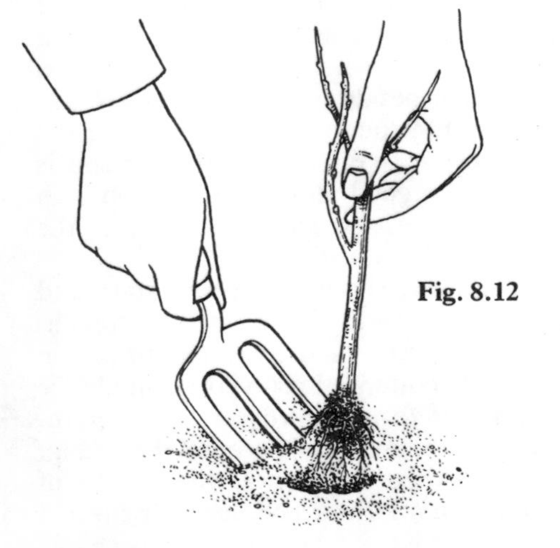

Fig. 8.12 When the cuttings are putting on lots of topgrowth they have rooted, and during spring they should be transplanted.

same technique is used for rooting in a garden frame.

The cuttings need practically no attention from now onwards. Check them occasionally, particularly after a period of hard weather and, if the action of frost in the soil has moved, or lifted them, push them back down into the soil and tread them in. Over the following spring and summer hoe between the cuttings regularly to keep down weed growth, and if it is necessary, spray them to keep pests and diseases at bay.

With any cutting, as we have seen, it is a race against time to get it to produce roots before it collapses from lack of water. Whereas softwood cuttings have a very short life span but tend to root quickly, hardwood cuttings are just the opposite. A piece of leafless hardwood takes a long time to dry out, but takes months to form a callous at the base and then go on to make roots. Those cuttings that take successfully will be ready to move to their permanent quarters the following autumn (Fig. 8.12).

Some of the more difficult plants root better if they are subjected to bottom heat. Therefore a greenhouse bench with soil-warming cables or a heated propagator should be used to house the cuttings. Insert them in containers of appropriate size; 12.5 cm (5 in) pots will hold six to ten cuttings without overcrowding them. A suitable compost is a mixture of equal parts peat and coarse sand. Just as if they were outdoors, insert the cuttings by pushing them into the compost, to about half to two-thirds their length.

A temperature at the base of the cuttings in the region of 18–21°C (64–70°F) would be ideal. Heat like this will enable you to propagate such plants as *Ficus carica* (fig), and laburnum. You will find that the top-growth is often quickly produced when the cuttings are rooted in heat. Do not be tempted to lift the cuttings too quickly, however. Wait until the roots are showing under the bottom of the container, for you then know that rooting has been successful and that a reasonable root system will have developed.

It will probably be mid-spring before the cuttings are ready for potting individually. Use a large pot, and you may use either a loam-based or peat-based compost – there is little to distinguish between the two for rooted hardwood cuttings. After potting, return the rooted cuttings to the greenhouse or frame so that they may become established in their containers before being taken outside for the summer.

Stem sections

The stem cuttings considered so far in this chapter refer to the preparation and rooting of tips, usually with a few young leaves attached. However, a wide range of thick-stemmed plants, particularly tender, exotic or house plants, may be propagated by rooting sections of their stem, or trunk.

Lower parts of stems will often root if the top of the piece is trimmed just above a node and the rest is prepared and treated as if it were a tip cutting. An offcut, pruning, or the remainder of a shoot that had its tips taken for a separate cutting all make good, suitable material for stem section cuttings.

Plants that can be propagated from woody stem sections include cordylines, dracaenas and dieffenbachias (leopard lilies, or dumb canes). Very often stem sections of the Polynesian ti plant, a variety of *Cordyline terminalis*, are sold in gift shops and by mail-order. They are packaged and prepared in such a way as to root fairly easily.

If the stem of your chosen plant is a thick one – say, thicker than your thumb – quite short pieces may be used for rooting. Each piece need be only about 5 cm (2 in) in length, provided it includes one or two nodes.

Fill a pot or tray with moist compost, either one prepared at home, or a proprietary brand, provided it is specifically designed for cuttings, and gently insert the short, thick pieces of stem horizontally (Fig. 8.13). Make sure they are half-buried, with a node or leaf-bud facing upward. In this way, the stem and leaves that develop from the buds will grow upright. An alternative is to plant such cuttings vertically, as long as they are planted with the bottom end down, as on the original stem. Water the cuttings and place the container in a warm, humid place, as if it contained ordinary softwood stem cuttings of tender plants.

You will be able to tell if the cuttings have taken root, because the nodes above the compost will develop the characteristic leaves of the plant; the nodes buried below the compost will sprout roots. When the stem cuttings have developed enough root to fill the allotted space in the container, pot them up individually, and treat them as you would any other rooted cutting or young plant.

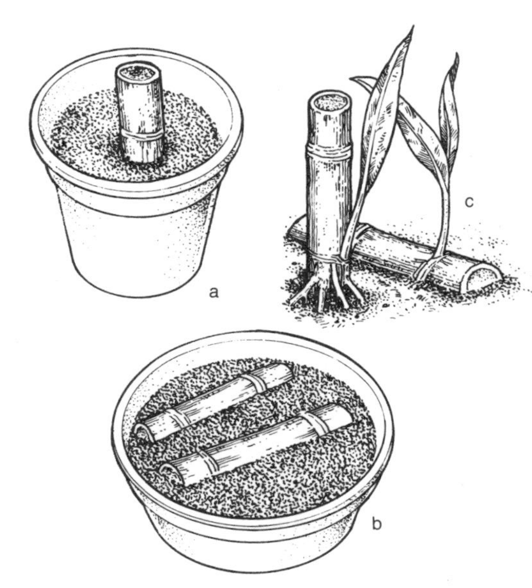

Fig. 8.13

Fig. 8.13 Stem sections – about 5 cm (2 in) long – may be used as cuttings, as long as they have one or two nodes. (*a*) potted vertically, (*b*) potted horizontally, (*c*) cutting with new shoot.

Eye-cuttings

This method of propagation is most often used for grape vines, but it can apply to many hardwoods; it is particularly useful if material for propagation is scarce. Eye-cuttings are taken from fully ripened wood when the leaves have fallen in autumn. Cut a sturdy vine stem which has grown the previous year. As always, use healthy, typical growth that is free of pest or disease. One good branch will provide plenty of cuttings. Check that the wood is firm, and all of the dormant buds are plump and healthy. Do not be tempted to use the thinner tip of the branch.

Fill a pot or tray with a good cuttings compost. Firm it, water it and scatter a thin layer of silver sand on top to improve drainage. There are two ways to prepare the cuttings (making sure that all cuts are made with a pair of sharp secateurs, and being

Fig. 8.14 Grapevine eye-cuttings should be 2.5–4 cm (1–1½ in) long, and have a bud in the centre or at the top.

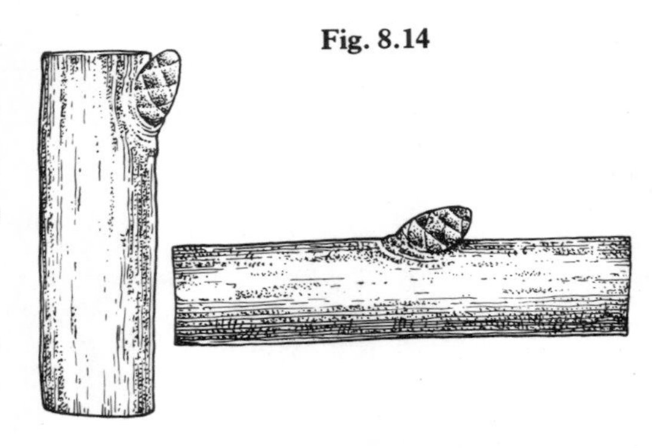
Fig. 8.14

careful not to damage the buds). Either make the first cut just above the bud, and the second about 2.5–4 cm (1–1½ in) below, or cut the vine so that the bud is in the centre (Fig. 8.14).

Dust the lower end of the first eye-cutting with hormone rooting powder formulated for hardwood cuttings. Then insert the cutting, upright, so that the base of the bud is just level with the surface of the compost.

The second type of eye-cutting can be inserted horizontally, but it will tend to root more quickly if a narrow sliver of bark is stripped from the side opposite the bud. Use a sharp knife for this, and dust the wound with rooting powder. This operation exposes the inner tissue, or cambium layer, from which the roots will emerge.

Whichever type of eye-cutting you choose, make sure it is well watered, and then place it in a warm, humid place. They root most rapidly in gentle heat in a propagator, but it is important to know that too much heat will force the buds into growth before there are any roots to support them.

The buds will sprout shoots during late winter and spring, and some may require some form of support, such as that provided by a thin cane. Once the eye-cuttings are well rooted they should be moved singly, into 10 cm (4 in) pots, using potting compost. They should be gently hardened off in a cold frame, and the young plants put into their final positions (in the greenhouse or a sheltered, open spot outdoors) the following autumn.

Pipings

These are a special type of softwood cutting used in the propagation of garden pinks (dianthus), both the perpetual-flowering border varieties, and the alpine or rock garden forms. Such plants start to decline in vigour after a couple of years, and the older they are, the less frequently they flower, so propagation to replace stocks should be carried out regularly. Pipings are very easy to root, and they are usually taken in mid- to late summer.

Hold the selected stem with one hand while pulling the tip carefully with the other (Fig. 8.15). Try to ensure that the cutting parts from its parent just above a node. There is no need to use a knife. The lower leaves are then removed, but retain three or four pairs of well-developed leaves on the cutting. The base is dipped into hormone rooting powder and the cutting is then ready for insertion into a sandy compost. Heat is not required, but shelter is! A garden frame is usually sufficient protection. Good ventilation is important, but keep them out of direct sunlight, especially in the first week or two. Roots will form within a month.

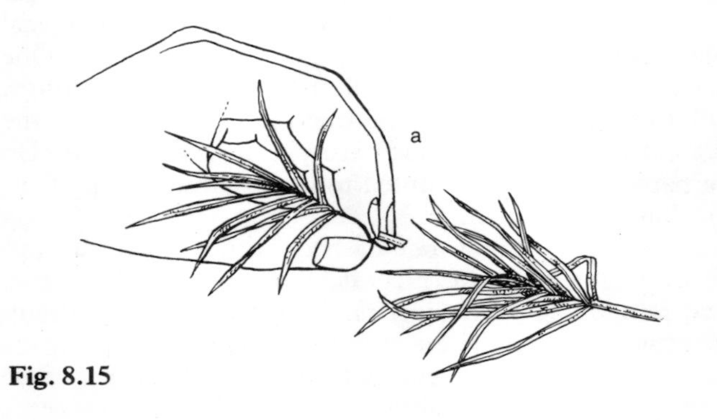

Fig. 8.15

Fig. 8.15 Pipings – a softwood cutting used in the propagation of garden pinks. (*a*) Carefully pull the tip away from the selected stem. (*b*) Cuttings potted up. Push the piping into the compost up to the base of the lower leaves.

CHAPTER 9

Root Cuttings

In propagation, we now come to an ironic twist. For in the same way that stems and upper plant parts can be encouraged to develop roots, some plant roots can be persuaded to send up shoots and form new plants that way. A wide range of trees, shrubs, climbers, bush and cane fruits, alpines, perennial plants and even some vegetables may be quickly and easily increased from root cuttings taken during the period from late autumn to late winter. (Root cuttings should not be confused with *rooted* cuttings; the former are taken from sections of plant root, the latter are the cuttings detached from any part of the plant, and which have 'taken' successfully.)

Botanically it is possible for roots to take because of the presence of 'adventitious' buds. Normally above-ground shoots originate from buds found at the nodes of stems, but buds may appear, almost randomly, on other parts of the plant, including internodes (the area between nodes), leaves and roots.

Generally, root cuttings are regarded as a means of increase for herbaceous plants which have fairly thick roots, although some fibrous rooted kinds (such as border phlox) can be propagated this way. Incidentally, if the phlox has become infested with stem eelworms (causing twisted, distorted or dead young shoots and unusually narrow leaves), then propagation from root cuttings, instead of stem cuttings, will usually produce healthy eelworm-free new plants.

Root cuttings can also be used to propagate trees and shrubs which tend to produce suckers naturally – indeed, from adventitious buds – and which therefore sprout rapidly from severed root sections. Many of these plants are, however, just as easily increased by lifting well-rooted suckers in autumn; so taking root cuttings from them is mainly of interest to gardeners only where natural suckers are scarce, or propagation on a large scale is being carried out.

As we will see in Chapters 15 and 16, it is not possible to propagate grafted or budded trees and shrubs from root cuttings, nor from suckers. Anything that grows from the roots of these will only reproduce the inferior rootstock onto which the top-growth was grafted. Lilacs (syringa) for example, are easily raised by this method, but instead of increasing your stock of, say, a favourite double-flowered white variety, root cuttings may well give you the single, purple-flowered rootstock of the common lilac, *Syringa vulgaris.* Similarly, root cuttings taken from variegated hybrid plants often revert to the green of the original plant, so these should be avoided.

It is also worth noting that certain plants, including those of the genus *Acanthus*, have a juvenile type of foliage during their early weeks and months. Root cuttings taken from young plants with the juvenile foliage will develop leaves of the same shape, whereas root cuttings taken from an older, mature plant will develop mature foliage characteristics. Generally a degree of maturity is required before the plant will flower, so it follows that root cuttings from a mature plant will bloom quicker.

Preparing the cuttings

The taking of root cuttings should, generally, be timed to when the plants are dormant and therefore least likely to suffer from the disturbance involved. At this time of year the plants are resting, but warmer spring days, when the cuttings can be expected to start sprouting, are not far off either!

Collecting the material

Collecting the cuttings material is not always straightforward, and whichever method you choose will depend on the type of plant to be propagated. Choose a mild day when the ground is nice and workable, ie. not frozen or waterlogged. The small herbaceous plants that are suitable for root cutting propagation should be lifted completely (Fig. 9.1). Wash most of the soil off the roots by

Fig. 9.1 Plants to be propagated by root cuttings should be lifted during the dormant season.

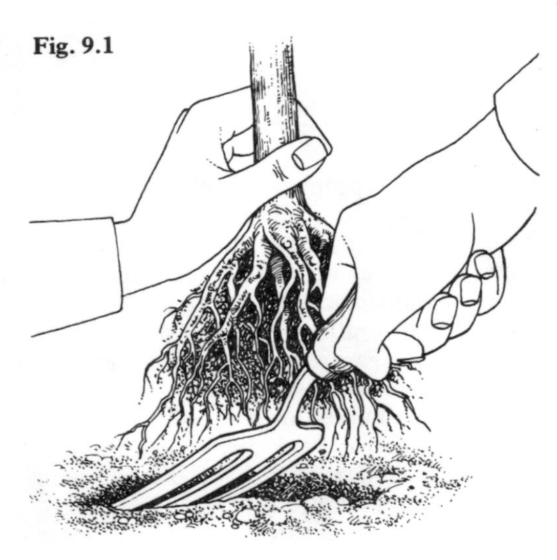

Fig. 9.1

Fig. 9.2 Wash the majority of soil from the roots. Choose only healthy undamaged material for the cuttings.

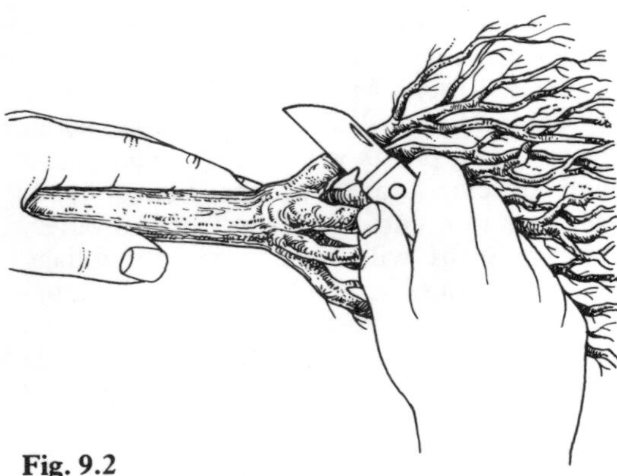

Fig. 9.2

gently dunking them in a bucket of water. Of course, you need not wash off every grain of soil, as the roots have only to be clean enough for you to see what you are doing. It will be either difficult or impossible to lift larger plants, shrubs or trees, so it is prudent with these to expose the roots by scraping away some of the soil. This also applies to those plants that resent too much disturbance.

Plants with thin, wiry roots give you little choice in terms of variety of thickness. With the fleshier rooted plants, such as *Papaver orientale* (oriental poppy) and *Rhus typhina* (stag's horn sumach), it is best to go for roots of about pencil thickness. However, in all cases try to choose young roots, as these will give the best success rate.

Method for root cuttings

Unless you are embarking on large-scale propagation, two or three roots per plant will usually give you sufficient cuttings, and the removal of a small number will have practically no effect on the parent plant. Make all cuts cleanly with a sharp knife, picking only sturdy, firm, undamaged roots (Fig. 9.2). If you are not going to prepare them and insert them straight away, pop them into a plastic bag until you are; this way they will not dry out.

The parent plant wants to be out of the soil for as little time as possible, so once you have taken off as many roots as you need, replant it. Make sure you firm the soil around the root ball, and if the soil is on the dry side, apply a little water.

Thick-rooted plants

The thickest retained roots are prepared by first trimming off any thin, whiskery side roots. Try to keep down the number of cuts and wounds, to minimize the risk of diseases getting a hold. Before making the major cuts, lay out the root lengths on your work surface, so that you can tell which is the top, that is the part of the root nearest to the parent plant's crown. Now cut the roots into sections 5–8 cm (2–3 in) long, the cut nearest the crown of the plant being made straight across, and the lower cut being on the slant, so that you will always know which way to insert them (Fig. 9.3). If they go in upside down, by mistake, the shoots will not be prevented from growing, but they will instead grow downwards and have to turn themselves round.

Using a proprietary (or home-made) cuttings compost mixed with a small amount of sharp sand, fill a medium-sized pot or a standard seedtray and firm it down lightly. Water the mixture and leave it to drain a while before inserting the roots.

It is not necessary to use a hormone rooting powder with these types of cutting. However, it is very worthwhile lightly dusting the cuttings with a fungicidal powder for disease prevention. Seed dressing compounds are ideal.

Insert the cuttings, vertically, into predibbed holes in the compost. Space them about 5 cm (2 in) each way. Push them down so their tops (with the flat cut) are just about level with the surface (Fig. 9.4). If you are in any doubt whatsoever concerning which is the top end of the root cutting, it is always a good idea to lay them horizontal on the surface of the compost, and then to lightly cover them with some of the same.

Thin-rooted plants
A similar technique to this last method is used for thin-rooted subjects. Here it is best to cut the roots into sections, perhaps a little shorter, and there is no need here to make straight and angled cuts. Lay the sections flat on the surface of a tray partly filled with a suitable compost. Then cover them with a shallow layer – perhaps 1 cm (½ in) – of the same compost (Fig. 9.5).

Starting into growth
Root cuttings of hardy plants do not generally require heat to start them into growth. The filled containers may therefore be placed straight away into an unheated greenhouse, garden frame, or even a sheltered place in the open garden. There are, as always, one or two exceptions to the rule. Some subjects tend to root a little better with heat. These include ceanothus (Californian lilac) and *Romneya coulteri* (fried egg plant).

The compost in the pots or trays should be kept damp, and if outdoors, protected from drenching rain. By spring the cutting will have branched out and buds will appear at the surface of the compost.

Potting up
Do not be tempted to lift the cuttings at this stage, for often shoot growth precedes adequate root growth. Occasionally roots will sprout only from the new shoots, and not from the original root cutting at all. Do not be in too much of a hurry to pot them up, or they may fail.

Conversely, do not keep the rooted cuttings in these containers too long: there are no plant foods in cuttings compost, so the young plants will quickly become deficient. The timing is just right if you pot them up when plenty of new, fibrous roots have grown out from the old root sections.

Fig. 9.3

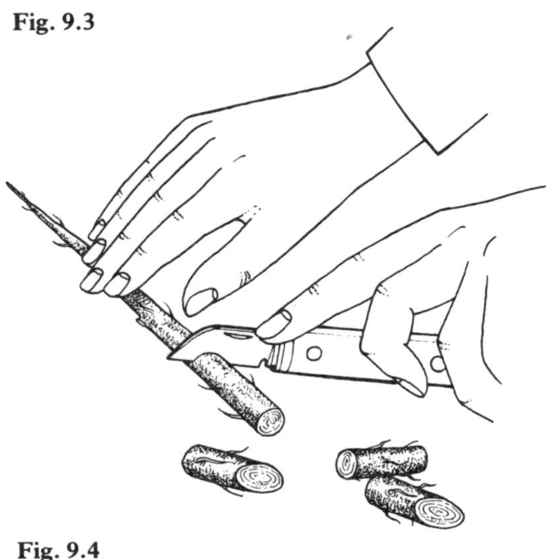

Fig. 9.3 Trim the root cuttings so that each has a flat top and a slanted base.

Fig. 9.4

Fig. 9.4 Thicker, fleshy root cuttings should be 'planted' in a pot containing potting compost.

Fig. 9.5

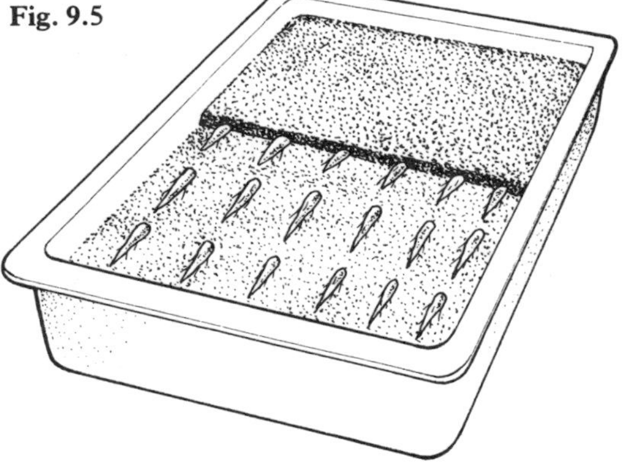

Fig. 9.5 The thinner rooted cuttings may be horizontally laid on compost in a standard seed tray.

CHAPTER 10

Leaf and Leaf-bud Cuttings

Stems and roots are not the only parts of plants that can be used in cuttings propagation: leaves and leaf stalks can also be used. When inserted these will form roots and young plants will develop. This type of plant increase is most often applied to greenhouse or house plants, although there are also a number of hardy plants that respond to it.

There is no dependable way of distinguishing plants that may be raised this way from others, but most do, in fact, have thick, fleshy leaves which often grow in rosettes. Included among the tender plant groups are streptocarpus (cape primrose), saintpaulia (African violet), sinningia (florist's gloxinia), peperomia, some species of begonia, and a number of succulents, including echeveria and crassula. Hardy plants that may be propagated from leaf or leaf-bud cuttings include clematis, mahonia and hedera (ivy), as well as small perennial or alpine plants such as ramonda, haberlea and lewisia.

Leaf and leaf-bud cuttings, for they are two completely different yet related methods, are generally taken in spring or summer. Even so, they usually require artificial heat to take root, so a greenhouse bench warmed by under-soil cables, or a heated propagating case are recommended. Both types of cutting can be prepared in a variety of ways, according to both the plant being propagated, and the gardener's own preference. Let us look at each in turn.

Leaf cuttings

There are a number of techniques open to us, some plants responding better to certain methods than others.

Cutting into sections

Perhaps we should start by looking at a technique of taking leaf cuttings that applies to two extremely popular house plants, cape primroses and mother-in-law's tongues.

The former, known botanically as streptocarpus, has rather long, fleshy, prominently-veined leaves that are easily damaged, and will readily turn brown at the edges and tips. However, masses of flowers are thrown out by healthy, well-grown plants, so this anguish is worth enduring: the cape primrose is one of the most popular of all house plants. Propagation by seed is not particularly difficult, but taking leaf cuttings is easier, and there is usually a ready supply from which to choose.

The mother-in-law's tongue (*Sansevieria trifasciata*) is a very different plant. The leaves are very long, hard and sword-like. The most popular form is the yellow-edged *S. trifasciata* 'Laurentii', but if you take cuttings of this, the new plants will not retain the attractive variegation: they will, instead, revert to the natural marbled grey-green colouring.

Preparation

With both types of plant, it is worthwhile preparing them some months in advance. In late winter, therefore, remove all yellow or brown leaves, and treat the plant to a good feed and give it plenty of moisture, so as to encourage strong, healthy growth. Spring and summer cuttings are most successful, although leaf cuttings may be taken whenever there is a plentiful supply of foliage!

Method 1

Before you tackle the cutting, prepare a tray of cuttings compost, and water it thoroughly. Due to the nature of the leaf cuttings from these plants, it is better to use trays rather than pots. Take your plant and, using a razor blade or sharp knife, cut off the leaf of your choice, as near to the base of the leaf stalk as possible. Lay it on a flat surface. Because the leaves are lengthy, each streptocarpus leaf can be cut into two, three or four sections (Fig. 10.1), and considerably more for the sansevieria (each section may be 5–8 cm (2–3 in) long. The lower part of the leaf, nearest the leaf stalk, normally roots more readily than the tip.

Brush a tiny amount of hormone rooting powder, formulated for softwood cuttings, along the lower edge of each section.

Take each piece of leaf and, holding it vertically (and the right way up!), push it into the compost until it is in about one-third its depth (Fig. 10.2). Place each subsequent section behind the one in front, so that you are making a row, or rows, of cuttings. The greenhouse is not the best place for these cuttings as they are likely to be scorched by the bright light. It is better to keep the tray of cuttings indoors, in a warm position, 18–21°C (65–70°F), but out of direct sun. If the greenhouse *must* be used to accommodate the cuttings, make sure they are well shaded.

Keep the compost moist, but do not splash any drops of water onto the foliage. There are a number of plants which have tiny hairs covering their leaves, including Cape primroses, African violets and gloxinias, and as a result they are prone to scorch when water is left on the upper leaf surfaces. The best way to water the cuttings, therefore, is to either dribble water onto the compost by holding a can carefully and pointing the spout under the leaves, or you can sit the tray in an inch or so of water for ten minutes, so that the compost sucks up a quantity of water like blotting paper.

These cuttings are often slow to root and produce plantlets. Be patient. Remove any cuttings that become limp and brown. In due course each section will develop new plantlets at the points where the veins meet the surface of the compost (Fig. 10.3). After a few weeks. remove the cuttings from the tray, separate them and pot them up, individually.

Method 2

An alternative way of propagating Cape primroses by leaf cuttings is to cut the leaf longitudinally. Take another strong, healthy leaf – perhaps for this method you should choose one that is as straight as possible. Lay it on the work surface, upper surface downwards. You will notice the thick midrib vein and the many smaller lateral veins; each vein *could*, in theory, produce a new plant. Run your knife along each side of the midrib so that you are, in effect, cutting

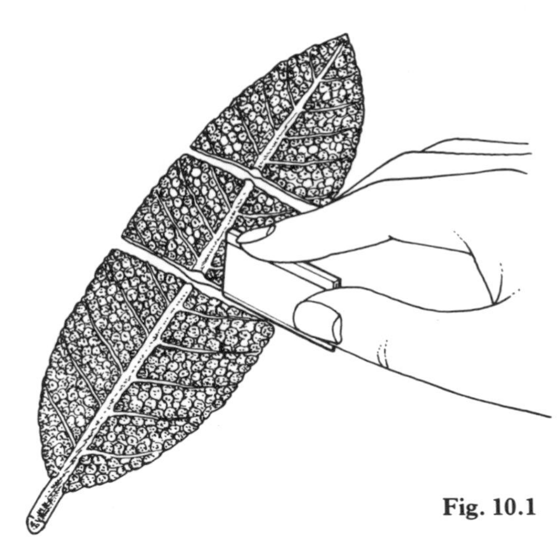

Fig. 10.1

Fig. 10.1 To take a number of cuttings from a streptocarpus leaf, use a sharp knife or razor blade to slice across its surface. Each section may be 5–8 cm (2–3 in) wide.

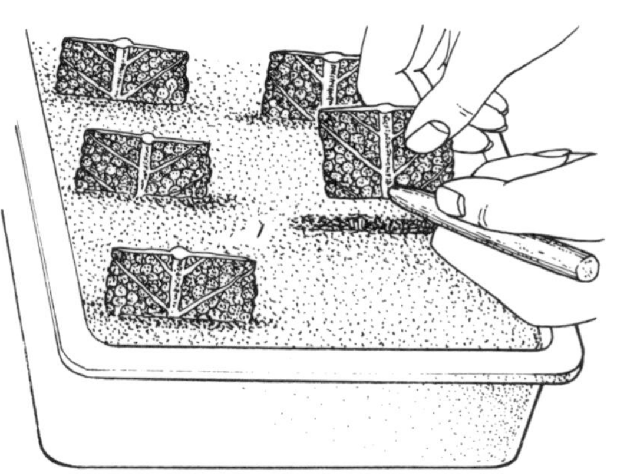

Fig. 10.2

Fig. 10.2 Each leaf section should be pushed, vertically, into a tray containing cuttings compost.

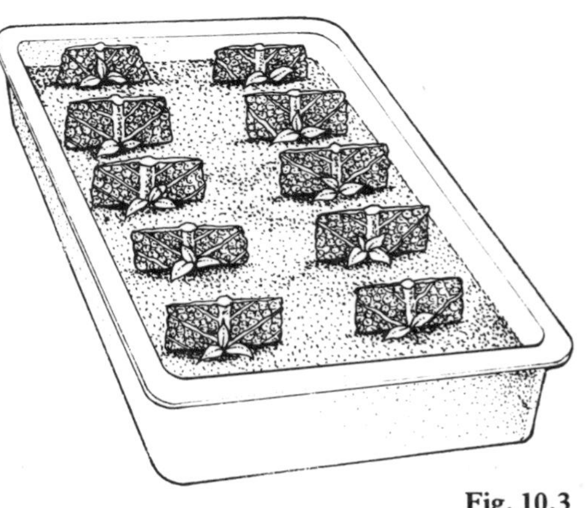

Fig. 10.3

Fig. 10.3 The cuttings may be slow to root and produce plantlets, but be patient. In due course each leaf section will develop new plantlets at the points where the veins meet the surface of the compost.

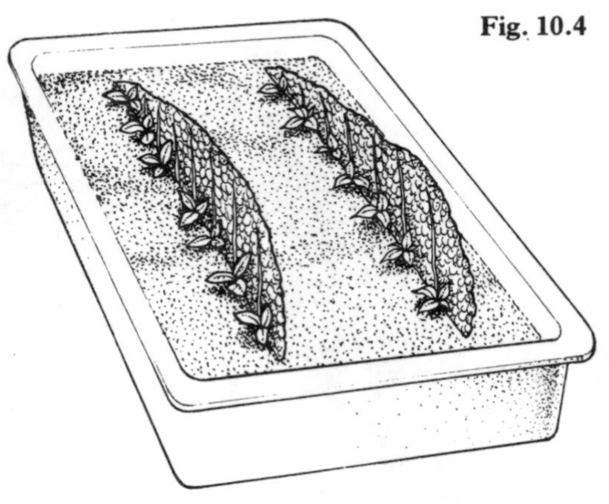
Fig. 10.4
Streptocarpus leaves may be cut longitudinally. Perhaps a dozen or more plantlets may develop along the cut base.

it out. Now throw it away. You are left with two leaf halves with exposed lateral veins. Dust some hormone powder along the cut edges and insert each half into the tray of compost, to about a quarter of its depth (Fig. 10.4).

These leaf cuttings are rather more difficult to handle than horizontally-cut ones – wide leaves may need some support at the back to keep them vertical – but if they are given the same conditions for rooting, they will generally provide a dozen or more plants, each. Once the small plantlets are well developed, they can be removed from the tray and potted up individually.

Leaf with stalk

Perhaps the most common house plant of all propagated by leaf cuttings with stalks, is the saintpaulia or African violet. This will produce several new plantlets from one leaf, and the procedure may be undertaken virtually any time of year, but spring and summer cuttings root most easily.

Method

Fill a small flower pot with a proprietary or home-made cuttings compost, and firm it down lightly. If you prefer, you may also use an inert material, such as vermiculite, which gives good results. Water the compost thoroughly. Then, using a dibber or a pencil, make a hole in the compost that is slightly off-centre. This is needed, because the single leaf, when it is inserted, will sprout new growth in one direction, and this should be aimed towards the centre of the pot. If a large number of cuttings are being taken, a standard seedtray may be used.

Now approach your mature plant. Any well-grown African violet may be stripped of a few leaves without harm, but do not overdo it; take no more than 10% of the foliage. Choose leaves that are fully open and healthy (Fig. 10.5).

Cut off the leaf as near to the base of the leaf stalk as possible. Lay the leaf on a flat worktop and trim the stalk so that it is about 2.5–4 cm (1–1½ in) in length, with a clean cut across the base (Fig. 10.6). Ensure that you cause as little harm as possible to the leaf blade as you do this; damaged leaves will rot.

Many gardeners prefer to use a hormone rooting powder developed for softwood cuttings, but quite good results may be expected without this. If you do use it, make sure you tap off the excess powder from the base of the stalk. Insert the saintpaulia leaf so that half to two-thirds of the leaf stalk is buried (Fig. 10.7). Firm it in gently with your fingers.

Rooting conditions, in every respect, are the same as for the Cape primroses. In only a few days roots will form at the base of each leaf stalk, but it will be several weeks before you see tiny new plantlets pushing up through the soil. How long they take to form and develop varies with the conditions and season, but six to eight weeks is about average. Once the top-growth appears, further development will be fairly rapid, and when the pot or tray has more than 80% of its surface covered with leaves, it is time to transplant the cuttings.

Most of the African violet leaf cuttings will produce a cluster of new baby plants (Fig. 10.8), and each of these must be separated at the potting up stage. First, turn your container upside down, and then tap the rim on a firm surface to loosen the compost and remove the plants. Once they are out of the pot, it should be fairly easy to recognize the separate crown of each plantlet, and tease them apart. Pot each one up in its own 9 cm (3½ in) pot, containing potting compost, and make sure you firm each plant in position. As always, give the new plant a drink of water to settle it in, but do not get water on the leaves.

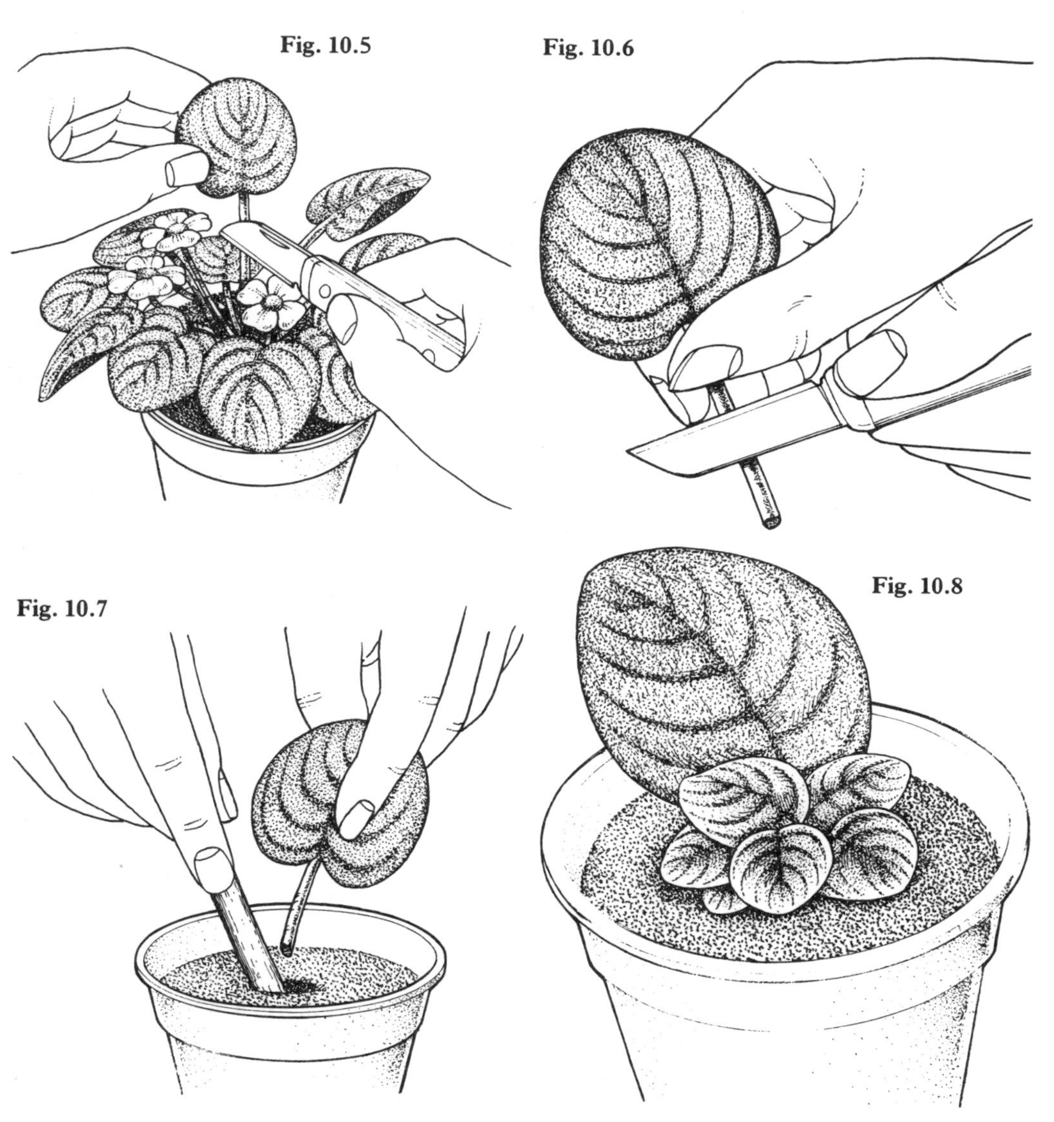

Fig. 10.5 When taking whole leaf cuttings of saintpaulias, choose only healthy, unblemished material.

Fig. 10.6 Trim the stalk so that it is about 2.5–4 cm (1–1½ in) in length, with a clean cut across the base.

Fig. 10.7 Insert the leaf so that half to two-thirds of the leaf stalk is buried, and firm it in with your fingers.

Fig. 10.8 Most of the saintpaulia leaf cuttings will produce a cluster of new baby plants. Each of these must be separated at the potting up stage.

If the original leaf cutting is still in a good condition, it may be used again but the chances of producing many new plantlets is somewhat reduced. This is particularly useful if the variety of saintpaulia you are propagating is uncommon. Cut off the base of the leaf stalk and repeat the process. If, however, the leaf has gone limp, or looks as though it is about to shrivel up and die, simply dispose of it. There is every chance that you will be able to choose a completely fresh leaf from the parent plant, and propagate from that.

There is, in fact, one method of leaf propagation that applies to both 'sections' and 'stalk'. Take, for example, begonias and peperomias; cut the leaves into small triangular sections with a piece of stalk at the apex of each. The triangles are inserted, apex down, into the compost, covered and treated as before. Each section should develop a new plantlet at compost level.

Pegged down leaf

There are one or two species of greenhouse begonia, such as *B. rex* and *B. masoniana*, that can be propagated from leaves, fairly large ones, and without the stalk.

Fig. 10.9 A whole leaf of *Begonia rex* should be pegged down on to the surface of the compost. Plantlets should develop from slits made in the leaf veins.

Fig. 10.9

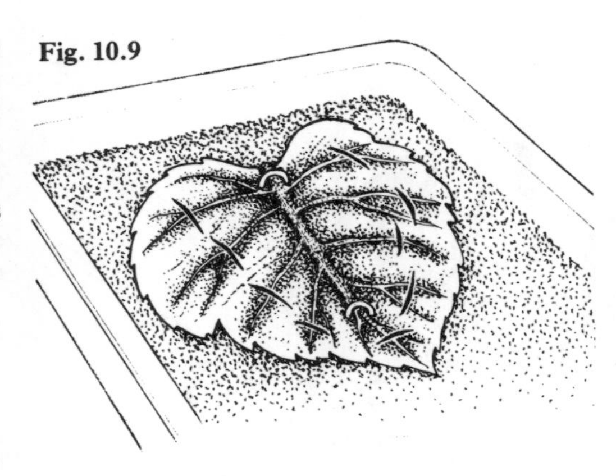

Method

Pick a healthy, unblemished, fully developed leaf and lay it face down on the worktop. First, cut off the stalk. Then, with your knife, cut across the larger veins, which will be standing out quite prominently. Make sure you cut right through the veins, with each slit around, say, 1 cm (½ in).

Take the leaf and lay it, this time face uppermost, on a tray of moist cuttings compost. Press it lightly into the surface so that the cut veins are in good contact with the compost. It will need holding in position, and there are a number of ways in which this can be done; a few pebbles or coins, short pieces of wire, hairpins (Fig. 10.9) – in fact anything that does not weigh too much.

It is a good idea to spray mist over the compost occasionally; this will help to prevent the compost from drying out. Place the container in a warm, humid place (as for the other leaf cuttings), and plantlets should develop on the upper side from each slit in the leaf veins.

Leaf-bud cuttings

We have seen that a number of plants have the ability to produce offspring from one small section of tissue, be it stem, root or leaf. But others need more base material with which to work. A leaf-bud cutting comprises a whole leaf, but this time we retain a portion of stem with it, accompanied by a growth bud in the axil of the leaf (where the leaf stalk and stem meet).

A number of tender greenhouse and hardy outdoor plants are quite easily raised by this type of propagation. However, as the technique for each varies slightly, it is best if we consider them as they pertain to specific plants, starting with the outdoor subjects.

Hardy plants

One of the most popular of garden shrubs is the camellia, yet it can be one of the trickiest plants to propagate. Therefore, it is worth trying any form of propagation open to us, leaf-bud cuttings being relatively successful. During late summer, cut a healthy shoot that has grown during the spring and which is starting to ripen at the base. This will provide you with enough material for several cuttings.

Method

Each cutting will consist of a leaf, complete with its axillary bud and a piece of stem around 2.5 cm (1 in) in length. Using a pair of sharp secateurs, make the top cut just above a bud. Now dip the base of the cutting into some hormone rooting powder formulated for semi-ripe cuttings (Fig. 10.10). Knock off any excess powder, and insert the cutting into a pot or seedtray containing moistened cuttings compost mixed with a small handful of silver sand (per 10 cm/4 in pot). The cutting should be positioned so that the leaf and bud are just above the compost surface. Use your fingers to gently firm it in place.

If the leaf is large, the cutting may require some support; a piece of split cane is useful. Now water the inserted cuttings to settle them in. Whether or not warmth should be

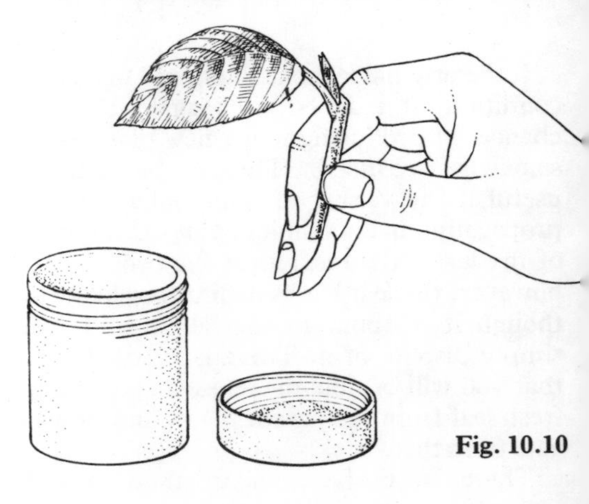

Fig. 10.10

Fig. 10.10 Use hormone rooting powder when taking a leaf-bud cutting of camellia.

provided for these rather difficult cuttings is a matter of some controversy. Some gardeners advocate placing them in a garden frame. However, the greatest success rate by far is achieved if you are able to provide a greenhouse propagating case with under-soil heating or a mist unit. In any case, be sure to shade the cuttings from continual direct sunlight. Keep the compost and atmosphere moist, and check the cuttings regularly for yellow leaves, removing them straight away.

Camellia cuttings may take two months or more to form a good root system, so be patient. At this point they should be potted individually into small pots, using an ericaceous compost (ie. specially prepared for plants requiring an acid-based growing medium). The cuttings should be overwintered in a cold frame or greenhouse.

Other hardy garden shrubs may be propagated in a similar fashion. Try, for example, clematis, hederas (ivy) and passifloras (passion flower). For these, use soft growth during the spring. The cuttings should, essentially, be prepared in the same way, but the need for hormone rooting powder is not so great. Root the cuttings in warm, moist conditions, as provided by a heated propagating frame and mist unit.

Mid-autumn is the best time to take leaf-bud cuttings of mahonia (Oregon grape), although some success can be achieved with mid-spring cuttings. Use leaves from healthy, green semi-ripe stems, not diseased and brown (this colour telling you that the stem is fully ripened, or hard). Mahonia leaves are quite large and heavy, so it is wise to reduce them in size, by about a third, and to have a slightly longer stem, say up to 5 cm (2 in). Again, rooting can be a slow process. Potting up should take place as soon as a good root system has been formed.

Species of rubus, namely blackberries and loganberries, may also be raised from leaf-bud cuttings taken in late summer. Root them in a garden frame, using ordinary cuttings compost. Planting out may take place the following spring.

Tender plants

A similar type of leaf-bud cutting is sometimes used in the propagation of *Ficus elastica* (large-leaved rubber plant). Select a suitable young stem during spring or summer, and with a pair of sharp secateurs cut it up into 1–2.5 cm (½–1 in) sections; each section containing one leaf and its bud. After dipping the base of each cutting in hormone rooting powder formulated for softwood cuttings, insert it in a 7.5 cm (3 in) pot of cuttings compost.

Fig. 10.11

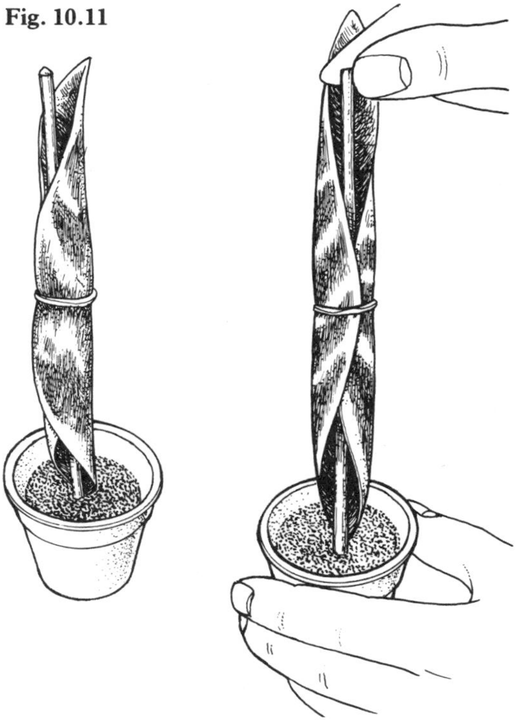

Fig. 10.11 Leaf-bud cuttings of *Ficus elastica* should be easier to handle if folded and held back with an elastic band.

An elastic band placed around the leaf will keep it compact, therefore saving valuable greenhouse or window sill space, and also make handling much easier (Fig. 10.11).

Rooting should take place within a month in warmth, and potting on a few weeks after that. Try to perform the above operations as quickly as you can, because when ficus stems are severed they lose 'blood' (the white latex sap).

Similar techniques are used in the propagation of other greenhouse and house plants, including *Aphelandra squarrosa* (saffron spike), epipremnum (devil's ivy), dracaena, *Monstera deliciosa* (Swiss cheese plant), philodendron (sweetheart vine), rhoicissus (grape ivy), and the more tender forms of hedera (ivy).

CHAPTER 11
Division

Propagation by 'division' is a term which hardly requires a definition. It really is self-explanatory, meaning specifically that some plants can be divided into several pieces, each of which will grow on into a self-sustaining new plant. It is probably the simplest and most basic way to increase any plant, and requires a minimum in the way of tools, equipment and conditions.

Importance of division

Strictly speaking all plant propagation, except by seed, is effected by division. If we propagate by cuttings or, as we shall see later, suckers, grafts or buds, we must, in all cases, divide the plant to obtain them. However, division as a separate distinction within the overall subject of propagation, is usually understood to imply the parting of tufted plants, so that each part has roots and, if possible, growing points.

The history of plant division is as old as Nature itself. Long before Man existed, and when the plant kingdom was just beginning to evolve, and flowers, fruits and seeds were setting for the first time, many plants had already developed a means of increasing and spreading by natural vegetative growths. These growths were invariably underground and it was probably some time before Man fully understood what was happening to the plant, and indeed how it managed to perpetuate.

Rhizomes

Many of the early plants that Man observed were spread by means of rhizomes and suckers. The former are actually modified food-storing stems, though they are often mistaken for roots. Each piece of rhizome with a bud attached will grow into a new plant. Some grasses have slim underground rhizomes which produce new plants wherever they are broken. The infamous weed couch grass is a prime example. More desirable rhizomatous plants include the bearded iris, bergenia (elephant's ears) and convallaria (lily of the valley).

Suckers

Suckers are familiar to all rose growers, but many garden plants have a tendency to produce them. Basically, a sucker is the name given to shoots arising directly from the roots, often at some distance from the parent plant, and often mistaken for seedlings! Although more often than not regarded as a nuisance, suckers can be used as a means of increasing stock, and are simply separated from the parent root, and replanted in a new position. Familiar suckering plants include raspberries and other cane fruits, roses, *Rhus typhina* (stag's horn sumach), syringa (lilac) and symphoricarpos (snowberry).

Nature is not always so straightforward and simple, however. Although Man noticed that wild plants had this almost magical ability to spread and increase, the new plants were usually crowded together, within close proximity to their parents. Consequently there was severe competition for the available nutrients from the soil, as well as light and air. This can be seen today, if a garden is left to its own devices for a period of a year or more; weeds and scrub take over, neat shrubs will expand and overcrowd smaller subjects, and the plant mass will treble quite quickly.

The lesson that Man has learned from all of this is that if we are to maintain many of our garden plants in optimum condition, and to prevent them from spreading too far and encroaching upon their neighbours, we need periodically to operate on them and make them smaller. This often takes the form of pruning, and it also occurs as 'division', but with the added bonus here that the pieces created from dividing a large plant, can be replanted, so increasing our stock.

Simple division

In Nature, the simplest form of division is shown by many algae and fungi. The threads from which they are composed may be broken into several pieces, and yet they continue to grow. Liverworts frequently produce tiny detachable buds, known as 'gemmae', in shallow cups. Splashes of rain will wash these gemmae out and they develop into new plants on the soil nearby.

In modern day gardening, splitting up plants into several smaller ones, each with roots attached, is known as 'simple division'. It is a method of plant increase that even the most inexperienced gardener can attempt and often achieve excellent results, if not 100% success. Many different types of plant lend themselves to this method, but they are invariably the clump-forming, tufted or matted types. A number of hardy perennial plants are included here, such as species of aster (Michaelmas daisy), rudbeckia (cone flower) and *Primula denticulata* (drumstick primula). Some alpines may also be increased by division, including species of aubrieta, arabis, dianthus, and veronica.

Let us examine each of the appropriate plant areas, and see how division is carried out within them.

Alpine and rock plants

Nurserymen regularly use division as a means of increasing stocks of alpines, and for a number of reasons. First, it is cheap; one mature plant may bear as many as six or more offspring. Second, no specialist equipment is needed, other than pots, compost and labour! Third, it is relatively quick; substantial plants of flowering standard may be produced in 9 cm (3½ in) pots within a year.

The following types of alpine, generally, may be propagated by division: mat-formers, cushions, carpeting types, rosette types, as well as those producing offsets (young plants growing out from a parent plant, and attached to it by a short stem). The best time of the year to undertake this is early to mid-spring, just as the plants are starting into growth. It is at this time of the year that the divisions establish most quickly. An alternative time would be late summer or early autumn, to establish the new growth before winter sets in.

It should be understood that most alpines take a long time to form large clumps, and it is therefore a pity to destroy the appearance of a rock garden by lifting whole plants and splitting them. The best method is carefully to dig out rooted pieces from the edges of clumps, pot them up and treat them as if they were young seedlings. An exception to this rule are plants such as arenaria, raolia and thymus (thyme), which have the ability to settle in again quickly if lifted.

One important point: *Gentiana sino-ornata* (the autumn-flowering gentian) should be lifted and divided every three years, and then replanted in a different part of the rock garden. Alternatively, lift it, replace the soil and replant it in the same position. If you do not do either of these things, the soil can become 'gentian-sick', and the plant will begin to deteriorate and eventually die.

Division technique

Lift alpines carefully with a fork. Using your fingers, gently pull the plants apart into pieces about 5 cm (2 in) across, making sure that each piece has some root growth attached (Fig. 11.1).

Fig. 11.1

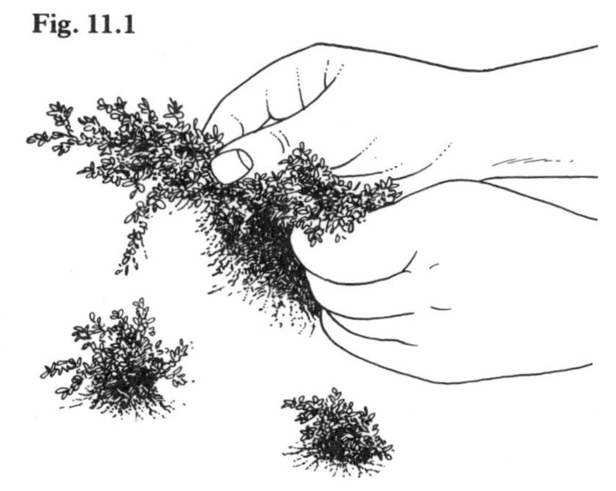

Fig. 11.1 Alpine plants may be divided with fingers. Pull them apart gently, as if you were pulling apart the two halves of an orange.

Certain alpines have the ability to reproduce themselves by means of offsets, such as sempervivum (houseleek) and *Androsace sarmentosa* (rock jasmine). Usually it is quite possible to remove these offsets without disturbing the parent plant. Simply ease them out of the soil with a small hand fork and pull them away from the parent plant – they generally 'give' readily.

Aftercare

Many gardeners replant their divisions straight back into the rock garden, and there

Fig. 11.2 Each division from the parent alpine should be potted singly.

Fig. 11.2

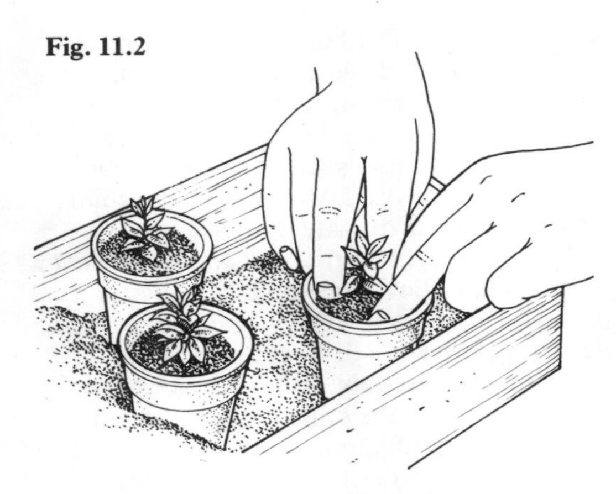

is nothing wrong with this. But it is often good practice to grow the divisions into larger plants before planting out. In this case, pot them into 9 cm (3½ in) pots, using a loam-based potting compost mixed with a small handful of grit or sharp sand (per pot) to improve drainage (Fig.11.2), which all alpines appreciate. Potting itself can be a problem with some species of alpine, for they are often flattish plants, and it is sometimes difficult to avoid getting compost all over the foliage. All experts can suggest is that you pot with care, and try to avoid this happening as far as possible.

Firm the compost around the small plants, and then water them in with a can and fine rose. Now place the pots in a sunny but sheltered spot in the open garden. It is essential that you keep an eye on the water requirements of these pots; they will soon dry out in hot weather, and some alpines will not tolerate being dry for too long. Sometimes it is possible to plunge the pots up to their rims in garden soil. This prevents them from drying out too rapidly. Weathered fire ash is even better, as it holds moisture well.

It is best to overwinter pots in garden frames to prevent them from getting too wet. Ensure good ventilation throughout winter, especially on mild or sunny days. Fail to do this, and the alpine divisions will go 'soft', and may even start to grow, as if it were spring. Practically all alpines will endure frost and ice, but they succumb quite readily in prolonged wet conditions.

The following spring, or a year or so from division, the plants will have developed a good root system within the pots, and will be ready for planting out onto the rock garden.

Aquatic and water plants

The basic difference between hardy aquatic plants and hardy border perennials is that the former live in water! This may seem frivolous, but the two types of plant are essentially the same in that they are both perennial, and often form clumps that may be divided to increase stocks.

There are several distinct types of water plant, so let's look at each.

Submerged oxygenators

These are also known as 'water weeds', for they are grown simply to provide some greenery in the pond. It is this greenery, however, that is the life-blood of most water gardens: it is essential for a well-balanced pond. It provides oxygen for fish and livestock (also providing useful breeding sites), and it helps to keep the water clear.

Division technique

Generally quite vigorous, these oxygenators will need lifting and reducing in quantity every couple of years. A plant may be scooped out of a pond by using a fork or rake, or you can use your hands. Be sure, though, not to bring with it any small fish or other water creatures. It will, undoubtedly, be a tangled mass. Gently pull it apart into small bundles, and tie these together with pieces of wire or elastic bands. Then replant several bundles, the number according to the size of your pond.

You will have quite a bit of weed left over. This can be given away to other pond owners or, more than likely, you will have to discard it. It makes a good addition to the compost heap!

Bog plants and marginals

Let us first understand the distinction between these two. Bog plants are subjects which thrive in the moist soil at the edge of a pool, such as ajuga (bugle), mimulus (monkey flower) and astrantia (masterwort). Marginal plants, on the other hand, grow in the shallow water at the edge of a pool. Included within this group of plants are juncus (rushes), acorus (sweet flag) and typha (reedmaces).

Division technique
This is exactly the same as for hardy border perennials (see page 83), the only difference being that the best time for dividing bog and marginal plants is late spring to early summer. When dividing, remove any dead or dying leaves, as these will only foul the water if left. In order to see what you are doing, it is a good idea to wash the soil and mud away from the roots of these plants. Replant the divisions straight away, in the same containers, and discard the unwanted pieces.

Water lilies
No one will argue that the water lily holds pride of place amongst aquatics; few plants have aroused such universal admiration. Botanically referred to as *Nymphaea spp.*, the water lily is closely related to the buttercup and delphinium, though few people can see the resemblance! Although water lilies may be raised from seed, this requires quite specific conditions, and it is more common for them to be propagated by division.

Division technique
Water lilies produce their top growth from tubers, and these are lifted in spring or early summer and cut into pieces with a sharp knife. Each portion, if possible, should contain a growing shoot (or 'eye'), a portion of the old tuber and a certain amount of fibrous root. Even small pieces the size of a thumbnail should grow!

The secrets of success are shallow water and not too much soil. Fill a pot with a suitable compost, such as heavy loam, finely screened, and a little powdered charcoal. The eyes are put into pots, firmed in with fingers and introduced to the pool so that the rim of the pot is just covered with water. Very small eyes should be kept in a garden frame until roots develop. These may be transferred outdoors during early to midsummer.

Avoiding weeds
If you are unfortunate enough to have blanket weed covering the surface of your pond, it is essential you do not propagate this at the same time you are propagating the desirable plants. Remove all traces of it before replanting the divisions; the tiny green leaves and filaments hide easily amongst larger foliage and roots.

Greenhouse and house plants

Many pot plants develop into clumps, and so division makes a most acceptable form of propagation. Examples are chlorophytum (spider plant), cyperus (umbrella plant), maranta (prayer plant), saintpaulia (African violet), sansevieria (mother-in-law's tongue) and many ferns. Some orchids may be propagated by division, and these are covered separately, later. Do not try to divide either woody plants or palms, even if they are clump-formers; such plants are unlikely to survive the treatment, althought suckers may be taken.

Division technique
Take the plant out of its pot, find a point where separation seems possible, and carefully divide the plant using your fingers (Fig. 11.3), or by using a sharp knife. It may be possible to pull apart certain clumps by hand. However, for plants such as ferns that make tight clumps with dense root systems, or for subjects with roots that are strongly linked together, a knife may be required. Try, wherever possible, to use the knife only to start with. If the division can be completed by pulling with the hands (Fig. 11.4), the stems and roots are less likely to be damaged.

Some gardeners advocate brushing or washing off much of the compost from around the roots, in order to see what they are doing, and to untangle the roots properly. This is not, however, generally a good thing to do as it may damage some of the very fine root hairs, and cause a slower settling-in of a division in a new pot.

Always plant a divided portion of a plant in a pot that is a little larger than the rootspread. Use a proprietary or home-made potting compost, and part-fill the pot with only enough soil to meet the lowest roots. Then put the division into the pot, still holding it with one hand and trickling more compost between and around the

Fig. 11.3 *Sansevieria trifasciata* (mother-in-law's tongue) will grow into a large clump.

Fig. 11.4 Offsets may be pulled off, or cut from the parent plant. Each offset may be planted into its own pot. It will soon establish itself and, in time, produce further offsets.

roots with the other. To prevent air pockets, settle the mixture by tapping the pot gently on a hard surface. Then top up as required. Soil or loam-based potting mixtures should be pressed firmly around the newly divided plant, whereas peat-based or soilless types do not need this treatment. Make sure you water the plant, to settle the compost around it.

Aftercare
If possible, keep the divisions in a warm place for the first couple of weeks to encourage them to form new roots. Good light, but shade from direct sunlight will help. Newly divided pot plants seldom require any special after-treatment. Wilting may occur with some subjects; if so, place the plant in a propagating case or cover it with a plastic bag to improve the humidity level.

Orchids
There are many different kinds of orchid, but the type most easily propagated by division are the epiphytic 'sympodial' species. These include species of cattleya (corsage orchid), odontoglossum (tiger orchid), cymbidium and paphiopedilum (slipper orchid). They usually have many stems, but all of which rise from a horizontal rhizome. In the wild, the rhizome runs over the surface of the support – a tree or branch – but in cultivation it sits on the surface of a potting mixture, sending feeding roots down into the compost. Thick stems, called 'pseudobulbs', rise from the rhizome at intervals. They are given this name because, like bulbs, they are the storage organs for food and water, and can help plants survive through short periods of drought.

The pseudobulbs are usually light green in colour, but they vary widely in size and shape, being either egg-shaped, cylindrical, spindle-shaped or round. They can appear like elongated bulbs or ordinary stems, or they can be more or less flattened out.

Division technique
As a first step in dividing such an orchid, cut through the rhizome about four weeks before actually dividing the topgrowth. This gives the cut ends time to heal before the upheaval of removal and repotting. In ideal circumstances, each section should carry two or three pseudobulbs (or clumps of leaves). After removing the plant from its container, disentangle the roots carefully and repot each part individually into a suitably sized container. Some keen orchid

growers also plant several pieces of detached rhizome together in a large hanging basket.

Aftercare
Water newly separated orchids sparingly; a daily overall mist-spray may be sufficient, at least until new growth indicates that additional rooting has occurred. Thereafter, treat each section as a mature plant.

Hardy perennials

Herbaceous border plants, including hardy ferns and ornamental grasses, grow quickly and soon make thick clumps. After a couple of years, the more vigorous plants are quite likely to start to deteriorate as the shoots become overcrowded and the centres of the clumps get woody. As we saw in the first part of this chapter, dividing plants into smaller pieces will rejuvenate them as well as increase one's stock, and this is particularly the case with perennials. As a general rule, the dividing of border plants can occur either as they are dying down in the autumn, or in spring, just as they are starting into growth. Spring division generally gives the best results, when the weather is getting milder and the whole of the growing season is ahead of us. Wait until the plants have plenty of sturdy young shoots.

The hardy garden species of both ferns and ornamental grasses are enjoying considerable popularity today. They are both clump-formers and so are suitable for propagation by division. Again, the best time of the year is mid-spring, just as the plants are beginning to grow apace. The majority of grasses and ferns benefit from division about every three or four years.

On the other hand, early-flowering perennials, such as *Primula denticulata* (drumstick primrose), some doronicum and pulmonaria, are best divided immediately after flowering. Also included here is the bearded iris, but the technique is slightly different (see pages 88-89).

It is also worth remembering that some common perennial plants do not respond well to being lifted and divided, due in the main to their resentment of root disturbance. Included here are species of paeonia (peony), helleborus (Christmas and Lenten roses), *Anemone* × *hybrida* (Japanese anemone) and alstroemeria (Peruvian lily). In addition, you will also find that some perennials cannot be divided because they are not clump-formers. In other words they are single-stemmed or tap-rooted, and do not have a fibrous root system. These include dianthus (carnations and pinks), *Papaver orientale* (Oriental poppy) and lupinus (lupin). All of these will need to be propagated by different methods (refer to the plant listings in Part II).

Fig. 11.5

Fig. 11.5 Dividing a perennial: carefully lift the plant with a garden fork.

Division technique
The sequence to follow is quite specific. Carefully lift the plant with a garden fork, retaining a good soil ball around its roots (Fig. 11.5). Once it is out of the ground, use your hands to remove most of the soil; this way you avoid doing too much damage to the delicate fibrous roots. Now lay the plant on a clear piece of ground.

If you are dealing with a large clump, insert two garden forks, back to back, through the centre of the plant so they are touching. Make sure they are pushed right through the clump, and that the prongs of the fork appear through the base (Fig. 11.6). Now pull – or push – the handles of the forks apart; this will divide the clump into two pieces. A very large plant can be further divided by the same method. Two handforks can be used in the same way on smaller plants, or those which require rather less force.

Plants such as, for example, some forms of primula (primrose, polyanthus and

Fig. 11.6

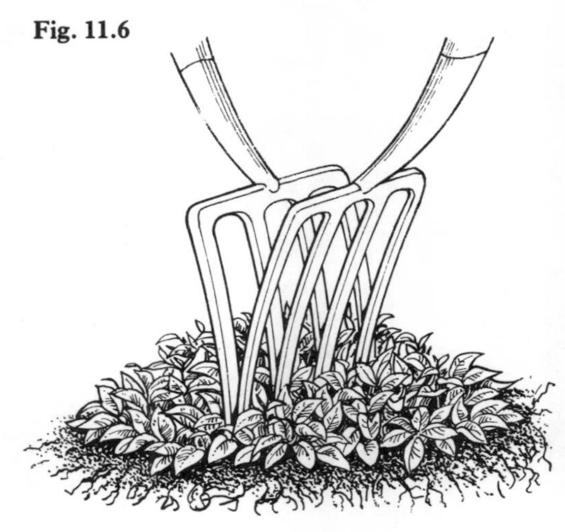

Fig. 11.6 If you are dealing with a large clump, split it by using two forks back to back.

auricula), can be tugged apart by hand, teasing the roots gently and being careful not to snap young shoots. Many species are very easy to pull to pieces; others need a little more patient work before they will come apart without damage to roots or stems.

Then, of course, there are some plants that are so tough, they will not divide without being cut with a knife or spade. Included here are some of the ornamental grasses, such as *Miscanthus sinensis* 'Zebrina' (zebra grass) and forms of cortaderia (pampas grass). Make sure each section has several strong shoots and plenty of fibrous roots.

The centres of perennial plants are usually hard and woody, and should be discarded for they will not produce good new plants (Fig. 11.7). The growths around the edge of the clump on the other hand should be strong, healthy and vigorous. These are the portions which should be retained for replanting. Ideally, each division should be about 10 cm (4 in) across the crown.

These divisions should give a good flower display the following summer – from either an autumn or spring divide. If you have something particularly precious that you went to increase as fast as possible, it can be split into smaller pieces. As long as each one has a shoot and some roots, it will survive. Tiny divisions of only 2.5 cm (1 in) or so across, should really be potted up, with a proprietary potting compost, to give them the best start possible.

It is important to avoid the air from drying out the roots of divided plants, so make sure you replant them as soon as possible (Fig. 11.8). Before you do, quickly dig over the place from where the original plant was lifted, removing perennial (and annual) weeds, and adding a handful of general fertilizer before replanting. Using a can with a rose fitted, water each of the divisions well, to help them settle in their new homes.

Rhizomatous border iris, with time, develop into a large circle with tufts of

Fig. 11.7

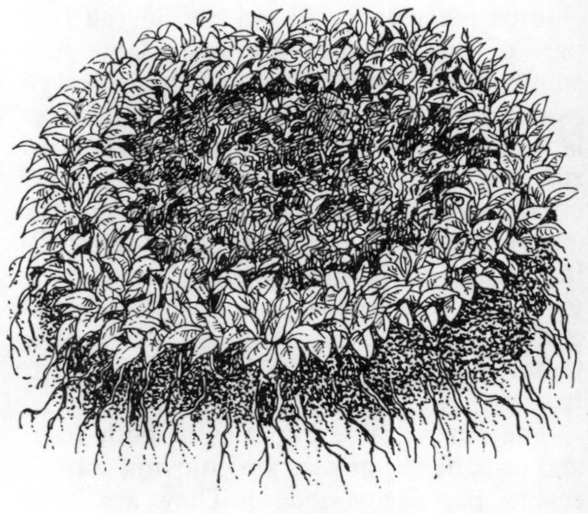

Fig. 11.7 The old, centre section of a large clump will not be worth retaining. It is the material at the edge that is fresh, and will grow away most vigorously.

Fig. 11.8

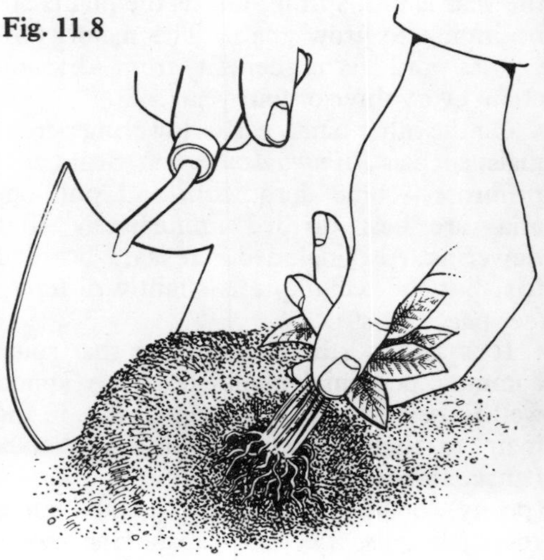

Fig. 11.8 Replant the small divisions as quickly as possible, to avoid them drying out.

leaves round the edge and a large dead patch in the centre, which usually becomes infested with weeds and is difficult to keep tidy. Like the ferns and grasses, these plants should be lifted and divided every three or four years as a matter of routine. These, however, should be lifted in summer (Fig. 11.9), once they have finished flowering.

Established clumps are very tough, and will require quite a bit of digging out. With a knife, cut off and retain all the healthy, vigorous rhizomes with fans of leaves emanating from them, and discard the unwanted woody portions from the centre. Cut through each portion of rhizome about 8 cm (3 in) in front of the fan of leaves (Fig. 11.10). When you have fully prepared them, each section of rhizome should have plenty of roots and a fan of leaves.

Before replanting, reduce the amount of leaf (removing the top two-thirds), to balance the plant (Figs. 11.11 and 11.12). This will also prevent the plant from being rocked by the wind. Tickle over the soil with a fork and add a handful of general fertilizer, then plant the rhizome sections shallowly, directly where you wish them to flower.

Bergenias

These are evergreen perennials, often referred to as 'elephant's ears', and flower from late winter until mid-spring. They produce thick fleshy rhizomes at soil level, and these can be used for propagation. In mid-winter, before flowering, carefully lift mature plants and cut off some of the rhizomes. Provided you do not cut them back too hard, and

Fig. 11.9

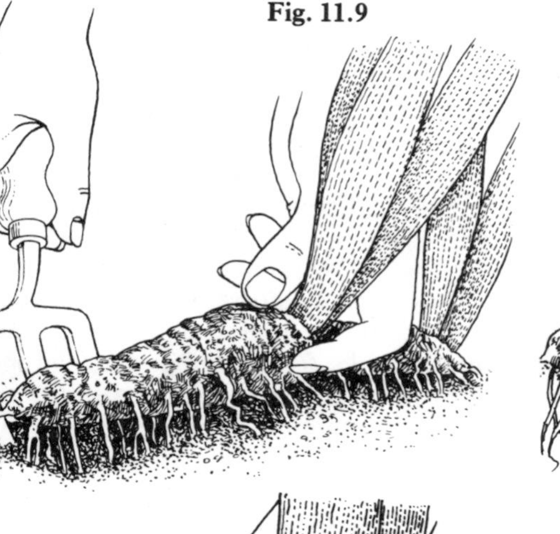

Fig. 11.10

Fig. 11.9 Border irises may be divided. First, lift the underground stem, or rhizome, with a fork.

Fig. 11.10 Cut the rhizome with a sharp knife. Discard unwanted, woody portions, retaining only the healthy, vigorous portions with leaves.

Fig. 11.11

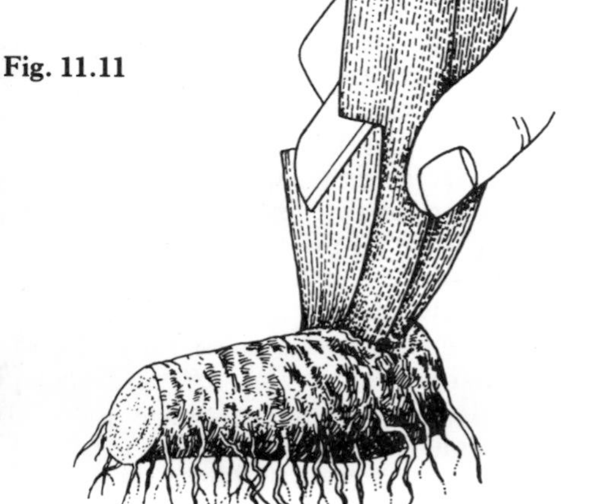

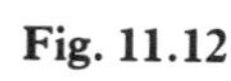

Fig. 11.12

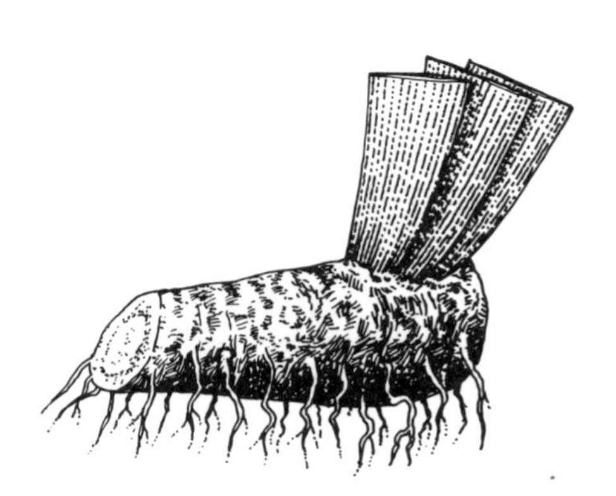

Fig. 11.11 Reduce the amount of leaf to balance the plant.

Fig. 11.12 A trimmed piece of rhizome, with some leaf attached. This is ready for replanting.

there is still plenty of fibrous root, the plant will not be damaged – except the check it will receive as a result of it being lifted.

Wash the rhizomes thoroughly and search for the dormant buds that should be present. Cut each section so that it has at least one bud, and is about 2.5–3.5 cm (1–1½ in) long. Put the sections in a polythene bag with a small amount of fungicide containing gamma-HCH, and lightly shake them together. This will help to prevent rotting.

Root the rhizome sections using ordinary, but sterile, potting compost. An inert substance, such as vermiculite or perlite may be used. Fill a seedtray with the medium of your choice, and then push the sections horizontally into it, so they are buried to about half their depth. Now water with a can and fine rose. Gentle bottom heat is desirable for starting bergenia rhizomes into growth, so place the tray on a greenhouse bench heated by soil-warming cables, or use an electrically heated propagating case. Set the thermostat for about 21°C (70°F).

Shoots and leaves will soon appear, but do not remove them from the compost yet. As with rooted stem cuttings, you should always wait until the first fibrous roots appear through the bottom of the tray. At this stage, pot them up undividually, using a soil-based potting compost. Gradually harden them off, and plant them outdoors during late spring after the danger of frosts has passed. Although a mature bergenia is an extremely tough and hardy plant, such a small specimen as a rooted rhizome, and one that has been nurtured in heat, will be soft and fleshy, and easily damaged by the cold weather.

Trees and shrubs

Certain trees and shrubs produce shoots from below ground level, and these are known as suckers. Very often, suckers are regarded as a nuisance, because they cause the plant to spread beyond its allocated space. With fruiting trees and shrubs, suckers sap the energy from the plant, which could otherwise be put toward further fruit or flower production.

Nevertheless, in the majority of cases suckers can be used as a means of increasing stock, and are simply separated from the parent root and transplanted to a new position. This is generally undertaken at the normal time for planting such plants – autumn or early spring.

Many plants can be increased in this way, including shrubby cornus, *Rhus typhina* (stag's horn sumach), raspberries, and passiflora (passion flower). However, it must be remembered that suckers rising from grafted plants such as fruit trees, roses, syringa (lilac), and so on, will produce plants identical to the rootstock, which is most unlikely to have the desirable characteristics of the grafted portion.

Division technique

The job is simple enough. With a spade, carefully dig around the roots of the tree or shrub, trying not to damage the parent plant. With a knife, or a pair of secateurs in the case of larger specimens, cut the sucker away from the original root, but preferably with a small piece of that root still attached to its base.

These suckers, once lifted, should ideally be grown on in a nursery bed, in a sheltered part of the garden. If the weather is unfavourable for planting them properly here, you can heel the suckers into a spare piece of ground, until you are ready.

Aftercare

Because some suckers are quite large when they are separated, they can be a little more difficult to get established. Make sure you water them well at each stage, both when they are planted in the nursery (and if they were heeled in first), and at the transplanting time. In the first few days in the nursery bed it is as well if the suckers can be shaded from the strong sunshine and sheltered from the drying winds.

The following spring, move the suckers to their final position in the garden. They will by now have an established root system. It is worth remembering that, in time, they will produce more suckers of their own.

CHAPTER 12

Layering

Propagation by layers is a very old but useful and much practised art, as many plants can be increased by this method that cannot, with reliance, be multiplied by others means. These include such common garden fruits as rubus (blackberry, loganberry) and fragaria (strawberry), and many ornamental trees and shrubs, such as rhododendron, *Liquidambar styraciflua* (sweet gum), hamamelis (witch hazel), some species of magnolia and tilia (lime). The basic outdoor forms of layering are used only for those plants whose shoots and stems can easily be brought down into contact with the soil.

Layering, and there are many methods the amateur gardener may adopt, often occurs naturally. The simplest form of layering is called, not surprisingly, 'simple layering', and it involves weighing down a stem or branch into the soil while it is still attached to the parent plant. The idea is for roots to develop at the point where the branch touches the ground, and at an appropriate stage the rooted part is severed from the main plant. Perhaps the best example is that of the bramble (species of rubus). The shoot tips of these plants will easily root into the soil once they come into contact with it, and due to the bramble's growth habit, a thick, thorny impenetrable jungle may soon develop.

In gardens a number of cultivated shrubs and plants may layer themselves naturally; these may include the spring-flowering forsythia (golden bell), the winter-flowering chimonanthus (winter sweet) and forms of the summer-flowering dianthus (border carnations). In the wild, hedgerows and woodland for example, branches of fagus (beech), juniperus (juniper) and some viburnum sweep down to the soil and root naturally.

The advantages of layering for the gardener are that it costs practically nothing to undertake. Generally speaking, no artificial heating or lighting is required, no special tools or equipment; in fact, all you need is the mother plant, something with which to peg down the stem(s), some string and a little peat and sand! What is more, as this is a form of vegetative propagation, the offspring from your endeavours will be identical in every way to the parent plant. With a little care, layering can result in 100% success. Let us look, in more detail , at each of the many forms of layering.

Simple layering

Numerous trees and shrubs can be propagated by simple layering. In Part Two, individual plants are listed, but layering is only recommended for those which may prove difficult for home gardeners to propagate by other means. In other words, if you have a garden shrub or climber, for example, that you would like to layer, but find little or no information available on the technique to adopt, try the method of simple layering as described below. There is a very good chance of achieving complete success.

Simple layering may, technically, be accomplished at any time during the year. However, it is generally considered to be more successful when the plants are actively growing (that is, mid-spring to mid-summer), when rooting should happen more quickly. Choose a mild day when the soil is not too wet. Good soil preparation is one of the most important ingredients in simple layering. With a border fork, dig over the soil around the plant to be propagated, being careful not to damage its roots. Do not leave the soil in clods, break it down into a fine tilth, almost as if you were going to sow seeds in it. Now the soil should be given a moderate firming with your heels. This should not be overdone, so that the ground is like concrete. All you need to do is firm and settle it in order to get rid of any pockets of air.

A light, sandy soil is ideal for layering, for three good reasons. First, it is well-drained; second, it is open and aerated; and third, it warms up quickly in spring. A word of warning, though. If your soil is very sandy, it perhaps has little or no water retention qualities, therefore a bucketful or two of moist sphagnum moss peat should be incorporated at this stage to help retain water during dry periods. If, on the other hand, you garden on a heavy, clay soil, it would be as well to incorporate a bucketful or two each of sand and peat to help open up the ground to encourage rooting. Always ensure that any added peat or sand is mixed thoroughly with the garden soil, and to a depth of about 30 cm (12 in). This will encourage the new roots to develop a good, large system, prior to being lifted.

Now you must choose the stems, or 'layers', you are going to bring to soil level. The age of the tree, shrub or plant is of no concern to the process (in fact, an old specimen can be ideal for propagation in this way, and so perpetuate its usefulness in a second generation), but the layers should comprise young wood that was formed in the previous or current season. If older wood manages to root at all, it would probably take an inordinately long time to do so.

A single plant may offer anywhere between one layer and a dozen or more. The only limiting factor is the amount of ground space available for accommodating the layers! In commerce, propagators often cut subjects down to the ground so they will throw out masses of long whippy stems, which will be ideal material for pegging down in a large circle around the parent plant. From the centre of the plant will often grow more shoots, giving the opportunity for a second batch of layering once the original ones have rooted.

To encourage rooting in the quickest possible time, it is important to wound the part of the stem that is to be in contact with the soil. This constricts the flow of hormones and nutrients. But there are one or two stages to get through before the wound is made. First, grasp the chosen shoot about 23–30 cm (9–12 in) behind its tip and bend it down to the soil (Fig. 12.1). If the bending is comfortable, with no noticeable straining, mark the place on the stem where you are holding it and which is touching the ground, for it is this spot that should be wounded, and subsequently buried. At the same time, mark the point on the ground where the stem touched, for it is here that the stem will be buried. Dig a small hole (Fig. 12.2), or make a depression in the soil.

Strip off the leaves and leaf stalks or stem for about 10 cm (4 in) either side of the point to be layered. This will help to divert energy into the formation of roots. Now make the wound. The best way to do this is to make an angled cut with a sharp knife (Fig. 12.3), cutting less than half way into the stem to avoid snapping it completely. A small stone, about the size of a pea, should be pushed into the cut to prevent it from healing.

Pull the shoot down to ground level, into the hole. The tip of the stem, with its leaves still in place, should be above ground, pointing skyward. Use a large stone as a weight, or a piece of thick bent wire as a peg, to secure the stem firmly into the soil (Fig. 12.4). Fail to do this, and it is quite likely that the stem will work itself loose, and out of the soil, particularly if it is windy.

Backfill the soil over the piece of buried stem, making sure that you create total darkness around the wounded section. Now gently tread the soil so that it is firmed in place, and water the area well. Many gardeners at this stage insert a short cane alongside the exposed stem tip, and tie the two together with twine. This serves the purpose of holding the stem in a more or less upright position; a form of pre-training for the newly layered plant!

Aftercare

Inspect the layers regularly during dry weather; remember, the soil should be kept moist at all times if you are to encourage good rooting. For this reason it is important to make sure that weeds do not grow to maturity around the parent plant, for these will take moisture out of the soil. Try to hand-weed only, as hoeing may damage the new root growth, and the use of strong chemicals could have an adverse effect on its development.

Roots will form during the following season, but it may be 12–18 months, depending on the plant and the conditions, before the layers are ready for separation. You will instinctively know that rooting has

Fig. 12.1

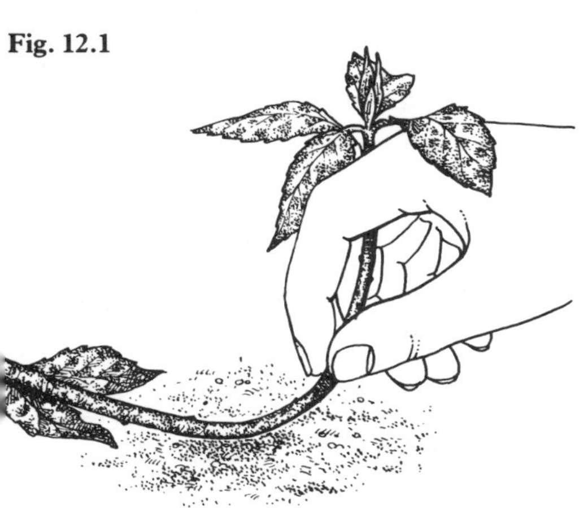

Fig. 12.2

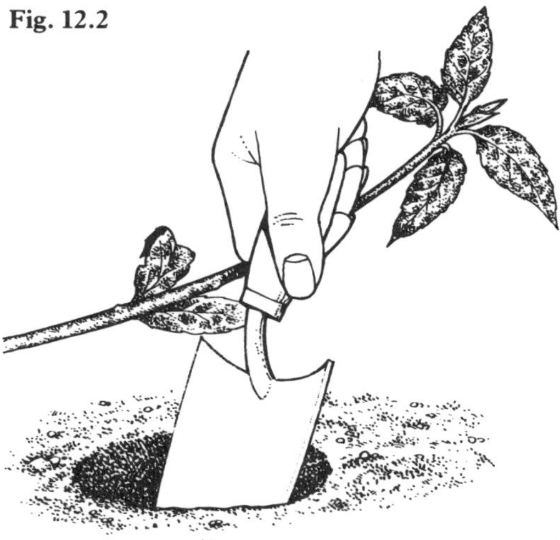

Fig. 12.1 In simple layering, begin by bending a flexible branch down to ground level.

Fig. 12.2 Using a trowel, make a hole about 10 cm (4 in) deep at the point where the branch came into contact.

Fig. 12.3

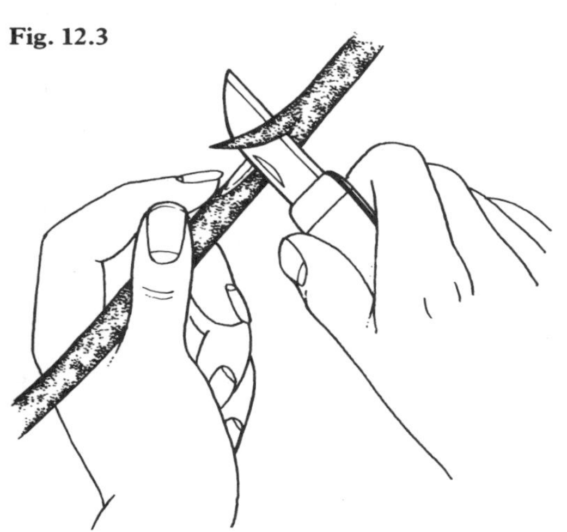

Fig. 12.4

Fig. 12.3 Slit the stem with a sharp knife and wedge it open with a small stone.

Fig. 12.4 Pull the branch down and, using a piece of bent wire, peg it firmly into the soil.

been successful, and this will more than likely be evident by quick spurts and development of new topgrowth. During autumn or spring, take a pair of secateurs and simply sever the connecting stem between the parent plant and the rooted layer. Don't be tempted to dig up (Fig. 12.5) and reposition the new plant straight away; leave it to get used to feeding off its own system of roots for a month or two.

Serpentine layering

This is a form of layering that is particularly suited to long stemmed plants – often creepers and trainers – such as lonicera (honeysuckle), clematis, wisteria, and jasminum (jasmine). The process is an adaptation of simple layering, and basically involves

Fig. 12.5

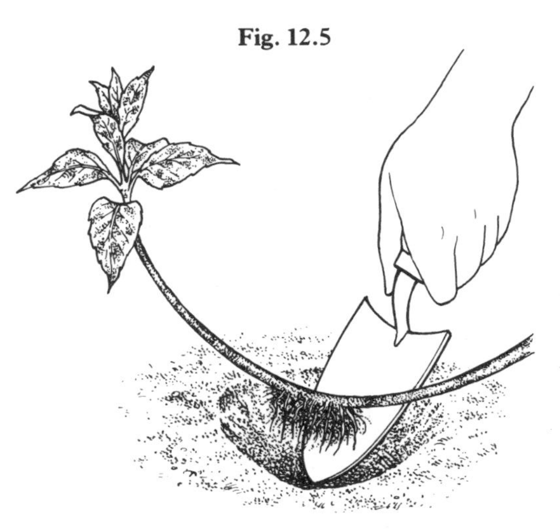

Fig. 12.5 Eventually, roots will come from the area around the slit. When digging it up, try not to damage any of the small roots.

Fig. 12.6 Serpentine layering produces a number of plants from one long, pliable shoot.

producing a number of new plants from one stem. Such a stem must, naturally, be pliable and long enough to carry several layering positions.

Serpentine layering can be carried out at the same time of year as simple layering, with an equally high success rate. As before, the age of the plant has no bearing on the propagation process. What you must do, however, is choose material that is suitable for the purpose: a vigorous, healthy shoot of wood grown during the current season.

Two or more parts of the stem will be submerged in the ground (the soil being previously prepared, as for simple layering), and roots will emerge from the buried leaf joints (or nodes) (Fig. 12.6). However, between each buried node, a length of shoot must remain above ground. An ideal stem should have nodes at regular intervals, perhaps every few inches. Choose as many nodes as you want layers, and carefully strip the leaves from them. Ensure that there are two or three nodes between each prepared one.

The stem should be wounded several times to encourage rooting: make a slanting cut half way through the shoot for a length of some 5 cm (2 in). This should be positioned immediately behind each prepared node. A small stone should be used to prop this 'tongue' open, and to stop it healing.

The parts of the stem that are wounded are then pegged, weighted or buried in the soil. Sometimes pots of compost sunk into the ground make acceptable vessels in which to root layers; it makes for easier transplanting later. The stem need not be laid out in a straight line; this would take up too much space. Instead you could consider winding it in a sweeping curve around the parent plant. When the planting of each section is complete, you will have a series of loops of stem above and below ground – something like the popular photographs of the Loch Ness monster!

Aftercare

As with simple layering, the soil should never be allowed to dry out, or become weed-infested. The layers should have rooted within a year or so, depending on conditions and species. At this stage the whole of the layered shoot should be carefully lifted and cut into several portions; each one possessing a length of stem with shoots and, hopefully, leaves, as well as a small but sufficient root system.

Each new plant should be repositioned as quickly as possible. This may be in a nursery bed for further development, or straight out into the open garden in its new, permanent home.

Tip layering

This form of layering exploits a natural method of propagation, and it is very simple to do . It is used widely for increasing stocks of cultivated blackberries and loganberries (rubus); we are emulating the natural propagation of wild brambles. Basically, the tips of strong shoots are pegged down or dug into the soil where they touch.

Fig. 12.7 Tip layering occurs naturally with members of the bramble family, and can be used to propagate many other shrubs.

Fig. 12.7

The best time to do this is mid-to late summer when there is plenty of new wood available. Take a supple shoot with a strong, healthy growing point, and bend it to the ground. Using a trowel, dig a hole some 10 cm (4 in) deep where the tip touches the soil. The soil should have been prepared, as described for simple layering.

Push the stem tip into the hole, so that the

extreme tip touches the bottom of the hole. Now bury it, replacing the soil around it and firming it with your fingers. You will need to anchor most of these flexible stems if the soil is light; otherwise they will spring out (Fig. 12.7).

Aftercare
Within a few weeks, especially if the soil has been kept moist, a new shoot will start to appear from the buried tip. If all goes well it will grow away strongly. However, it is best to wait until mid-autumn before you lift the rooted portion, for it is only by now that the roots will have developed a system of sufficient density to support the plant.

Sever the new layer close to its roots, and transplant it to its new quarters. Alternatively, if you prefer, you can sever the plant at this time of year, but then leave it where it is until spring, and then transplant it. This extra time will allow it to develop an even better root system.

Mound layering
A relatively unusual form of propagation, but quite useful, this is also referred to as 'stooling'. Gooseberries and blackcurrants are easily propagated by this method, as well as a number of ornamental shrubs, particularly those with upright, rigid stems, such as deutzia and philadelphus (mock orange).

First of all, it is necessary to establish a permanent row of plants, each with a single stem, and spaced about 30 cm (12 in) apart. During winter cut back the stems to a height of about 45 cm (18 in). During the following year they are allowed to grow freely, but next winter they are cut to a mere 2.5 cm (1 in) above the ground.

Aftercare
This is quite a lengthy form of layering, taking place over several seasons. It is therefore important to keep the area free of weeds. In dry, warm periods, make sure the soil is kept moist. When shoots appear in spring they are left to grow to a height of around 15 cm (6 in), and then they are earthed up, as you would earth up potatoes (Fig. 12.8). Eventually the soil should be 'mounded' up around the new growths, to a depth of 20–25 cm (8–10 in). However, never completely cover shoots. Each new shoot will also have its own small root system, so when they reach this height you can dig up the new stems. Sever them from the main plant and reposition them in a nursery bed where they are allowed to develop naturally.

Fig. 12.8

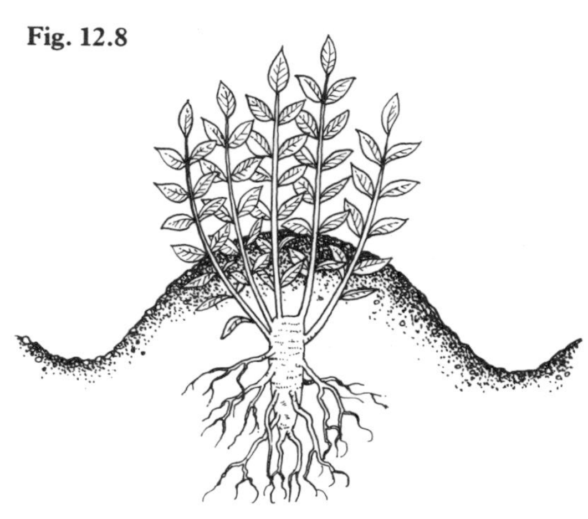

Fig. 12.8 Mound layering. A rootstock is allowed to develop a good root system, and then it is cut to within 2.5 cm (1 in) of the ground. The shoots coming from this are 'earthed up' (illustrated), and the following winter each shoot is detached from the parent; it has its own root system.

Continuous layering
This is used mainly in the propagation of fruit tree rootstocks, but certain shrubs with long, slender stems may also be tackled. These include *Cotinus coggygria* (smoke bush), *Hydrangea arborescens* and *H. paniculata*. It is a method not generally practised by amateurs, for reasons which will become apparent. Continuous layering (also known as French or etiolation layering) may be carried out any time of the year but, naturally, spring or summer, when growth is at its most rapid, are the best times. First, you must have a row of single-stemmed plants, spaced about 60 cm (2 ft) apart, that are growing, healthy, and with their own root systems. Over a period of a year or more these plants – let us call them 'stems' – must be pegged down on the ground to form a continuous line of layered shoots (Fig. 12.9). This will need to be done in stages if you are to avoid snapping them.

Immediately prior to the stems finally being laid horizontally, take out a trench 5 cm (2 in) deep. The stems should now be laid along the trench, and the soil should be replaced over them. Weight or peg them in place, and gently firm the soil over them.

Fig. 12.9

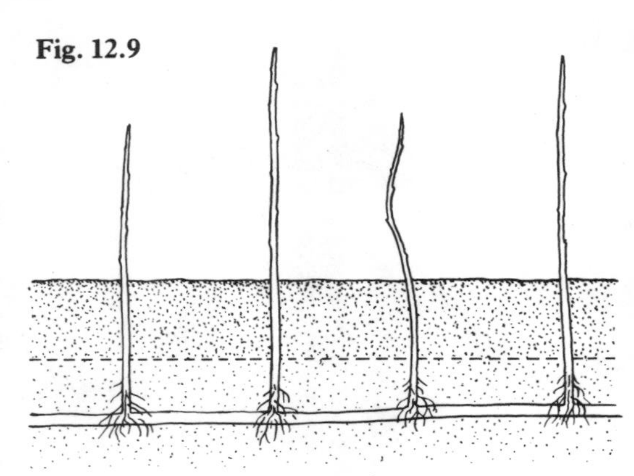

Fig. 12.10

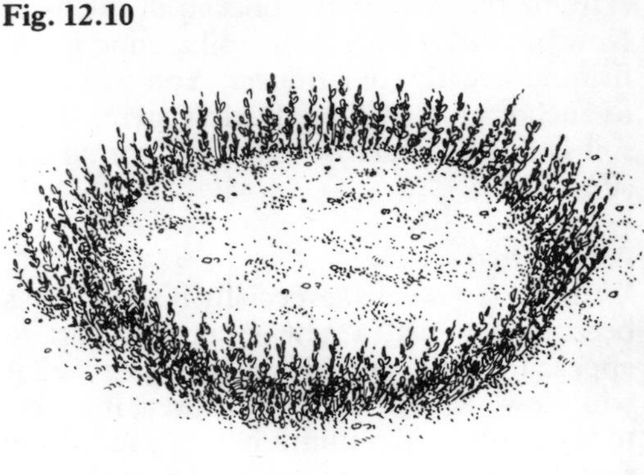

Fig. 12.9 Continuous layering. A rootstock is planted at 45 °, and the following winter is laid down, horizontally, in a shallow trench, and covered with soil. Shoots will grow from the buds on the rootstock, and these are 'earthed up' as they grow (illustrated).

Fig. 12.10 Heathers may be propagated by 'dropping'. One method is to lift and replant the heather deeply, so that 2.5 cm (1 in) of the shoot tips are showing. Backfill the soil in and around the plant. Many of the shoots will form roots at the soil level.

Aftercare

In a month or so young shoots will arise from buds along the stems. These new shoots will also carry small roots. As the top growths appear through the soil, earth them up, as for mound layering. During autumn, rooted shoots are severed from the original plant and repositioned in a nursery bed.

Dropping

This is a type of layering adopted for ericas and daboecias (heaths) and callunas (heather or ling), but also with larger shrubs, like pernettya (prickly heath) and vaccinium (blueberry, amongst others). It is ideal for the amateur who wishes to raise small stocks and has no specialized equipment for propagation by cuttings. If it is carried out in mid-spring, the rooted shoots may be lifted the same year (during autumn). It may also be carried out early to mid-autumn. Strong, healthy plants around three or four years old, are best. There are two different types of dropping.

Method 1

The more involved method requires the plant to be lifted and transplanted deeply, so that only 2.5 cm (1 in) of the stem tips are showing above ground level. Fill in the hole, and around the plant, with a peat, sand and soil mixture. Firm it with your fingers. It is important to spread the shoots out, and to work some of this mixture amongst them (Fig. 12.10).

As always with layering, make sure the plants are well watered during dry weather. During the autumn (with a spring drop), or the following spring (with an autumn drop), lift the plants again and cut off all of the rooted stems.

Method 2

Several branches around the plant's perimeter can be selected and pegged to the ground so that a few centimetres of the stems are buried.

After about a year, sufficient roots are established to enable new plants to be cut from the parent plant. Ideally they should be transplanted to a nursery bed for about six months before putting them in their permanent homes.

Layering carnations

The layering of certain species of dianthus (border carnations and pinks) is not really difficult. It should be carried out in mid- to late summer, when the flowers have faded, but the plants are still growing freely. Before the actual layering, lightly fork over the soil around the plants to be propagated, and then a mix of loam, peat and sand should be spread around the plants to a depth of 5 cm (2 in).

Select some of the strongest young shoots and with each one remove a few of the lower leaves. Now use a sharp knife to make a 'tongue' or slit near the base of the shoot. This slit should be through a node or leaf-joint. Penetrate the stem about half-way, and in an upward direction. Next, bend the stem so that the slit is opened, and use a small stone to ensure that it stays open. This

Arguably the most vivid garden colour is obtained from summer-flowering annuals, and all may be sown and grown by the amateur gardener. Hardy types may be sown outdoors – during autumn or early spring – and the frost-tender types are sown under cover from mid-spring, for planting outside when the weather is suitable.

Just a few of the many aids available for the home gardener who wants to raise his own plants. Unheated propagators are the most economical way of providing high humidity for seed germination.

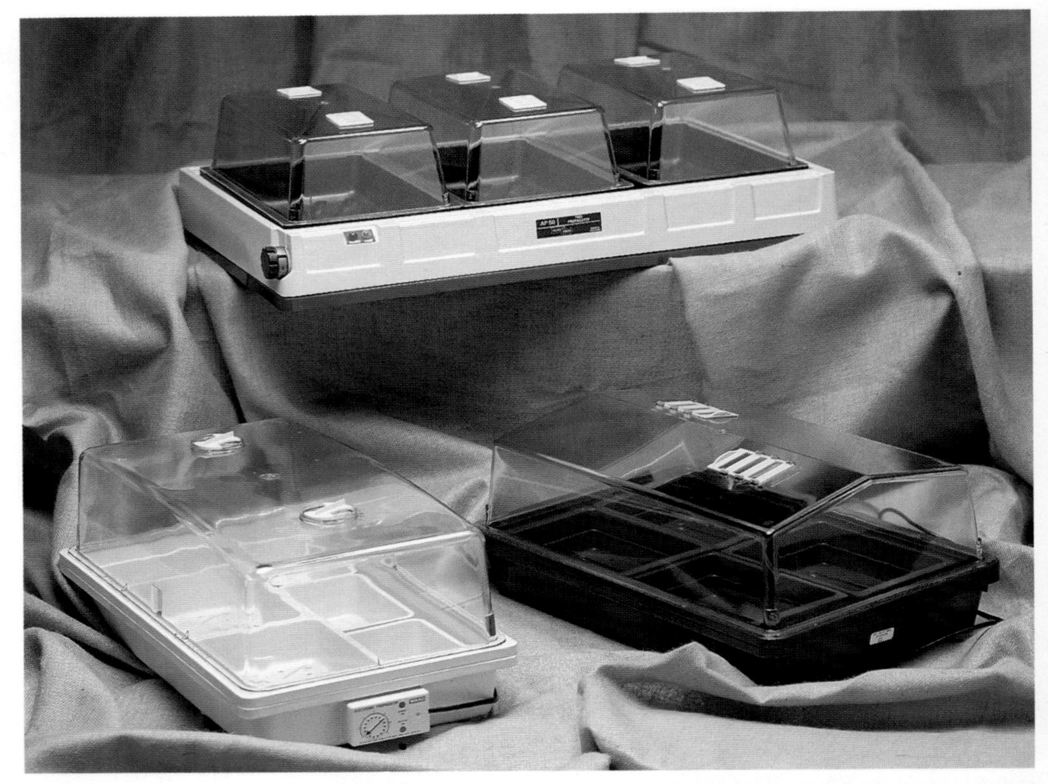

The compost into which seedlings and cuttings are expected to flourish must be of good quality. The soilless or peat-based types are more open and have better drainage, whereas the John Innes, or other soil or loam-based composts, are heavier, and often need lightening to improve drainage.

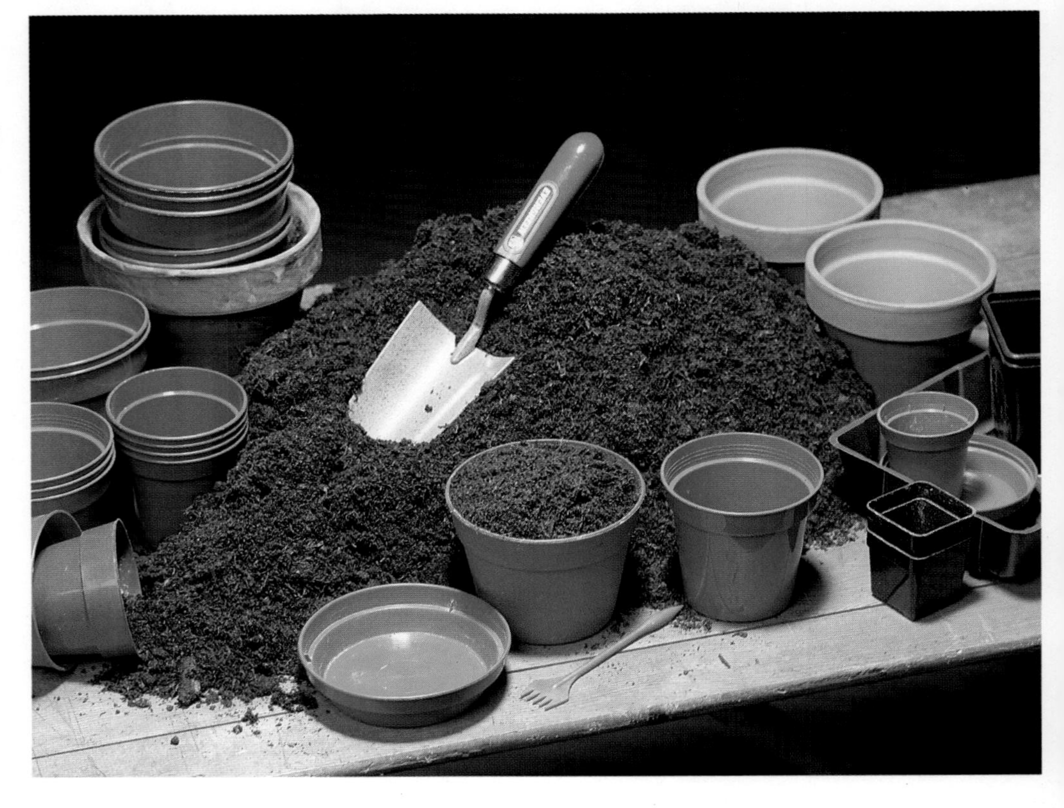

Damping off: a generic term for a number of similar fungal diseases that will affect seedlings of many plant types. Whole 'patches' of seedlings way wilt, some individuals will develop stem lesions, while in others the roots rot away.

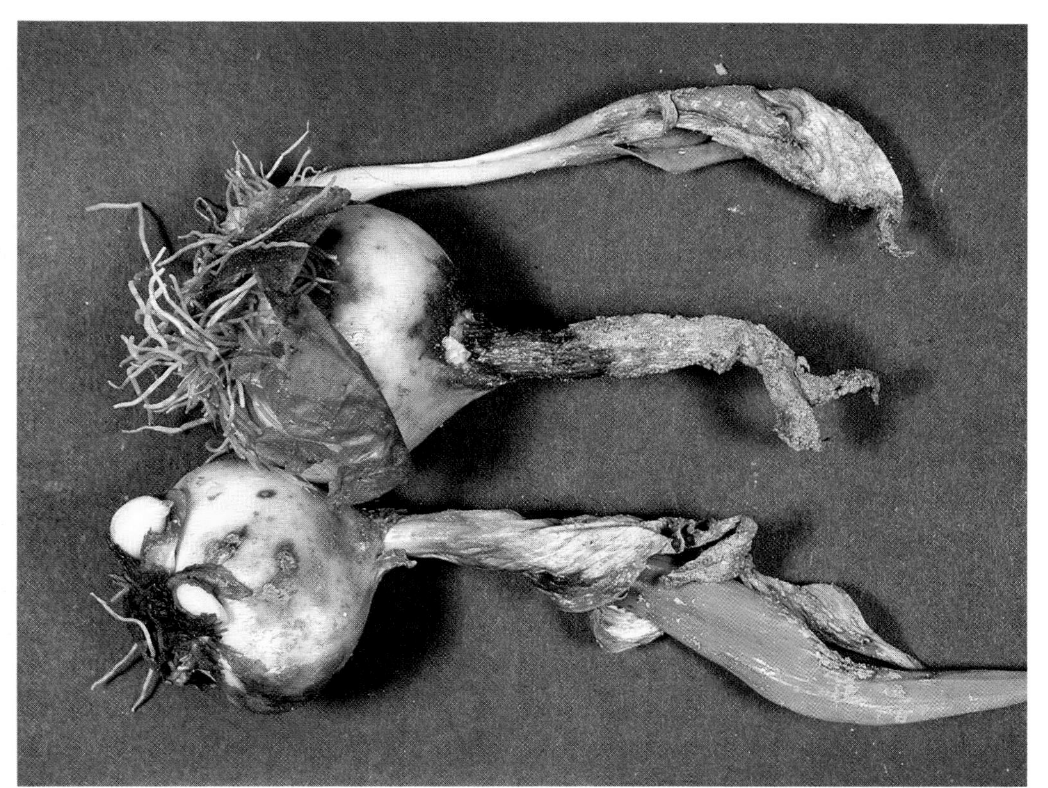

Tulip fire: a fungal disease of tulips that is encouraged by cold, wet weather, and is often spread by planting infected bulbs or bulblets. The disease shows itself as a grey mould on certain tissues above and below ground.

Alchemilla mollis (Lady's mantle), an attractive border perennial, much beloved of flower arrangers. It releases its seeds during late summer and autumn. If saved and dried, they may be sown the following late spring or early summer; germination takes place freely.

F_1 hybrids – flowers **and** vegetables – are specially bred by commercial growers for superiority in one or more of the following: colour, scent, yield, taste or resistance to either disease or damage by adverse weather conditions, such as rainfall. Consequently F_1 hybrid seeds are much more expensive to buy. Here we see Aquilegia 'Music'.

There is nothing quite as satisfying as to see a tray of healthy seedlings. However, seeds tend to germinate erratically. Therefore, when sowing, always try to spread the seed 'evenly', and resist the temptation to over-sow.

There are about 10,000 species of ferns, spread throughout every part of the world that can support vegetation. In the civilized Western world, ferns are more likely to be used as pot plants or garden decoration. Spores, rather than flowers, are the reproductive units, and keen amateurs can raise new plants from them.

Rooted cuttings will be identical to their parents in all respects, which means that great care should be taken when selecting the parent: only use sturdy, healthy, typical plants.

Papaver orientale (Oriental poppy) is just one of the many border perennials that may be propagated from root cuttings. These should be taken during early to mid-winter, and rooting allowed to take place in a garden frame.

Potted orchids may be propagated by division. Cut through the rhizome about a month before separating the topgrowth, to give the cut ends time to heal before the upheaval of repotting.

Saxifraga stolonifera (Mother-of-thousands) is regarded as being the only member of a massive plant genus that requires warm, frost-free conditions. It produces attractive trailing shoots that carry miniature plantlets. These may be removed, once well-developed, and rooted in individual pots.

is now pegged firmly into the soil, covered with 2.5 cm (1 in) of the mixture.

Aftercare
As always, keep the compost moist. The layers should be rooted by the autumn, at which time they may be severed from the parent plant, and potted up or transplanted. The new plants should reach flowering stage by the following summer.

Air layering

So far we have seen how plants may be layered by bringing stems and shoots into contact with the ground. However, certain plants have stems that are so rigid and upright, it is impossible to bring them down without breaking them. Fortunately we have developed a way of bringing the ground to the stem, so to speak, and it is called Chinese or air layering.

A number of hardy trees and shrubs may be propagated by this method, but where amateurs are concerned, it is with greenhouse and house plants that air layering comes into its own. These plants include various species of ficus (rubber plant), philodendron, dracaena, dieffenbachia (dumb cane) and codiaeum (croton). One can try air layering any hardy outdoor shrub or tree which may have no suitable branches for bending to the ground. Often included here are species of magnolia, Hamamelis (witch hazel), fatsia (Japanese aralia) and tilia (lime).

The principle of air-layering is the same for both hardy and tender plants. The most advantageous time of the year to undertake this propagation, for both types of plants, is between mid-spring and late summer, when growth is rapid. Let us take *Ficus elastica* (rubber plant) as a prime example. After several years these plants usually become lanky and top-heavy with a cluster of leaves near the ceiling. The ideal would be to make a new plant from the top section, and to cut the main plant back so that it will bush out with new growth.

First, choose the point on your overgrown specimen at which you wish to root a new plant. You may well find that an ideal place is around 30 cm (12 in) down from the top. If possible this should also be around 13 cm (5 in) below a leaf. The stem should be sturdy but young, as old, hard wood will not root easily. Remove a leaf, and with a sharp knife make a diagonal cut upwards through the node (Fig. 12.11), taking care not to sever the stem completely.

Insert some hormone rooting powder – the type formulated for semi-ripe or hardwood cuttings – into the cut, trying to spread it evenly over the cut surface. A matchstick may be placed inside the cut to keep it open.

Damp sphagnum moss should be pressed around the cut area, and inside it (Fig. 12.12). This will provide the humid conditions which will encourage the plant to form new roots, If you cannot obtain moss then moist coarse peat (or even a peat-based compost) would do.

Now the mass of moss needs to be wrapped in polythene. The self-clinging film used in kitchens does the job well and is easy to use (Fig. 12.13). Otherwise cut a strip of clear polythene 20 cm (8 in) wide, and wrap it over the moss several times so that it will not dry out. Tie the polythene firmly into a neat package, but not so tightly that the stem is damaged.

Aftercare
Keep an eye on the plant, and if the moss seems to be drying out, loosen the top tie and give it some water.

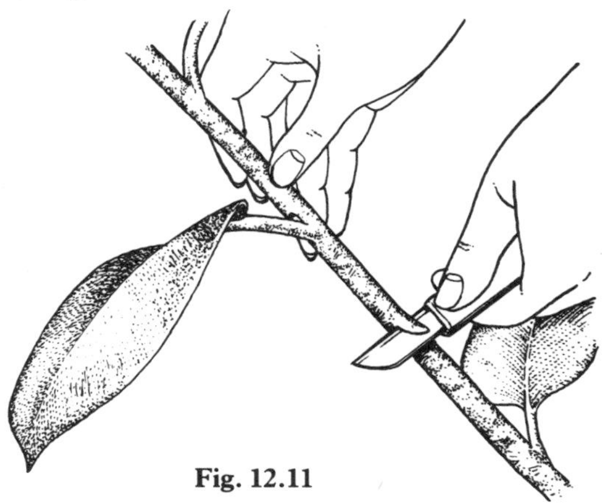

Fig. 12.11. Air-layering. Wound the stem with a sharp knife.

After six to eight weeks, roots should have formed from the cut, and be visible through the polythene (Fig. 12.14).

Shoots of trees and shrubs will take much

Fig. 12.12 Apply a small amount of hormone rooting powder to the wound, and then pack damp moss all around this part of the stem.

Fig. 12.13 Seal the moss with polythene, perhaps self-clinging kitchen film.

Fig. 12.14 Eventually roots will begin to show through the clear polythene.

Fig. 12.15 When you can see a mass of white roots inside the polythene, cut the stem neatly below the rooted area, and above a leaf.

Fig. 12.16 Unwrap the bundle carefully, taking care not to damage the roots, and pot up the new plant.

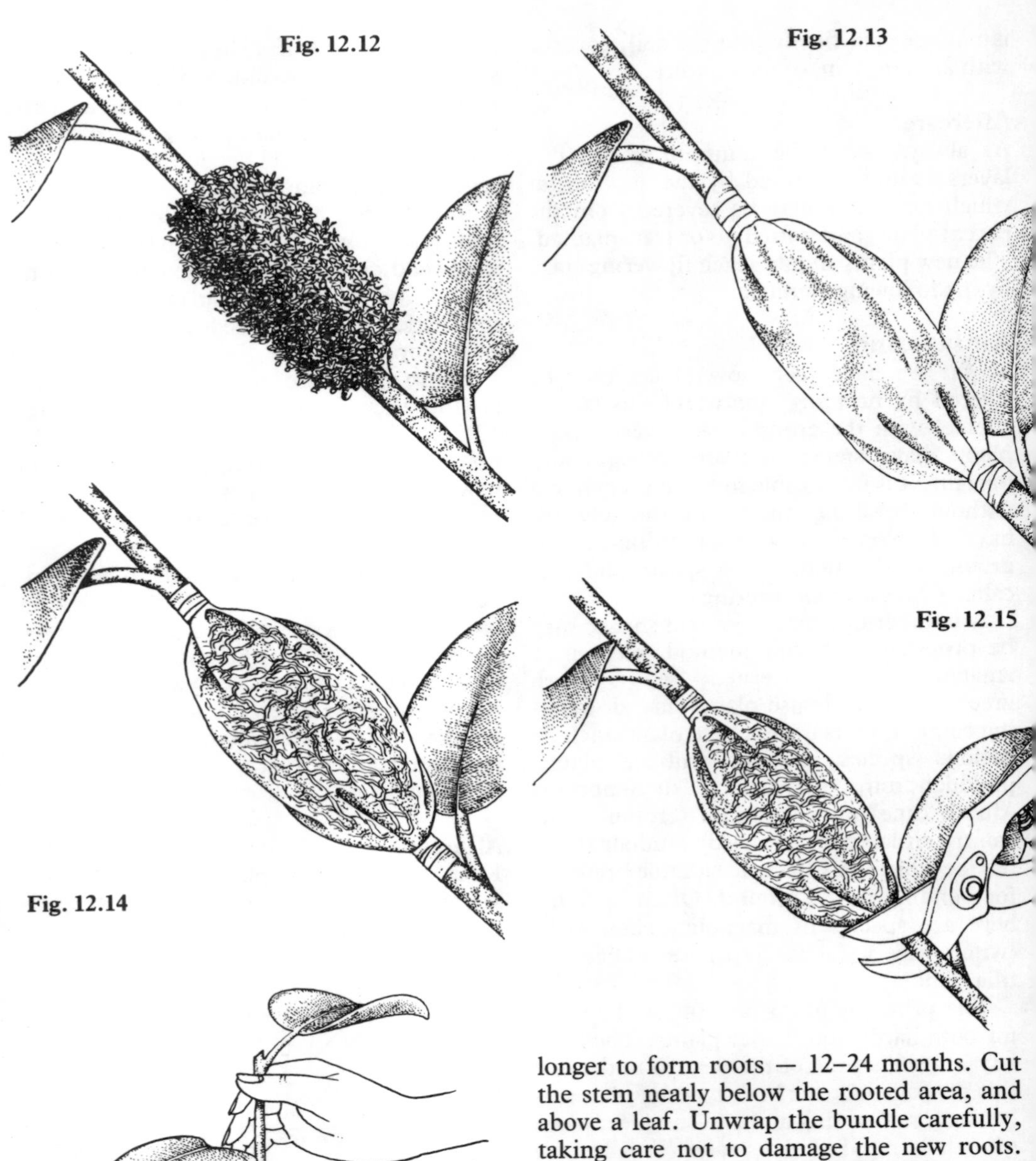

longer to form roots — 12–24 months. Cut the stem neatly below the rooted area, and above a leaf. Unwrap the bundle carefully, taking care not to damage the new roots. Remove any loose moss, but leave any whose removal would cause damage to the roots. Cut off the length of stem beneath the rooted area (Fig.12.15), as this will only rot away if left in place, which could cause the death of the young plant.

Fill a suitable-sized pot with a peat-based potting compost, and pot up your new plant (Fig.12.16). For the first few weeks, indoor plants should be kept in as humid a place as possible, and hardy plants kept in a garden frame or unheated greenhouse.

CHAPTER 13

Plantlets

We have seen that Man can propagate certain plants vegetatively, by carrying out techniques known as division and layering. A small group of plants have the natural ability to propagate themselves in a similar fashion but, it has to be said, in a more curious way: they produce their own plantlets, while still attached to the parent. Sometimes the activities of these plants will resemble Man's attempts at plant propagation, but some plants throw out new, baby replicas from long shoots, from within the leaf-structures, or from the stems. Let us look at specific examples of these 'ready-made' plants.

Layering plantlets

Probably the most widely grown plant that offers the facility of naturally formed plantlets is *Chlorophytum comosum* (spider plant). It carries lance-shaped leaves of green and cream, and during the active growing period pale yellow stems up to 60 cm (2 ft) long arch upward between the leaves. These carry small, six-petaled white flowers, which are succeeded by little plants appearing either singly or in groups (Fig.13.1). These plantlets weigh the stems down as they grow, making them arch in a more defined and, some feel, graceful way. Roots are also produced as the plantlets develop; in their natural habitat these would simply grow into the soil around the plant.

Fig. 13.1

However, we can take advantage of Nature's own method, and allow the plantlets to root into pots of compost, and so provide us with new plants. This may be carried out at any time of year, but rooting will be much quicker during the warmer months. It is best to use a 9 cm (3½ in) pot filled with a suitable potting compost – either loamless or soil-less. Stand it alongside the parent plant and, if necessary, raise it to a suitable level to meet the length of stem carrying the plantlet. Then peg the plantlet onto the surface of the compost, using a piece of wire (or a hairpin). Position the peg an inch or so along the stem from the plantlet, so as not to cause constriction once it starts to get bigger.

You may find that thick fleshy white roots have started to form before you peg the plantlet down. Whether this is the case or not, after six weeks or so, the young plants will be well rooted and can be severed from the long stem and the parent plant. At this stage the new plants are growing independently. They will need watering and feeding to establish them in their new homes.

Another tender plant that is propagated by the method described above, is *Saxifraga stolonifera* (mother of thousands). It produces many plantlets on the end of its long, thin stems; hence its common name.

Hardy decorative plants increased this way include *Ajuga reptans* (bugle), geum (avens), and *Viola odorata* (sweet violet) and its varieties.

An alternative method of increasing chlorophytums while still making use of the plantlets on the stem is to sever them from the parent plant once the leaves are 5–8 cm (2–3 in) long. These severed portions may then be potted up and kept in humid conditions for rooting. Or you can try to root

Fig. 13.1 *Chlorophytum comosum* (spider plant) is, perhaps, the most popular of indoor plants, and produces naturally formed plantlets at the end of long, arching stems.

them by simply putting them in a jar of water. This can sometimes have amazing success. In either case it is a good idea, beforehand, to strip off the small lower leaves to prevent possible rot.

The process of layering plantlets is widely used in the propagation of strawberries (fragaria). These popular fruits begin to deteriorate after three years or so, therefore regular replacement of plants will ensure consistently high yields. Like the spider plants, strawberries produce new plants on long stems, and these are referred to as runners. Often such an abundance of plantlets are created that if you are *not* wanting to propagate from them, they constitute a nuisance!

Fig. 13.2

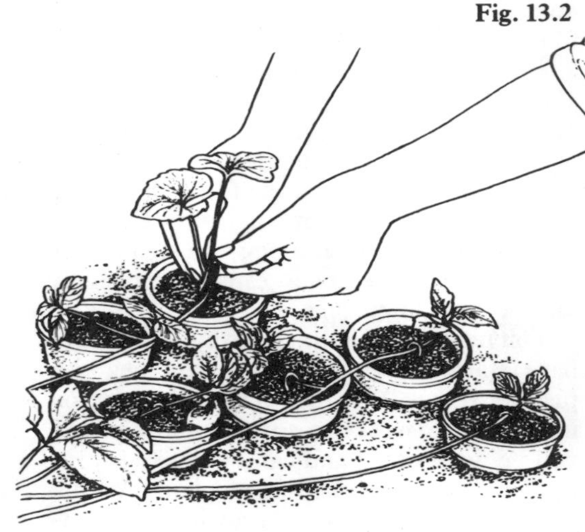

Fig. 13.2 Strawberry runners produce plantlets easily, and these can be potted up outdoors.

In early or mid-summer, select four or five well-spaced runners on each of the sturdiest plants (Fig.13.2), and remove the remainder, as close to the parent plant as possible. The runners may now be encourage to root – but only one plantlet per runner! This is because each runner carries two or three plantlets, with the smallest towards the tip. The one nearest the mother plant is the strongest, so this is the only one worth layering. It can be allowed to root directly into the soil surrounding the plant (it will establish itself here well enough if left alone), or you could consider sinking a pot into the soil and pegging the runner into it. Fill the pots with a loam-based potting compost, or good garden soil. This process of layering plantlets in pots makes lifting the new plant easier, and prevents unnecessary damage to the fibrous roots.

In most respects you can treat the strawberry runner in the same way as the chlorophytum layer, except that this is likely to take place outdoors. However, greenhouse-grown strawberries in pots may also be propagated by their runners. For this you will reqiure a network of connected pots on the benches!

Using a piece of wire, peg the plantlet down into the pot so that it makes good contact with the soil. Water the pot, compost and plantlet thoroughly, making sure that over the next month or so it is never allowed to dry out. Strawberry plantlets soon form new roots, and may even be precocious enough to send out their own runners. These should be cut off immediately, otherwise the plant will spend valuable energy needlessly.

Once the plant is well established – in other words, putting on lots of top growth, and perhaps with white roots showing through the bottom of the pot – sever the runner close to the new plant's crown. Mid-autumn is a good time to plant out new strawberries.

Rooting plantlets

Not all plantlet-producing plants do so with the use of runners or long arching stems. *Tolmiea menziesii* (Fig.13.3) (known variously as mother of thousands, piggy-back plant, pick-a-back plant, or youth on age),

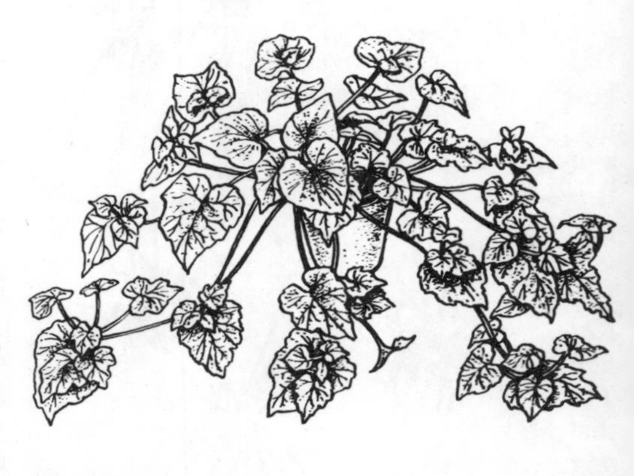

Fig. 13.3

Fig. 13.3 *Tolmiea menziesii* has the curious habit of producing plantlets on many of its older leaves.

has the curious habit of producing plantlets on several of its older leaves. Each plantlet grows on the upper surface of a leaf at its junction with the stalk. This extra weight bears down the slender leafstalks, so that a number of them appear as if they are trailing, thus making the plant particularly attractive when viewed from below.

Propagation takes place between mid-spring and late summer. Cut off a leaf that has a well-developed plantlet on it, leaving about 2.5 cm (1 in) of its stalk attached. Replant this in a small pot containing moist potting compost, which is preferably peat-based. Make sure that the stalk is buried and that the leaf carrying the plantlet sits on the compost surface. Keep the pot in bright light, watering only enough to make the compost damp. After about three weeks new growth will indicate that rooting has occurred. Thereafter, treat the plant as you would a mature one. The original leaf that carried the plantlet may remain green for several months, but it will evenually turn brown and shrivel. At this stage, carefully remove it from the new plant.

If you prefer, you could layer the plantlets into pots of compost while they are still attached to the parent, as for chlorophytums. There really is no preference. Tolmieas do grow rapidly, so an owner will be in possession of an attractive house plant only five months or so after the start of the propagation process.

Some ferns, including species of dryopteris, polystichum and asplenium, have a curious way of producing plantlets. Occasionally bulbils – in this context these are leafy plantlets with smaller bases – appear on the fronds. These will grow into small plants, even if left on the parent . Therefore, a useful method of increase is to detach them and grow them on as described for tolmiea.

Finally, leaf plantlets take the form of 'embryos' with the popular succulent *Kalanchoe tubiflora* (also known as *Bryophyllum tubiflorum* or the devil's backbone). These embryos are really tiny baby plants that form naturally, and in great numbers, on the edges of the parent plant's leaves. Each plantlet has a pair of leaves and a couple of tiny, hair-thin roots. When they reach a certain stage of maturity, these plantlets drop off: they will do so, anyway, if you brush past them. In the wild, of course, they will root in the ground where they fall. All we have to do, however, is to prick them out into pots of compost, and we have practically ready-made plants!

Offsets

These are somewhat different to leaf embyros and plantlets on stems and runners. An offset is a complete new plant which develops alongside its parent, and is easily detached from it, along with its own system of ready-formed roots. There are several kinds: bromeliads (including billbergia) produce offsets around the old flowering shoot; bulbs produce them freely (see Chapter 14), while some cacti and succulents produce a cluster of them around their bases. Sometimes they are attached to the crown of the plant, often by short underground stems; or by above-ground stalks, as with echeverias and sempervivums (houseleeks) (Fig.13.4).

Young offsets, that is, those which have not yet developed roots, should be removed cleanly from the parent and potted into a container filled with free-draining cuttings compost. Firm them in with your fingers, and then keep them moist and out of direct sunlight for the first week or so. Soon after this they will develop their own roots.

Fig. 13.4

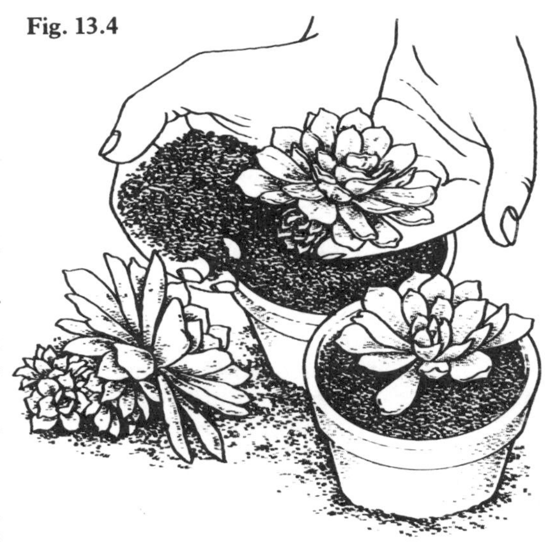

Fig. 13.4 Young *Sempervivum* offsets should be removed, cleanly, from the parent, and then potted up.

CHAPTER 14

Vegetative Propagation of Bulbs

So far in this book we have only briefly discussed the propagation of bulbous plants, and specifically the increase of stocks by sexual means, that is, seed reproduction, (see Chapter 5). This, of course, occurs naturally, and for many species is the main method by which plants are perpetuated. Vegetative reproduction is, however, an alternative form of bulb propagation. It is broken down into a number of headings, and we shall examine all of these in this chapter.

As we have already seen in previous sections of this book, 'vegetative propagation' is the method most often used for multiplying plants that are identical to their parents. The phrase encompasses all aspects of propagation that are non-sexual, ie. cuttings, layering, division, and so on.

For the sake of simplicity it is probably best to think of the vegetative multiplication of bulbous plants as being similar to simple division and taking cuttings, even though the plants appear to possess nothing that could remotely be called a cutting. Whether or not you succeed in the task depends on your understanding of the differences between the structures of tubers, bulbs and corms, and the way that their component parts work.

Bulblets and cormlets

Let us first look at the vegetative reproduction of bulbous plants by 'bulblets'. True bulbs (as opposed to rhizomes, corms and tubers) are very much like the buds of herbaceous plants or shrubs. They are built up by a number of leaf-like layers which are attached to a disc of tissue called the 'basal plate'. With shrubby plants the stems elongate as they grow, so the leaves grow more or less widely spaced apart. A bulb stem does not elongate like this, but instead grows sideways to form the basal plate. It is this that the 'leaves', separated (although almost imperceptibly), are attached.

It is important to understand this when you wish to propagate a bulb. In most subjects of the plant kingdom, cells capable of division (and so able to propagate their host plant vegetatively) are found at places where stems and leaves join. With bulbs, it is from these small dividing cells, where the scales join the plate, that 'offsets' originate.

These take the form of bulblets, or simply small bulbs growing around the base of a parent bulb (Fig.14.1). If any true bulbous plant that has been growing in the ground for a few years is dug up, you will immediately see a cluster of tiny young bulblets. Depending on the type of plant, these will grow into a suitable size for flowering in one to three years.

Fig. 14.1

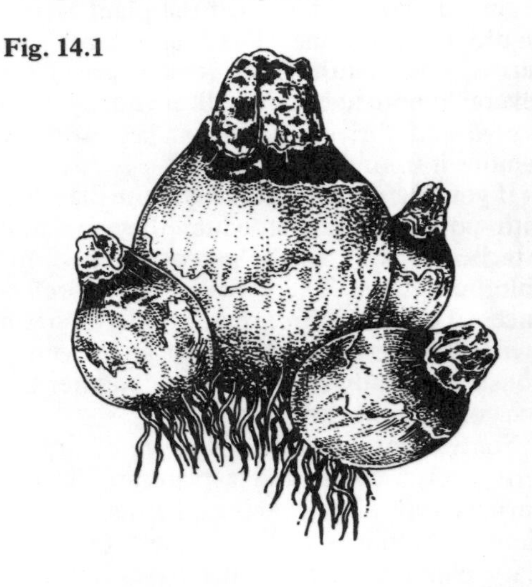

Fig. 14.1 Bulb offsets are easily visible, coming from the side of the parent.

To propagate from these, wait until the parent bulb is dormant, that is when the foliage has completely died down. The only common exceptions to this rule are species of galanthus (snowdrop). These are best lifted and separated immediately after flowering, while the leaves are still green. Replant the bulbs immediately. Generally, if you lift or disturb bulbous plants while the leaves are green, you are preventing the bulbs and bulblets from receiving food that has been manufactured by the foliage.

Once the dormant bulbs have been lifted, gently detach the bulblets with your fingers;

invariably they will come away with a simple flick or rub, so you will not have to use force. Keep the bulblets dry and cool until such time as you want to plant them, that is, the usual planting period for mature bulbs of the same species. The planting technique will differ slightly from that recommended for mature flowering bulbs, mainly because of the size of the bulblets. They are of course much smaller, so ensure that they are not planted as deep. They may be planted straight into the garden, in their flowering positions, but you may have to wait a few years for the first flower. Alternatively, plant them in a nursery bed, when they can develop and grow at their own speed. You will need to transplant them during a dormant period when they have reached flowering size.

There is always confusion over what is and what is not a true bulb. Briefly, the following are a few of the more common examples of true bulb, and may be propagated by the above method: allium (ornamental onion), cardiocrinum, chionodoxa, erythronium (dog's tooth violet), fritillaria (fritillary), galanthus (snowdrop), hyacinthus (hyacinth), leucojum (snowflake), lilium (lily), muscari (grape hyacinth), narcissus (daffodil), nerine (Guernsey lily), scilla, sternbergia and tulipa (tulip). There are, of course, many others.

Fig. 14.2

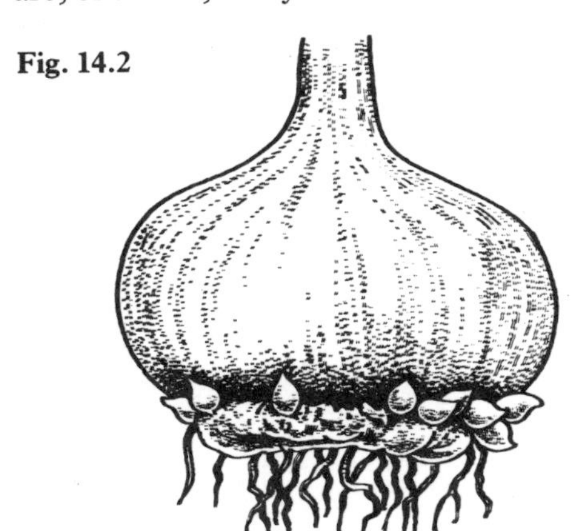

Fig. 14.2 Tiny cormlets are produced in large numbers, from around the side and base of the parent corm.

Corms (and rhizomes) often resemble bulbs, but are formed in quite a different way. Like bulbs, they are made up of stem and leaf tissues, but almost their entire bulk consists of the former, while the leaves exist as mere remnants, reduced to dry scales. In most instances these form a protective covering, several layers deep, around the corm.

In rhizomes the remnants are even less conspicuous, being reduced to insignificant papery tissues making the nodal positions. They still count as leaves, however, and as with corms, the places where they meet the 'stems' can be relied on to provide the appropriate tissues for cell division and therefore plant multiplication.

As would be expected, cormlets are small corms which are produced around the base of the parent (Fig. 14.2). To propagate from them, lift, detach, store and plant in exactly the same way as described for bulblets.

The following are all corm-producing plants and may be propagated from cormlets: colchicum, crocus and gladiolus.

With bulbs and corms, wounding the storage organ, such as cutting or slicing it with a knife, or pulling leaf scales off the basal plate, will stimulate the dividing cells to produce small bulblets or cormlets. At the least, it will enable the damage that has been done to be repaired. This will not work with tuberous plants.

Bulbils

Unlike bulblets which are offsets produced from the storage organ underground, bulblets are small bulbs which are found on the stems of some bulbous plants, usually in the leaf-axils. They will take two or three years to reach flowering size. Gently remove them from the stem (Fig.14.3), just prior to the foliage beginning to die down. If they are wet, they will require drying, in which case lay them out in a warm place, on some dry newspaper. After a couple of days they should be stored in a box or tray containing dry peat.

Keep them in a cool, frost-free and dry place until such time as they should be planted out, that is the normal time for planting out mature flowering bulbs of the same species. It is usual to plant them into nursery beds rather than straight into the open garden. As the bulbils are tiny, you may as well treat them as large seeds and 'space sow' them about 5 cm (2 in) deep in

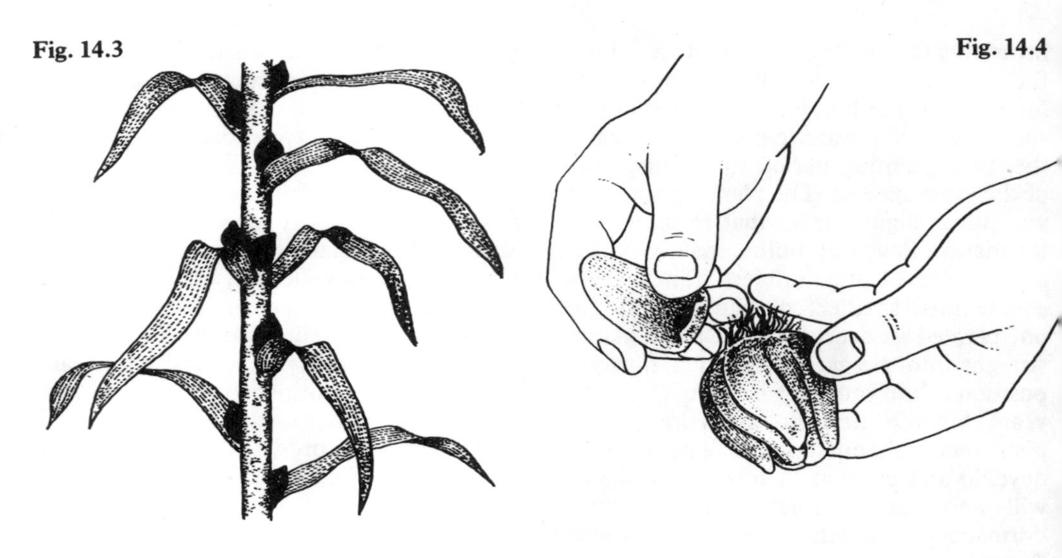

Fig. 14.3 Many bulbous plants, including species of *Lilium*, produce tiny bulbs – bulbils – in the leaf axils, along the stem.

Fig. 14.4 Many *Lilium* spp. may be propagated by their leaf scales. Use only clean, healthy scales. Pull them gently away from the parent bulb.

drills. However, lilium bulbils should be 'sown' in pots after collection and placed in a garden frame to grow on for a year. Lilies do not relish being dried off. After a couple of years or so, the bulbils will have grown into flowering-size bulbs, and these should be transplanted during their dormant period, when the foliage has died down.

The most commonly-grown plants to produce bulbils are the summer-flowering species of lilium (lily) and allium (ornamental onion), as well as culinary members of the onion family, like *Allium cepa* (tree onion) and *A. sativum* (garlic).

Lilies from scales

Like the storage organs of most bulbous plants, bulbs of the genus *Lilium* (lily) are made from masses of overlapping scales. However, these are slightly different in structure from other scales, in that they may be easily detached from the bulb, without damage, and are then used for propagation.

The art of propagating lilies from scales can be carried out any time of the year, but it is most convenient (and success is readily achievable) when the bulbs are dug up for replanting, and when new lilies have been purchased for planting – generally during spring

Take your bulbs and remove and discard the very old, semi-rotten remnants of scales on the outside. You are now left with a clean bulb comprising healthy scales. Some of these will need to be pulled off, but unless you wish to reduce the bulb's prospect of flowering, you should leave at least two thirds of them intact. Gently pull each chosen scale sideways (Fig. 14.4) and downwards so that it brakes off at the point where it joins the basal plate.

Provided the scales are kept alive and they continue to function, the cells along the broken edges will divide and repair the damaged tissue, and some will go on to produce bulbils. Often as many as four bulbils will develop along the base of one scale.

Fig. 14.5

Fig. 14.5 It is a good idea to dust the scales with a fungicide to protect them from moulds.

Fig. 14.6

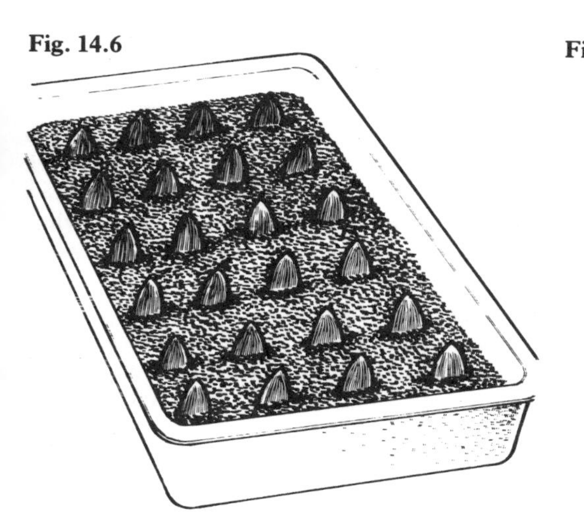

Fig. 14.7

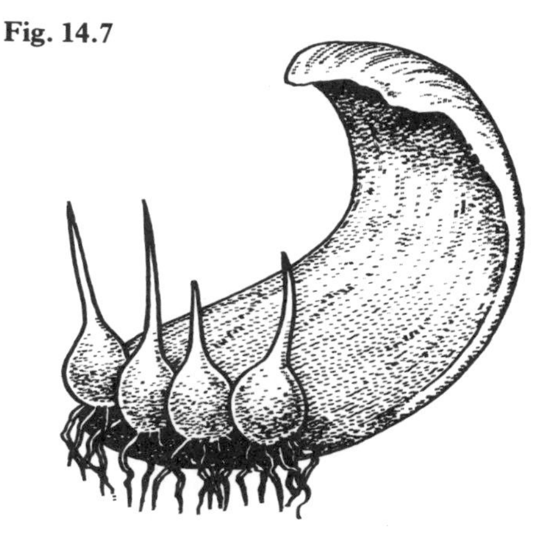

Fig. 14.6 Each scale should be inserted into a seedtray filled with moist compost.

Fig. 14.7 After a few weeks, tiny bulbs should have developed at the base of the scales.

As soon as the individual scales have been removed, they should be put into a polythene bag containing a small amount of powdered fungicide – such as benomyl – to protect them from moulds (Fig.14.5). Gently shake them around so that the dust coats them, and then remove them. Each scale should then be inserted into a seedtray or pot filled with a moist compost of equal parts sharp sand and peat (Fig.14.6). Make sure that the scales are 'planted' the right way up, and that they are two-thirds buried.

Leave the containers in a greenhouse or garden frame, where they are sheltered from the worst of the weather. Heating is not essential for encouraging the scales to root, but the prospects of success are greater if you can maintain a temperature of 21°C (70°F). Water sparingly throughout the rooting period; overwatering may cause the scales to rot.

Eventually the scales will produce roots, at which time a small bulb or bulbs will be produced at the scale bases (Fig. 14.7). At this stage watering may be increased. You should not be in a hurry to pot up the individual scales. Many gardeners prefer to leave the scales growing in the seedtrays over winter – while they are protected from frost – to be potted up singly during the following spring. Use a loam-based potting compost, and keep the pots well-watered during the summer and autumn.

The following spring, that is two years from initial scaling, the bulbs may be planted out into the garden. Do not expect flowers this year, however. It generally takes three or four years for bulbs of flowering size to develop from scale cuttings.

Tuber division

Tubers are simply swollen roots produced by plants to store food reserves, but they are very often confused with bulbs or rhizomes, particularly the latter (which are, botanically, swollen stems).

Our most familiar tuberous plants include alstroemeria (peruvian lily), arum, some begonias, the house plant caladium (angel's wings), cyclamen, dahlia, hemerocallis (day lily), the aquatic plant sagittaria (arrowhead), sinningia (florist's gloxinia) and, of course the humble *Solanum tuberosum* (potato).

Let us take one of these, the tuberous begonia, and look at a common method of propagation, division of the tuber. Each year in the life of the plant, the tuber increases in size, and after three years or so, the flower display begins to deteriorate, and the plant becomes more susceptible to disease. So, during the early spring potting period, it is a good idea to take the largest tubers and divide them.

Before you start work with the knife, it is important to know where the growth buds are, and this will take a little time to determine. Begonia tubers are generally saucer-shaped (although in a more exaggerated form), the side with the depression being the 'top' from where the shoots will arise. Take the whole tuber and press it, depression uppermost, into a tray of moist peat. Keep it in a warm greenhouse or on window sill indoors. The compost must be constantly moist, but do not splash water around as it will collect in the tuber hollows and may cause rotting.

Soon, clusters of plump buds will appear in the depression. At this point lift each tuber out of its container and with a clean, sharp knife divide it into as many pieces as you like, as long as each piece has at least one bud on it. Remember though, the smaller the piece, the less food reserves it has and the longer it will take to reach flowering size.

The cut edges will be prone to fungus attack, so some fungicide powder should be applied to these. Leave the tubers on a flat surface for a few hours to dry them out. Then pot up each section individually, using a peat-based potting compost, making sure that the cut sides are buried beneath the surface. Do not bury the whole thing, however; the shoots need some light. Water the pots carefully from around the edge, or stand the pots for a couple of minutes in a saucer of water, which is then sucked up though the compost.

Fig. 14.8

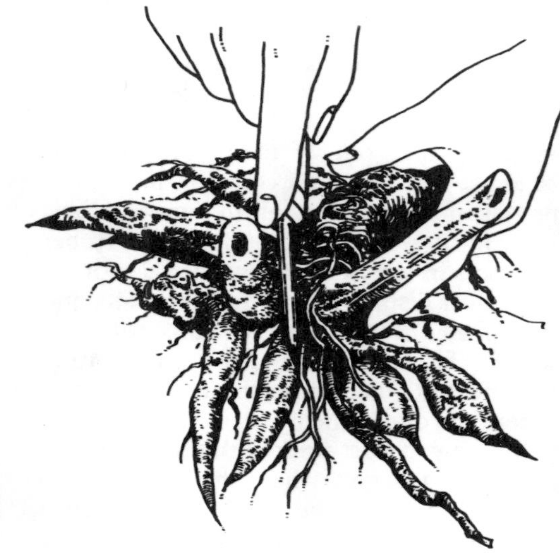

Fig. 14.8 Dahlia tubers should be divided every few years.

Dahlias are different in that they produce clumps of tuberous roots. Shoots arise from the crown, which is situated at the top of the clump, just beneath the soil surface. Each year the clumps will expand, so it is worth dividing them every few years (Fig. 14.8), carried out during the dormant period when the tubers are in winter storage, or when planting outdoors.

Each cut portion, or division, should consist of a piece of crown containing one or more growth buds, and at least one tuber. Tubers on their own do not carry growth buds, so they will not develop new shoots.

Some types of dahlia will divide easily, using your fingers. Others, however, are much tougher and will require a knife to cut them. Again, remember to dust the cut surfaces with a fungicide. The divisions should be planted outdoors in their flowering positions during mid-spring. Until this time, keep them dry and frost-free.

Division of congested clumps

We have seen that many bulbous plants multiply themselves very efficiently by developing small bulbs around the parent bulb, and that these are called 'offsets'. In specific terms, we have also seen how these offsets develop, and how they should be treated to perpetuate the individual plant (see page 102).

Garden bulbs, however, often grow *en masse*, particularly if they were planted to give a bold splash of colour in a border, or naturalized in grass. Daffodils, for example, increase quite rapidly and after a couple of years will develop into large clumps. By lifting the clump when the foliage had died down, and separating the bulblets (which are themselves almost at flowering size) and replanting them, you will help to keep the clump healthy and free-flowering. The same comments apply to galanthus (snowdrops) which are best separated after flowering.

Be careful how you lift the clump. Do not insert the fork at random. Think about the positions of the bulbs and avoid spearing them with the fork prongs. With your fingers, prise the bulbs out of the soil and shake any loose soil off. Remove the bulblets as already described and plant them as you would any newly-bought bulbs.

This disturbance and upheaval should not adversely affect flowering the following year. However, you should help the process by giving the bulbs – particularly flowerless or 'blind' ones – a liquid feed after the flowering period and while the leaves are still green.

Some types of bulb, including tulipa (tulip), tend to fade to nothing after flowering, relying solely on the development of its bulblets for perpetuation. When digging up these clumps then, discard the hollow, scaley old bulbs and retain only the healthy separated bulblets.

Fig. 14.9

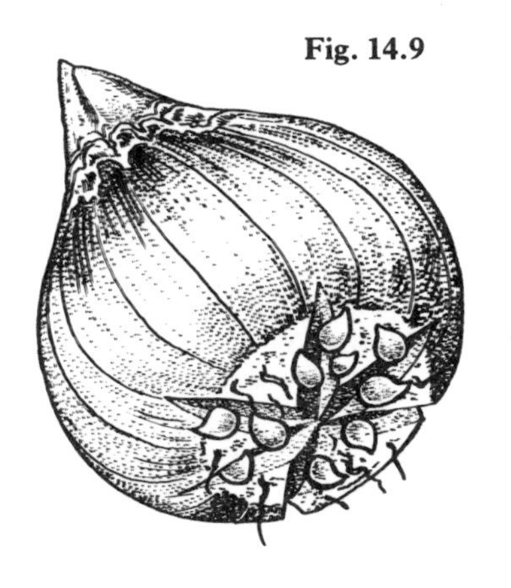

Fig. 14.9 The basal plate of some bulbs may be scored, from which many small bulbs will develop.

Scoring and scooping

Some bulbs, such as hyacinthus (hyacinths) and the large fritillarias (crown imperials), produce offsets very slowly. There are two methods by which the bulbs can be encouraged to produce bulblets, and these are 'scoring' and 'scooping', and either may be carried out during the dormant period, perhaps late summer or early autumn. In the case of scoring, take a healthy bulb and hold it upside down, so you can see the basal plate. With a knife, take two small wedges of tissue from this base, to make the shape of a cross, and then dust them with a fungicide. Now rest the bulb, cut side up, in a seedtray of sand, and leave it in a warm, dry place – the bottom of an airing cupboard is ideal.

After a couple of months some small bulblets, varying in size and number, will have grown from the cut surfaces (Fig.14.9). Use your fingers to remove them carefully from the parent bulb, and plant them in a seedtray filled with moist peat-based potting compost. They should now be left in a greenhouse or garden frame (no artificial heating necessary) until they reach flowering size, which may take a year or two depending on their size at the time of division. At this stage they may be planted outdoors in the normal way.

Scooping involves the very careful cutting out of the whole basal plate, using a sharpened teaspoon, grapefruit knife or, preferably, a curved scalpel. Be sure to cut through only the extreme end of each scale. This operation causes the tissues at the base of the leaf scales to be exposed, and it is from here that bulblets will form. Thereafter, treat the scooped bulb in exactly the same way as you would one that is scored.

CHAPTER 15

Grafting

Grafting is undoubtedly one of the most skilled and, some would say, laborious methods of propagation, but we should quickly add that without it – in its many forms – many of our garden plants would not give the cropping results they do. Grafting can increase both the potential yields of most fruit trees, and the quantity and quality of blooms on numerous shrubby flowering plants. Many home gardeners are deterred from grafting because of the skilled knifework required, and the time it takes for the benefits to show themselves. However, as with most things, practice makes perfect, and it is certainly worth having a go!

The art of grafting is as old as history itself, but we do not know which race first practised and succeeded with it. The ancient Greeks, Romans and Chinese tackled it. This much we know, and we can only guess that they attempted it after having observed the types of grafting found in Nature. It is not uncommon in wild woodland to find two branches which have grown into each other as a result of initial touching and subsequent rubbing together. The bark of one is abraded by the other, enabling the layers of cambium from each branch to come into contact and fuse together. Note that it is essential for the cambial layers to touch and connect: wood, pith and outer bark will not merge in this way.

Essentially grafting, as far as the amateur and propagating plants are concerned, involves placing two cut surfaces from two separate plants under conditions which will enable them to unite and grow together. The part of the plant with roots to be retained is called the 'rootstock', and the piece of plant that is to be grafted onto this is called the 'scion'. Typical rootstocks are made from trees and shrubs which have strong, vigorous and healthy root systems, but which may not flower or fruit particularly well, and which are not valued for their garden decoration. Typical scions, meanwhile, are from species or varieties that flower well, fruit well or have decorative stems, but which may have a weak root system.

Grafting is carried out for a number of reasons, although it is fair to say that not all of them can strictly be considered acts of propagation!

1. Many plants, including syringa (lilac), acer (maple) and juglans (walnut), are difficult to raise from cuttings or layering, and there are several which do not, in fact, grow well on their own roots. Included in these problem areas are fruiting species such as malus (apple and crab apple), pyrus (pear) and prunus (cherries, plums, peaches, almonds, apricots and so on). These all require special rootstocks to enable them to grow sufficiently well for fruiting; grafting is therefore the only practical means of increasing them in large numbers. Incidentally, such trees bought from nurseries and garden centres will have been grafted.

2. Grafting is carried out to obtain the advantages of certain rootstocks. Many plants grown on their own roots are susceptible to certain soil-borne problems. For example, tomatoes are often attacked through their roots by destructive organisms like root knot eelworm and wilt disease. Grafting them onto a rootstock resistant to these pests and diseases will overcome the problem. Similarly, apple varieties are grafted onto carefully selected rootstocks, this time in order to determine the eventual size of the tree. In other words, by grafting you can control a tree's size and speed of top growth. With speed in mind, grafting will often hasten growth and shorten the period to flowering and fruiting.

3. Many types of 'ordinary' tree may be converted to a weeping habit by being carefully grafted with a scion from a weeping variety.

4. Grafting will also aid pollination. For

example, an ilex (holly) will be either female or male, and so will only carry flowers of a single sex. Before the berries form on the female, pollination must take place. If you only have a female plant, one or two male branches can be grafted onto it, to overcome the problems of distance.

5. Trees with damaged trunks may be repaired by grafting. Provided the roots are sound, it is sometimes possible to effect a repair by bridging the injured region with new shoots grafted in.

6. Grafting will enable some trees to change variety. Cut back a chosen tree very hard, leaving just the trunk and a few healthy branches. A new variety can then be grafted on!

A large number of ornamental plants may be grafted, particularly where characteristics of a worthwhile named variety – such as double flowers, or good leaf variegation – need to be transferred to the plain 'type' species. In addition to those already mentioned, other decorative garden plants that may be grafted include many conifers, crataegus (hawthorn), hamamelis (witch hazel), rhododendron, and rosa (rose).

The type of graft you use is purely a matter of a personal preference. Roses are virtually always budded (see Chapter 16). Fruit trees are generally grafted by the whip and tongue method, and failures can be shield budded later in the same year. Saddle grafts are recommended for rhododendrons, whereas spliced side grafts are particularly useful with conifers. All of these will be described in this chapter, but we should start by examining the fundamental requirement of grafting: a good rootstock.

Rootstocks

Rootstocks for fruit trees are usually chosen because they will control the growth and ultimate size of the tree. The most commonly grafted plants are apples, and some rootstocks will produce large trees. Others will develop into medium-sized trees, whereas many modern rootstocks have been designed to produce dwarf or bush trees (and those for training in small areas, like cordons and espaliers).

Apple rootstocks were first developed at East Malling Research Station in Kent, and were given the prefix 'Malling', followed by a number in Roman numerals, to distinguish between them. More recently stocks have been given the 'MM' prefix (representing Malling and Merton, the latter place at the time being the headquarters of the John Innes Horticultural Institute, which worked closely with Malling in developing fruit tree production). Nowadays, the following rootstocks are available: M27 (very-dwarfing), M9 (dwarfing), M26 and MM106 (semi-dwarfing) and MM111 (vigorous).

Pears are usually grafted onto rootstocks of quince. The most readily available types are Quince A, which will produce a medium-sized tree and is semi-vigorous, and Quince C, which produces a tree that is vigorous at first, but which declines after it begins cropping. The latter is generally a good stock for small trees, cordons and espaliers. There are two common plum rootstocks: the first is the dwarfing Pixy, which makes a much smaller, compact tree than ever before; a mere 2.4 m (8 ft) or so, after pruning. The second is the semi-dwarfing St. Julian A rootstock, which provides a medium-sized tree, with slightly larger fruits.

Peaches, nectarines and apricots may be grafted onto St. Julian A, with the same expectations as for plums, or the semi-vigorous Brompton rootstock. This will provide medium to large trees with heavy crops.

The most recent developments in fruit tree rootstock have been with cherries. Only in recent years has it become sensible to recommend growing sweet cherries in smaller gardens, because no dwarfing stock was available. Now they are propagated onto the new Colt and Stella rootstocks, which restrict the natural vigour of cherry trees. They can be grown in a fan-trained shape, or they can be left in the open as a pyramid-shaped tree, where the height can be restricted to about 2.7 m (9 ft).

It is usually necessary to ensure that the plant being propagated is closely related to the rootstock onto which it will be grafted. Indeed, the stock may actually be the common counterpart of the hybrid or cultivar scion.

How do we obtain the rootstocks?
Most of the stocks specified above are available only to the nursery trade, and they will be difficult for an amateur to acquire. There is no harm, however, in talking to a local nurseryman to see if he can let you have, or sell you, a few stocks for experimental grafting. The better option is more likely to be trying the rootstocks of ornamental trees that you can easily raise from seed, such as any of the following: *Acer platanoides*, *Crataegus monogyna*, *Fagus sylvatica*, *Laburnum anagyroides*, *Prunus avium*, *Rhododendron ponticum*, *Robinia pseudoacacia*, and *Sorbus aucuparia*. See Chapter 5 for details on the sowing of trees and shrubs.

Before you undertake any grafting, the rootstock should have reached the right size. Ideally, it should have been planted out as a one-year-old seedling and left to grow in a nursery bed for another year. The grafting will take place on this site and after a further two years, your tree should be ready for transplanting to its permanent flowering and fruiting position.

Whip and tongue grafting

An ideal graft for the beginner, this is probably the simplest method of all. It is carried out during early spring, just before the trees start into growth. The first thing to do is prepare the rootstock by cutting it back. Use a razor-sharp knife as this is essential for producing a good clean cut and it also makes the job easier!

The point on the rootstock where you should make your cut is about 10–15 cm (4–6 in) from the ground. Make a slanting, upward-pointing cut about 5 cm (2 in) in length, and remove completely the top of the plant (Fig.15.1). Now make a nick, a short downward cut, in the exposed face on the rootstock in order to form the tongue (Fig.15.2). The length should be equal to about one-third of the way down from the top of the sloping surface.

The scion should be prepared at this point. From the plant you wish to propagate, cut off some sturdy, healthy shoots that grew the previous year. In an ideal world, the diameter of the scion should be as close as possible to the diameter of the rootstock. It is only the lower part of the scion that is required and needs to be prepared. This is done by making a sloping cut which ends just below a bud (Fig.15.3). It is important that the cut surface is perfectly flat, and it's a good idea to make several practice cuts beforehand on some spare lengths of wood.

To make sure that the rootstock and scion stay firmly together, a similar tongue is cut into the cut surface of the scion (Fig.15.4). This cut should be about one-third of the way up from the lowest point. You only need three good buds on the scion, so there is no harm in cutting off the remainder, just above the top bud.

Fit the scion onto the rootstock so that the two tongues interlock (Fig.15.5). As far as you can tell, ensure that the two cambium layers are touching. If you can see vast amounts of daylight between the two cuts, then the contact will not be good enough to make a perfect union, and you will really need to prepare another scion.

The tongues should ensure a good join between the two plant portions, but the grafting process is not complete until the point of contact has been sealed. It needs wrapping to protect it from drying out, or moving out of alignment. Where the stock and scion are the same size, polythene tape can be wrapped tightly around the area (Fig.15.6). (Years ago gardeners used raffia, and this is still an excellent material, particularly where the stock and scion are slightly different sizes; it makes wrapping easier!) Having bound the graft firmly, the whole area can now be covered with proprietary grafting wax, or a bituminous pruning paint, to protect it from drying out, as well as from fungal attacks. The tip of the scion, which is also an open, cut surface, should have a dab of wax or paint applied to it as well.

Over the next six to twelve months, a callus tissue will develop over the points of contact, and the two cut surfaces will unite. Shoots will, eventually, grow out from the scion's buds. The areas above and below the point of union will begin to swell and, before it does any damage, you should cut through the tape or raffia, to avoid constriction. Occasionally a shoot or two may sprout from previously dormant buds on the old

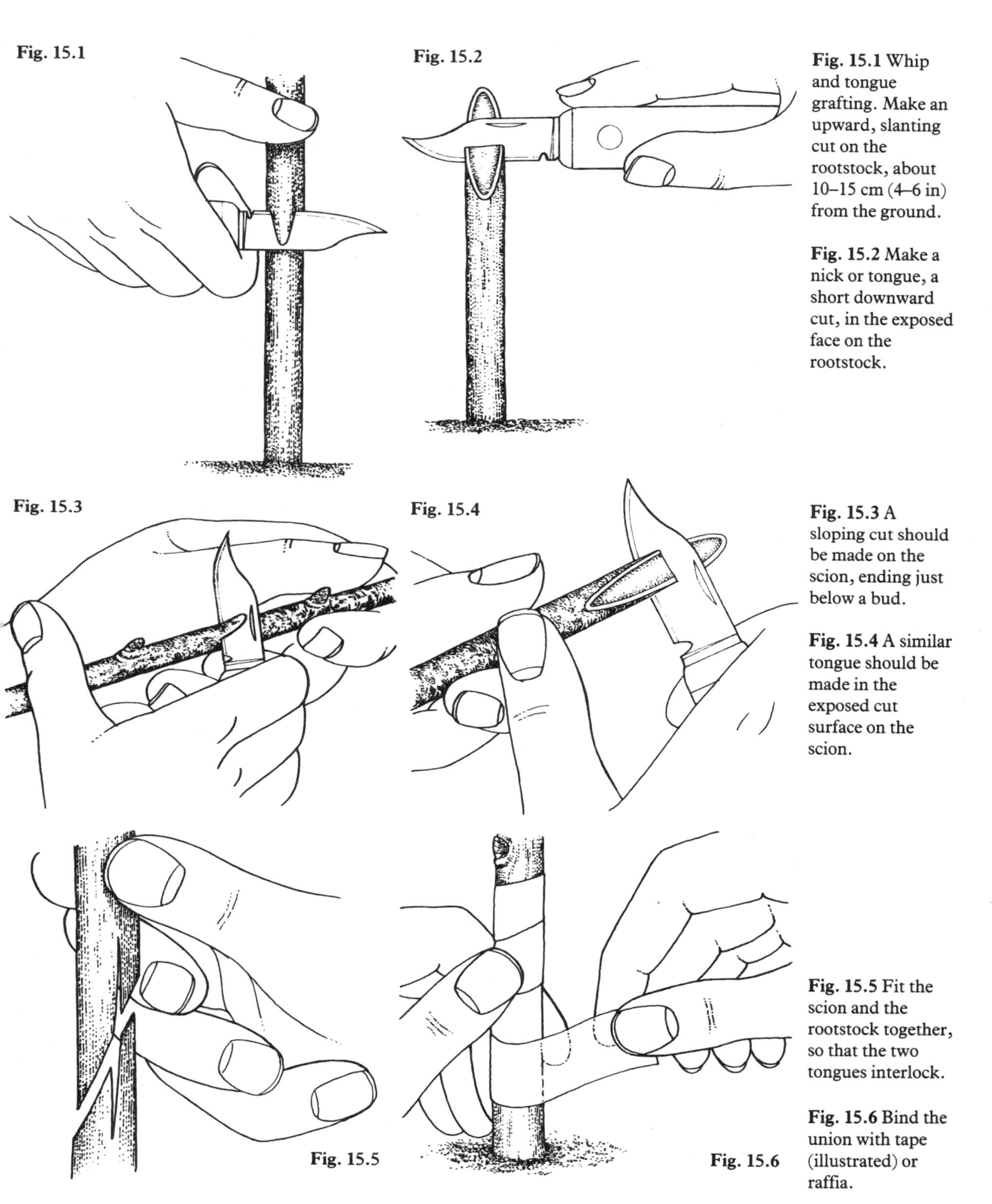

Fig. 15.1 Whip and tongue grafting. Make an upward, slanting cut on the rootstock, about 10–15 cm (4–6 in) from the ground.

Fig. 15.2 Make a nick or tongue, a short downward cut, in the exposed face on the rootstock.

Fig. 15.3 A sloping cut should be made on the scion, ending just below a bud.

Fig. 15.4 A similar tongue should be made in the exposed cut surface on the scion.

Fig. 15.5 Fit the scion and the rootstock together, so that the two tongues interlock.

Fig. 15.6 Bind the union with tape (illustrated) or raffia.

rootstock. These must be removed with a sharp knife immediately, otherwise they will grow vigorously and may soon overtake the scion growth.

Saddle grafting

This is a rather more specialized form of grafting, and is used mainly in propagating hardy hybrid rhododendrons. However, it may be used for almost any subject where the rootstock and the scion are the same size.

Let us take the rhododendron as the example. The saddle graft should take place in early spring, just as new growth is beginning to show. The rootstock should be a two or three-year-old seedling of the wild purple-flowered *Rhododendron ponticum* or R. 'Cunningham's White'. Keep the rootstock growing in a container, and a few weeks before grafting takes place, bring it into a greenhouse, to start it into growth.

With your razor-sharp knife cut the rootstock as far down as you can, to a wedge shape with long, even, sloping sides. The scion, which should be as close as possible in size to the rootstock should now be prepared. Cut the base of the scion to a V-shaped wedge, to match exactly the cuts made on the stock. Now fit the scion over the stock like a 'saddle', as always making sure that the layers of cambium are in contact with each other (Fig.15.7). Then bind the union with tape or raffia to keep it dry and firm.

This type of graft will only succeed if it is given some warmth and humidity. A propagating case is therefore the ideal home for the next six to eight weeks. When new growth is apparent binding can be removed, and the grafted plant can be gradually hardened off. It may be planted out in the garden during the following autumn, but it will be some two or three years before the plant reaches flowering size.

Spliced side grafting

Rhododendrons and conifers are regularly propagated by this method, which is actually slightly easier to do than saddle grafting. The type and size of rootstock, and the time of year, are exactly the same as for saddle grafting.

Prepare the rootstock by removing the top, making a clean, horizontal cut about 5 cm (2 in) above the compost level in the container. Now, make a short, downward cut at an angle of some 45°, about 2.5 cm (1 in) above the soil. It should be no longer than 6 mm (¼ in) or so. Take your knife to the flattened top of the rootstock and make a downward cut to meet the first. You should now be able to remove a sliver of wood, so exposing an inch or so of rootstock cambium, at the base of which is a lip.

The scion now comes into the picture. It should already have been removed from its parent plant, and then shortened to around 10 cm (4 in) in length. At its base, make a cut of the same size and length as the exposed wood of the rootstock; you will need to remove a complete length of wood

Fig. 15.7

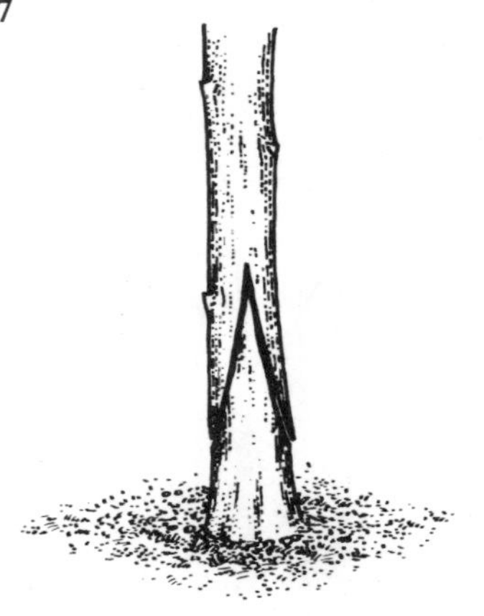

Fig. 15.7 Saddle grafting: the scion sits neatly on the pointed rootstock.

Fig. 15.8

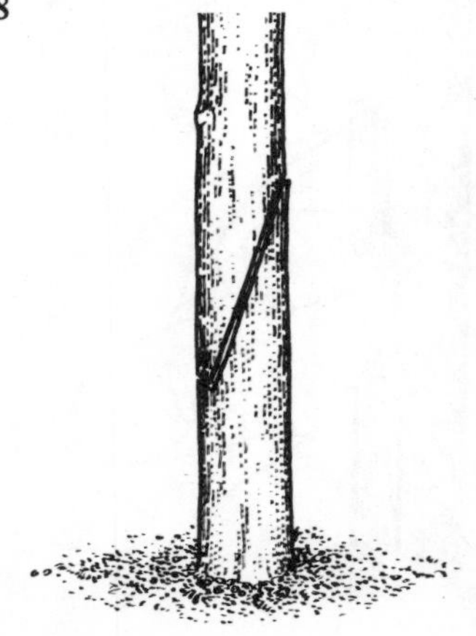

Fig. 15.8 Spliced side grafting. The base of the scion rests on the lip on the short rootstock.

from this scion. You will also need to make a tiny wedge at the extreme base of the scion – sufficient for it to sit comfortably within the lip of the rootstock (Fig.15.8).

Fit the two together, so that the cuts match all the way around. Bind with tape or raffia, and therefore the same conditions as for saddle grafting should be provided.

Veneer grafting

Many conifers are propagated from grafting, particularly those varieties which are very dwarf and slow-growing, and which may be difficult to raise from cuttings. Rootstocks will need to be grown from seed, or cuttings, in much the same way as other tree and shrub stocks. The following are vigorous conifers, and therefore suitable for growing as stocks: *Cedrus deodara* (deodar) *Chamaecyparis lawsoniana* (Lawson cypress), *Cupressus macrocarpa* (Monterey cypress), *Juniperus virginiana* (pencil cedar), *Larix decidua* (common larch), *Picea abies* (Norway spruce), *Pinus sylvestris* (Scots pine) and *Taxus baccata* (common yew).

As with saddle and spliced side grafting, the rootstocks should be two-year-old containerized wood, and should have been brought into the greenhouse a few weeks prior to the graft. The veneer graft merely involves making a vertical cut into the stock, as low down as possible and about 2.5 cm (1 in) long. Slice the knife in just underneath the bark. You should now have a flap of bark, with exposed wood underneath. Do not, as with the previous methods of grafting, remove the top portion of the rootstock; this should only be removed after the successful uniting of the stock and scion.

The scion should be prepared in the same way as for the previous graft methods. However, the length of it will depend on the growth habit of the conifer being propagated – some will not offer suitable scion wood longer than, perhaps, 5 cm (2 in), but this should not be a problem. The base of the scion is cut into a thin wedge, by making angled cuts on either side. The length of the cuts should be the same as the length of the flap of the rootstock.

Insert the scion into the stock, and make sure the contact is good (Fig.15.9). Binding should now take place, and then put the grafted plant into a warm greenhouse, or preferably a propagating case with bottom heat. The graft will unite, hopefully, within two months, at which time the binding material should be cut away. It is only now that the top of the original rootstock should be cut off, just above the union. The grafted plant should be hardened off, ready for planting outside.

Fig. 15.9

Fig. 15.9 Veneer grafting. The wedge-shaped scion is pushed into the flap of bark on the rootstock.

Spliced side veneer grafting

This form of grafting is used on a wide range of trees and shrubs, and is best carried out during early spring. It is similar to the spliced side graft, but the top of the rootstock is left intact, until the union has taken place.

The first step is to prepare the rootstock. Again, the graft should take place as low down on the stock as possible. Make a short downward cut, some 6 mm (¼ in) in length and at an angle of around 45°. Bring your knife up the stem about 2.5 cm (1 in) or so, and carve into the stem a downward cut to meet the first. A sliver of wood may now be removed, and there should be a lip at the base of the wound.

Your scion should be taken from one-year-old wood, and be around 10 cm (4 in) in length. On one side of the scion, at the base, use your knife to strip off a piece of wood the same length as the wound on the rootstock. Remove a tiny portion of wood at

Fig. 15.10 Spliced side veneer graft. The base of the scion rests on the lip of the unshortened rootstock.

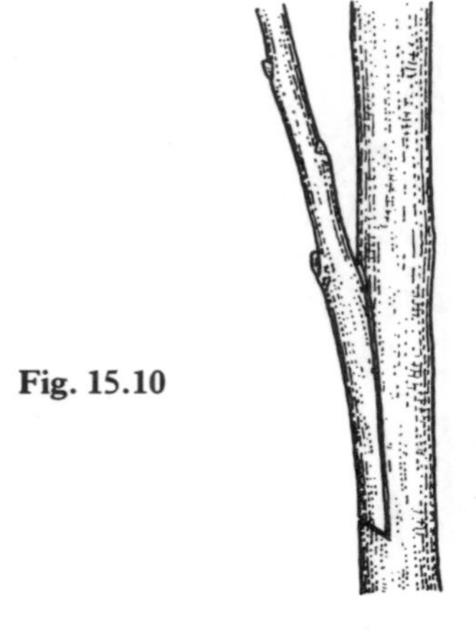

Fig. 15.10

the extreme base of the scion, to correspond with the lip on the rootstock (Fig.15.10).

Now fit the two portions of plant together, so that there are good points of contact. Bind the graft tightly, and place it in a propagating frame with some extra heating. After six to eight weeks, the graft should have succeeded in uniting. At this point cut away the binding, remove the top of the rootstock cleanly, just above the union, and begin to harden the plant off.

Other methods

Although there are a number of further grafting methods applicable to trees, shrubs and fruit, most are either loose variations of the above, or are so specialized that even professionals tend to avoid them if at all possible. Several forms of grafting are also not really to be considered acts of plant propagation, and so therefore do not come within the scope of this book.

There are, however, two forms of grafting that the amateur can consider, both of which take place under glass, and both of which are carried out on plants that could not, by any stretch of the imagination, be considered trees or shrubs. They are cacti and tomatoes! Let's look at both.

Tomato grafting

As we saw at the beginning of this chapter, tomatoes become susceptible to attack from soil-borne pests and diseases, particularly if they are grown in greenhouse border soil which has not been changed for some years. Obviously, the tomato crops will suffer if they are attacked in this manner. If soil change, or sterilization, is out of the question, you could try putting your favourite tomato varieties onto new, disease-resistant rootstocks. These may be grown from seed, and are available from some seedsmen. By far the best stock variety goes by the initials 'KNVF', and it is resistant to damage by corky rot, root knot eelworm and the very destructive wilt diseases.

Sow the rootstock variety and the fruiting variety at the same time, but keep them labelled and well apart, to avoid confusion. Move them on through the pricking out and potting stages, but when potting the plants for grafting, place them at the extreme edge of the pot, not in the middle,which is the usual position. The reason for this will become apparent. When the young potted seedlings have about four true leaves they are at the right stage for grafting.

Remove the seed leaves, if these are still attached to the plant, and use your knife to slice off a small piece of stem. Do this to both plants, making the cuts exactly the same height and length. Gently bring the two stems together, so that the cut surfaces are in contact. If necessary, you could lift the tomato plants out of their respective containers and plant them together in one. However there is less root disturbance and less chance of mixing the varieties if they are kept separate. Now bind the two stems together with a single thickness of sticky tape (Fig.15.11) and place plants in warmth and humidity.

After about three weeks the plants should

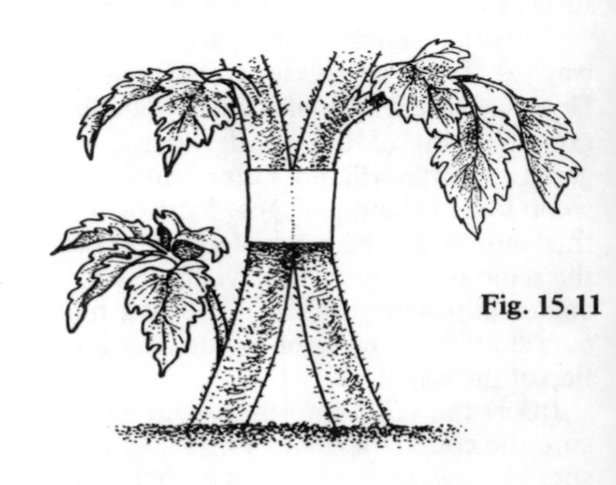

Fig. 15.11

Fig. 15.11 Tomato grafting. The cut surfaces on the two plants are held together with binding tape.

have united, and the tape can be removed, gently. Take the united plants out of their container(s), and cut away the roots of the fruiting variety, and the top of the rootstock variety. Replant the single tomato, making sure that the graft union remains well above the surface of the compost, or soil, and then treat it as you would any other greenhouse tomato.

Cactus grafting

Unlike tomato grafting, having a go with varieties of cacti should be considered for fun rather than for profit. It is not generally carried out by the amateur, but some professional growers graft scions of slow-growing varieties onto stocks of fast-movers.

Apart from simply seeing if it can be done, the amateur may wish to graft a variety which is difficult to propagate from any other method. This might include *Gymnocalycium mihanovichii* (red cap cactus) which contains little chlorophyll and so cannot manufacture its own food easily. Grafting this onto a normal, healthy green cactus, will enable it to lead a happy, if parasitic, existence.

The most common species used as rootstocks are from the *Echinopsis*, *Hylocereus* and *Trichcereus* genera. The grafting techique is simple. In early summer, slice off the rootstock, some 5 cm (2 in) from the top, leaving a flat cut surface. The edges of this should then be pared to round them slightly. The scion should be of a similar size. Cut a slim sliver off the base, to expose the inner tissue and to level the area.

By examining the cut surfaces you will notice the dark rings of tissue that are the equivalents of the cambium layers. Place the scion on top of the stock, trying to match the rings as closely as possible. Now use elastic bands to hold the scion down, stretching the band over the top of the scion, and down underneath the pot. Several bands should be used to provide even pressure (Fig.15.12).

Fig. 15.12

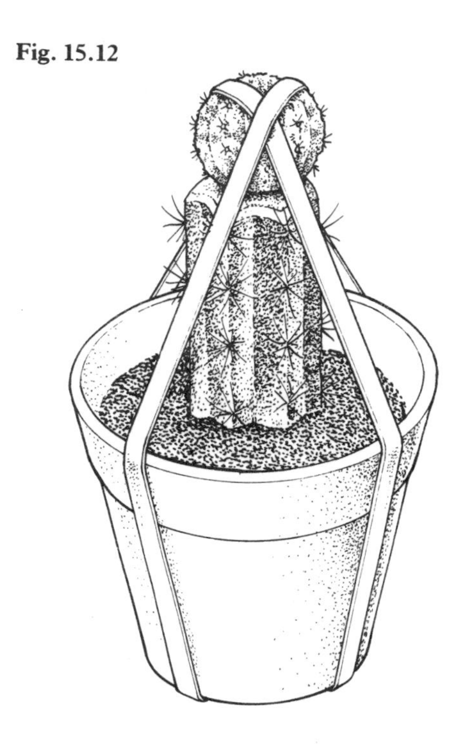

Fig. 15.12 Cactus grafting. The scion is held on to the stock by the use of elastic bands.

At this stage the grafted plant should be stood in a warm place out of direct sunlight, for several weeks. After this time has elapsed it can be moved to a sunnier spot. It will be a month or so before it is possible to tell whether or not the graft has taken.

The beauty of cactus grafting is that, if it does fail, all you need to do is separate the two portions, clean up the cut edges with a knife, and repeat the process.

CHAPTER 16

Budding

Budding is really a form of grafting, but one which is generally given its own definition in propagation terms. It is most commonly applied in the large-scale production of certain fruiting and ornamental trees and shrubs, particularly cultivars of *Rosa* (rose). Commercial growers use budding to mass produce the large-flowered and cluster-flowered bush roses, as well as climbers and ramblers. However, budding may be carried out by the enthusiastic amateur, often with complete success and the satisfaction that goes with it of raising new varieties.

Budding is carried out during the summer months, generally in the first half, when the buds on the scion have developed fully.

The fundamental difference between budding, and the sorts of grafting so far discussed, is simple. Usual grafting techniques require a piece of shoot to be grafted onto a rootstock; in the case of budding, it is only a growth bud from a desirable plant that is used, and this is then 'inserted' in the rootstock. This bud will, in time, unite with the stock, and the growth developing from it will become the new plant. The same generalizations concerning the use of rootstocks and scion material may be applied to budding.

In some cases, budding may be a useful alternative to grafting; many of the subjects grown by, say, the whip and tongue method may, instead, be budded. Certainly amateurs might regard the budding process as being much easier to carry out.

Rootstocks

For the reasons already explained, we shall consider the budding of roses throughout this chapter. As with all grafting methods, the amateur may find the availability of rootstocks a problem. Traditionally, the stock used is *Rosa canina* (dog rose or wild briar), althought this has one or two disadvantages. For example, the dog rose is found growing wild in woodland and hedgerow, and it is its ability to sustain these difficult conditions by being vigorous that make it such a good rootstock. However, the vigour also manifests itself in the production of suckers, the effects of which can be seen in gardens everywhere!

Nevertheless, the dog rose will make the best stock for the amateur who wants a bush rose. If a standard rose is desired it should, instead, be produced by budding the variety onto a rootstock of *Rosa rugosa* 'Hollandica'. This has a large coarse root system, and will produce a large, prickly maiden bush. Dog rose stocks are best raised by sowing seed that has undergone the process of stratification for at least a year (see Chapter 5 for techniques). The stock for a standard rose, being a variety and not a natural species, should be raised vegetatively, from hardwood cuttings. In either case, one-year-old seedlings or rooted cuttings should be planted outdoors in a nursery bed during autumn. The following summer they will be at the right stage for budding.

The buds

As in the case of grafting, you must choose carefully the material of the variety you want to propagate. This is gathered immediately prior to the budding taking place. Select a few well-ripened rose stems of the current year's growth which have, perhaps, already flowered. Make sure that in at least a few of the leaf axils there is a healthy dormant bud. Using a sharp budding knife, cut the shoot off the parent plant. The length of this shoot is immaterial, as it will be required only to offer a few buds.

At this stage, the shoots are called 'budsticks'. They should be prepared before budding takes place; cut off the soft tips of the shoots; remove all the leaves, but let the leaf stalks remain, and stand the sticks, the right way up, in a bucket of water, to keep them from drying out.

Having prepared the budsticks, before you go any further you need to know which of the two common methods of budding you are adopting. These are termed shield or T-budding, and chip budding. Both are simple to do.

Shield budding

This is the most widely used budding technique. Standing by the rootstock, take a suitable budstick from the bucket (which should also be at your side), and cut out a healthy bud with your knife. There is a proper and safe way of doing this. Inserting the knife just below the bud, shallowly scoop it out, with a piece of rind, shaped like a tiny shield (Fig.16.1). Make sure you keep the bud clean and moist while you prepare the rootstock. Indeed, many experienced budders develop the habit of collecting a bud, and popping it into their mouth for the few seconds it takes to make the cuts on the stock, at which time they retrieve it and insert it. This is done to prevent the delicate tissue at the rear of the bud from drying out during the process. An alternative is simply to drop the bud into a saucer of water. If you inserted the knife shallowly, there should be only a thin sliver of wood behind the bud. This sliver is not required – in fact it will hinder the budding process – so it should be flicked out carefully with the tip of the knife.

Now that you have cut out the bud and exposed its layer of cambium, you must tackle the rootstock in a similar fashion to expose *its* corresponding layer. About 2.5–5 cm (1–2 in) up the stem from ground level, make a shallow cut with your knife, in the shape of a 'T' (Fig.16.2). The horizontal top cut need be only around 6–12 mm (¼–½ in) in length, whereas the downward cut should be around 2.5–3.5 cm (1–1½ in) long. Two things will help you to make a clean cut. The first is to water the rootstock well a few days beforehand, to help the rind lift easily from the wood. The second is to bend the stem backwards slightly, so that the bark you are cutting is under tension.

True budding knives have flattened spatula-like handle ends, designed specifically for the next action, which is to slip the handle into the T-cut, so that it raises the two flaps of rind (Fig.16.3).

Carefully hold the bud by its edge, or by using any small remnant of leaf stalk which may exist, push it down into the T-cut on

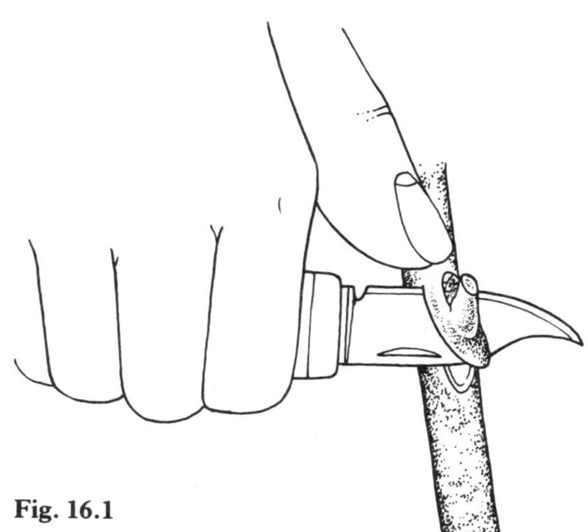

Fig. 16.1

Fig. 16.1 Shield budding. With a sharp knife, slice out a healthy bud from the variety you wish to propagate.

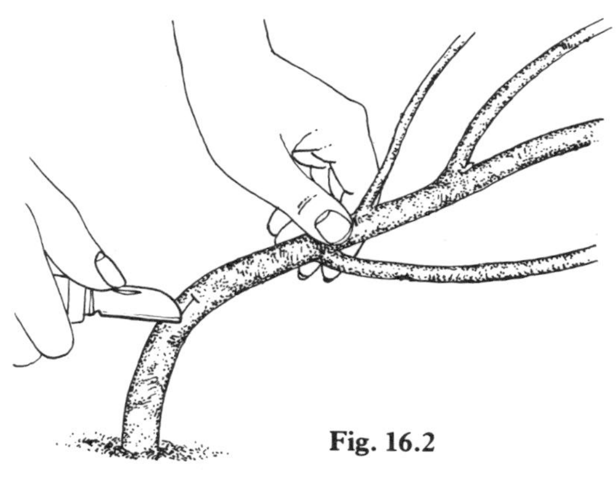

Fig. 16.2

Fig. 16.2 Make a shallow, T-shaped cut on the rootstock, about 3–5 cm (1–2 in) from the ground.

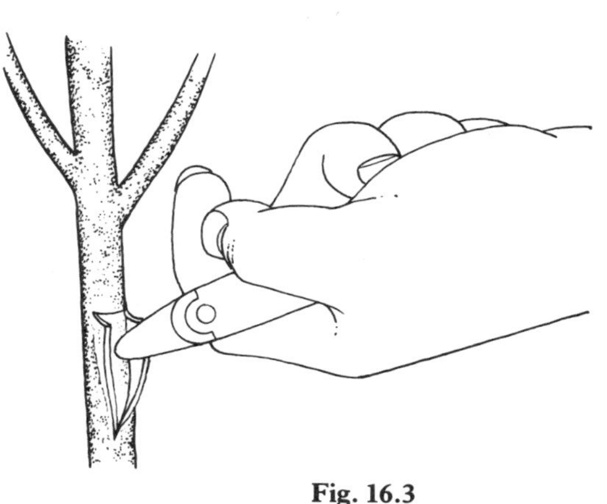

Fig. 16.3

Fig. 16.3 With the flat blade of the budding knife, or another flat object, peel back the flaps of bark.

Fig. 16.4 Push down the bud, into the T-cut, until it fits snugly.

Fig. 16.5 Bind the bud with a proprietary budding patch (illustrated) or raffia.

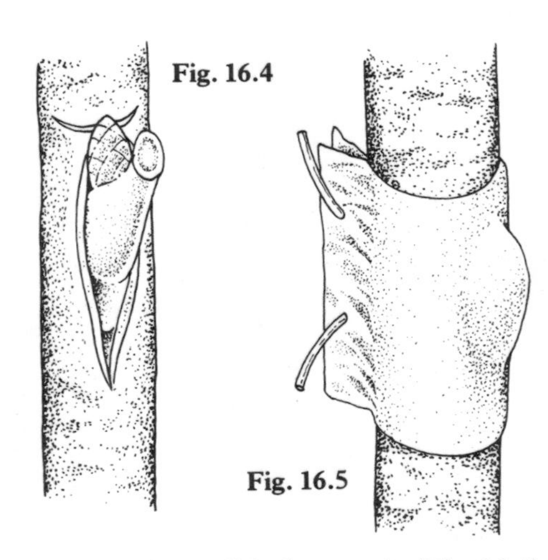

the rootstock, until it fits snugly (Fig.16.4). Do not be over-zealous, as the bud can be easily damaged. The idea is to enable the two layers of cambium to be in firm contact, so once the bud is in position, it will need binding. Raffia may be used, in which case try not to cover the actual bud, only the wood above and below it. Wrap the raffia around the stem several times and tie it tightly, but not so that it causes a disruption to the flow of sap within the stem.

Instead of raffia, some gardeners prefer to use proprietary budding patches (Fig.16.5). These are thin pieces of rubber, each one designed for stretching over a placed bud, so that the two ends meet behind the stem. They are then held together with a metal staple. The rubber will perish in a few weeks to a month, which is usually long enough for the bud to 'take'.

The bud will probably start throwing out new growth during the autumn months; slow takers may not move until after the cold winter. Whenever it starts, the rubber or raffia will need to be removed to prevent hindrance and constriction. In early spring, cut off the top growth of the rootstock, just above the new bud. The budded variety will then grow out as the only shoot, and it will become the plant's new topgrowth.

If you wish to grow a standard rose by budding, you will need to make a minimum of three T-cuts on the rootstock, and these sited towards the top of the stem, at the height from where you would like the branches to emerge

The successfully budded plant, bush or standard, may be lifted and transplanted to its permanent flowering position the following autumn.

Chip budding

This lesser-used method of budding has an advantage, in that it may be carried out during the dormant season. However, it is not as easy to complete. The best results occur in early spring, just prior to growth starting.

Prepare the rootstock and budsticks exactly the same way. The bud will need to be cut out, and this is done in two actions. First, insert the budding knife about 6 mm (¼ in) below the base of the bud and make a slighty, downward-sloping cut, travelling into the stem to about a third of its thickness (Fig.16.6). Withdraw the knife cleanly, and make the second cut. This should be started about 1 cm (½ in) above the bud; insert the knife and make a slicing cut down through the stem, and behind the bud, to meet the first cut. With luck, the bud should now be cleanly detached from the stem, and with it an arch-shaped piece of rind, wedge-shaped at the base.

Fig. 16.6 Chip budding. Cutting out the bud by first inserting the knife at an angle below it, and then by making a second insertion and slicing down behind it.

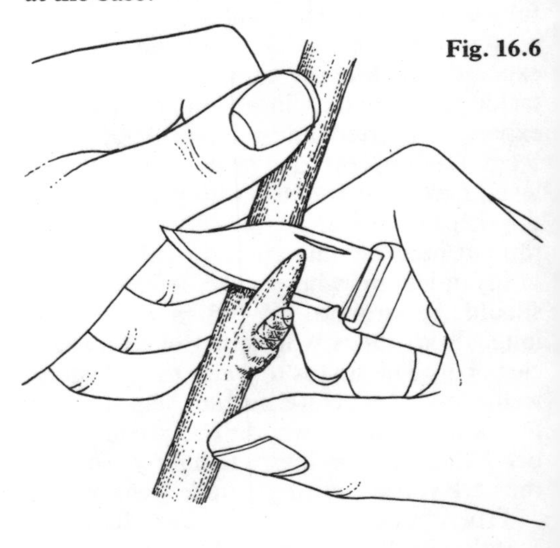

Keep it moist while you prepare the rootstock, as you would with the shield bud. Simply make the cut on the stock so that the bud will fit exactly into the space (Fig.16.7). Fit the bud onto the stock, and ensure good contact of the cambium layers by pushing

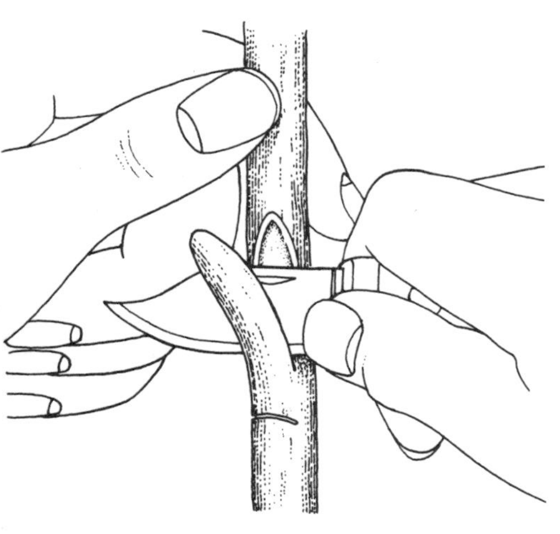

Fig. 16.7

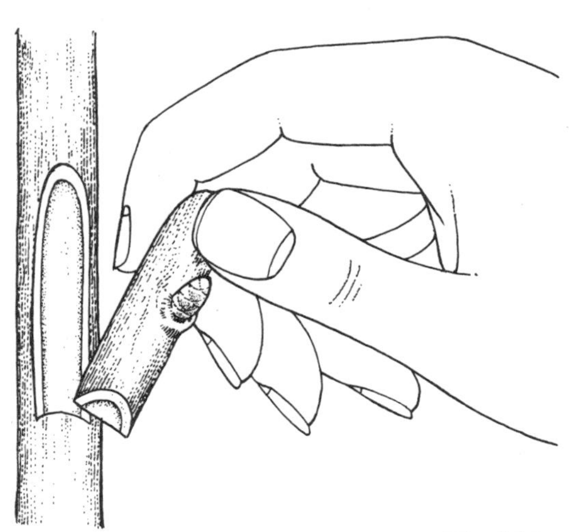

Fig. 16.8

Fig. 16.7 Now make a similar cut in the rootstock, so that the bud fits exactly.

Fig. 16.8 Fit the bud onto the rootstock by pushing the wedge-shaped base down into the lip.

the wedge at the base of the bud into a wedge-shaped hole cut into the stock (Fig.16.8).

The chip bud should now be tied firm, using either raffia or polythene tape. Within a month there should be signs of a successful 'take'. When it is obvious that growth is proceeding, cut the top of the rootstock, at a point just above the new bud. The plant may be transferred to its permanent flowering place the following autumn.

Chip budding may be carried out during the growing season, when the rind lifts more easily from the wood, but there seems little point when shield budding is an easier process.

Budding trees

Part II of this book comprises a comprehensive encyclopaedia of plants and their propagation. It should be noted that any woody plant that is recommended for raising by whip and tongue grafting, may also be raised by budding. Therefore it is not necessary to provide a large list of suitable subjects for budding here.

In all cases, the same techniques of shield or chip budding may be adopted; the only difference being the positioning of the new bud on the rootstock. With trees, the bud should take place some 15–20 cm (6–8 in) from the base of the stem, above soil level. The usual period for budding is from early summer to early autumn.

After the buds have successfully taken, and the old rootstock topgrowth has been removed, a stake (or cane) should be provided for supporting the new, often whippy, developing growth. Training may also be an important factor at this point (see Chapter 18).

CHAPTER 17

Micropropagation

Throughout this book we have concentrated on Man's discoveries in plant propagation, from the very earliest times. We have seen how seeds, cuttings, division, layering and grafting have all been used to multiply plant stocks, and that over the centuries advancements have been made within all of these techniques. One would imagine that we have now reached a stage whereby there is little new to learn. How wrong this would be!

In the last 60 years or so we have learnt how to reduce plants by cutting back and stripping leaves, even within growth buds, until we are left with a tiny unit of plant energy: the 'meristem', which can, too, be kept alive, encouraged to develop, and eventually be used to reproduce an exact copy of the parent plant from which it was taken

Further, Man has discovered that a plant can be reduced to its smallest working component, a single cell, and that even this could grow into a normal healthy seed-producing adult plant. This particular form of horticultural surgery, loosely termed 'totipotency', is practically impossible for the amateur to conduct, and should really be left to the white-coated laboratory technicians who have the facility of working in a near-sterile environment.

In a broader sense, the multiplication of plants from minute fragments, using modern technology, is termed 'micropropagation', and is now widely used by many nurserymen. Few people truly appreciate the impact that it has on modern-day plant production yet around 300 million plants are produced annually by various forms of micropropagation. It is in the commerical world of orchid, rose and fruit tree production that we see micropropagation being used at its best, and many homes will have foliage house plants which started life in a test tube!

This is particularly the case with modern plant varieties which originated from a sport (a genetic mutation) of an established variety. An example of this *Choisya ternata* 'Sundance' (yellow-leaved Mexican orange blossom), which was launched with a flurry of publicity in 1986. Other, more recent introductions similarly developed from sports, and subsequently micropropagated, include *Betula* 'Golden Cloud' (yellow-leaved birch) and *Dicentra eximia* 'Snowflakes' (white-flowered bleeding heart).

Micropropagation is a vegetative method of plant production using tissue culture techniques. This means we take small pieces of plant, called 'explants' such as a bud, section of leaf, stem or root, or a meristem and ensure that they are sterile. These tiny fragments are then provided with a complete growth medium containing sugar (for energy), vitamins, minerals and growth-regulating hormones.

Method

As can be seen from Fig.17.1, the micropropagation cycle, the explants (an apical meristem and an axillary bud) are first taken from the stock plant. Each would then be placed on a growth medium – usually a gel – which contains the hormone 'cytokinin', to stimulate multiple shoot formation. When enough shoots have been produced, they are then treated as micro-cuttings, and rooted on a fresh medium containing the hormone 'auxin'. The speed with which large numbers of identical plants can be produced is impressive. For example, one stock plant may easily give 1000 rooted offspring, from 100 explants, within three months.

Benefits of micropropagation

One can immediately see the advantages of micropropagation, that of mass production, so often bringing down the price of plants at retail level. The classic example of this is orchid production. At one time new varieties were prohibitively expensive, because conventional propagation was slow. Now, new forms can be multiplied quickly into thousands, and sold at reasonable prices. Enthusiastic amateur growers can

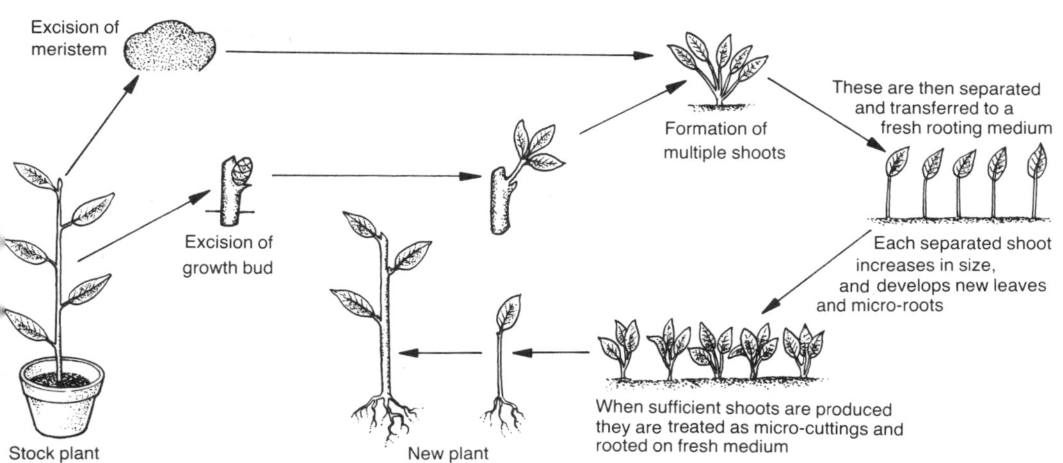

Fig. 17.1

Fig. 17.1 The micropropagation cycle. A simplistic view of the way millions of plants are propagated in commerce.

save even more by purchasing the micropropagated plantlets while they are still tiny, perhaps even in their culture gels.

The amateur also benefits in other ways. For example, plants are often 'rejuvenated', the offspring being of better quality and having greater vigour than conventionally propagated material. This often manifests itself as a significant improvement in their ability to root, when ordinary cuttings are taken from them at a later stage.

Soft fruit has particularly benefited from micropropagation in that they have been 'cleaned up'; that is, freed from viral or bacterial diseases. You can now buy rhubarb, gooseberries and tayberries that have been developed from disease-free stock plants. These are usually sold as 'virus tested'.

Another benefit for the amateur is that micropropagation is less seasonally dependant, because of the controlled conditions in which the plants are grown. Therefore plants may, generally, be raised all-year-round, increasing availability. Quality is improved, too. Huge numbers of plants are able to be grown at the same uniform high standard. This is particularly true in the case of house plants like the new miniature *Saintpaulia ionantha* (African violet), as well as begonias, marantas (prayer plants) and *Nephrolepis exaltata* (Boston fern).

Micropropagation will also enable new varieties to become available much more quickly than would otherwise be possible. We have seen this already in the worlds of chrysanthemums, strawberries and roses. The miniature rose 'Benson and Hedges Special', launched in 1983, was one of the first to be multiplied in laboratory conditions. Some plants, like the *Rhododendron yakushimanum* hybrids, for example, take so long to raise from conventional methods, these are prime candidates for investigation by micropropagation pioneers.

The future

Even newer approaches are currently being tried out; obsolescence, it seems, is not restricted merely to new technology in industry! One method undergoing trials, and which shows promise for use by the amateur, is the Swiss 'Sorbarod' system. This uses liquid growth media and cellulose plugs to support the plants. The plugs come in sterile trays, one type for the multiplication stage, and another for the rooting stage. The latter can be gradually vented to 'harden' the plants. They can then be planted into conventional cuttings or potting compost, still in their plugs, so minimizing disturbance to the young roots.

Although the amateur is currently restricted as to what he or she can do in micropropagation, the sensible action seems to be to stand back and take advantage of the benefits outlined above. One day it will be within the scope of all of us to use it. One thing is for sure, micropropagation is here to stay.

CHAPTER 18

Training Plants

In decorative and commercial horticulture, newly propagated plants of a woody nature, whether raised by seed, cutting, or any of the other accepted methods of increase, are unlikely to develop into near-replicas of their parents without some form of initial training and coaxing. In Nature, most young plants do not have the benefit of Man's encouragement, and so have to fight for their space: for moisture and nutrition from the soil; for light, and so on. As a result, all energies are put toward survival and procreation. Displaying their full potential, in terms of flower, foliage and fruiting ability, does not always take precedence in natural conditions. In other words, the plants grow tall and spindly (and generally grow in a most unsightly and unbalanced way), or they are dwarfed by lack of light, and they produce a minimum in terms of flowers and fruits.

A sad tale indeed! So, this is where the gardener must pay some attention to the formative training and pruning of woody plants. Of course, there are a number of excellent specialist books that cover this subject in much greater depth than we can give it here. However, it is essential for all who propagate their own plants at least to have an understanding of the subjects, and to give their plants every opportunity to develop properly in their early years.

Let us therefore consider a few of the subjects that will benefit most from initial training.

Training young trees

The first point to realize is that you must be patient with your plants. For example, it takes from three to five years to produce a tree from grafting. Throughout this period you will need to prune it and train it, so that it develops into a tree of the right shape and size for its space.

The most commonly grown fruiting tree is the apple, and the most easily managed form for today's average-sized garden is a dwarf bush tree, usually grown on M9 or M26 rootstocks. Open-centred bush trees are most commonly seen with about 60 cm (2 ft) of stem, with branches radiating up and out. There is a specific training operation that should be carried out in each of the first three years after planting.

Maiden trees should be planted between late autumn and mid-winter, at which time each one should be shortened to 60–75 cm (2–2½ ft). The place where the top portion is served from the base should be just above a bud (Fig.18.1). A year later the tree will

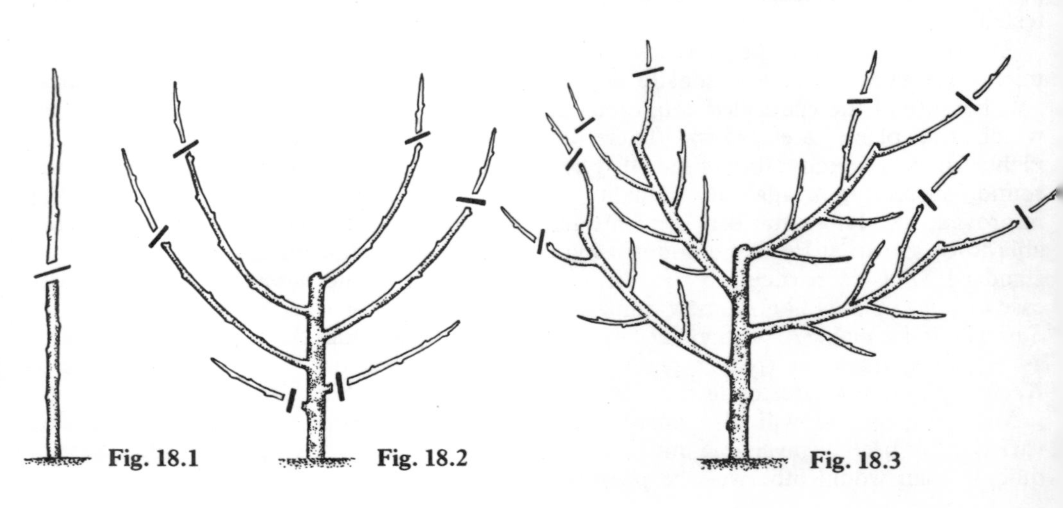

Fig. 18.1 Training a bush tree. A maiden tree should be planted between late autumn and mid-winter, at which time it should be shortened to 60–75 cm (2–2½ ft).

Fig. 18.2 A year later the tree will have formed a small framework of primary branches. Select four, and remove all others.

Fig. 18.3 Two years from planting. Prune the secondary branches back to about half their length.

have formed a small framework of healthy, strong primary branches. Select four of these which are coming out wide from the trunk, but remove all others at their points of origin, and cut back the selected four by about a half (Fig.18.2). Always cut to a bud, and preferably one that is pointing out from the centre of the tree.

A year after this, secondary branches will have formed. Measure the length of these from their tips, down to the point where they originate from the primary branch. Prune them at roughly half way (again, to outward-facing buds) (Fig.18.3). At the same time any lateral branches growing toward the centre of the tree should be cut back; it is important to keep the centre of the tree 'open', to enable air to circulate freely. A well-shaped bush tree is now becoming established.

Standard and half-standard trees are similar in overall shape to bush trees, except that the length of clear trunk beneath the lowest branch is around 1.8 m (6 ft) and 1 m (3 ft) respectively . These two tree forms were, for many years, the most popular styles. However, with modern garden sizes they have become less practicable. The formative training of standards and half-standards is exactly the same as for bush trees, except that it takes place much higher up, and the gardener will require a ladder!

There are some trees, both decorative and fruiting, which do not need much in the way of training, and these are left to grow quite naturally. Examples include certain weeping trees (although staking is required to ensure that the stems grow upright during the first few years), and fastigiate trees (which produce erect, closely-spaced branches from as near the ground as possible).

Staking

From the time a tree is first planted, and until it is well established and has made sufficient fresh roots, it will require staking. Ideally the stake should be put into position before the tree is planted, to avoid the possibility of damaging the roots, after they are hidden by soil, as you drive the stake down. The size of the stake depends on the size of your tree, but remember that at least 45 cm (18 in) of the base of the stake will be buried. Whatever the length of your tree's stem, you will need an equal amount of stake showing above soil-level. The top of the stake should be more than a few inches beneath the lowest branch. If it comes higher, some of the branches may be damaged by it as the tree rocks backwards and forwards in windy weather.

Your stake should be treated with a timber preservative, to give it a long life, before it is used to support the tree. But be careful here: all wood preservatives, unless specially formulated for horticultural use, damage young plant growth when freshly applied. So, do not use the stake unless the preservative was applied at least a week or two beforehand.

Ties are as important as the stakes. If they grip the area too tightly they will restrict the flow of sap, and interrupt growth. If they are too loose they will let the tree rock, with a resultant chafing of the stem.

Cordon training

The term 'cordon' usually refers to a single-stemmed fruiting plant, trained vertically or at an angle; 45° is recommended because it enables the plant to receive maximum direct sunlight. Cordons are particularly suited to small gardens where space is at a premium. A distance of 1 m (3¼ ft) is all that is needed between one cordon and another, allowing for example 10 apple trees to grow quite satisfactorily along a 9 m (30 ft) row. A sunny fence or wall running from north to south, with the trunks pointing northward, is ideal. To facilitate even growth and development, cordon rows (if there is to be more than one!) should be no more than 1.8–3 m (6–10 ft) apart.

Cordon trees require a supporting framework of posts and wire, which should be set up before the trees are planted. The posts must be sturdy; remember that wooden posts will need renewing eventually. The wires are best spaced at 60, 90, 120 and 150 cm (2, 3, 4 and 5 ft) from the ground. To support the actual tree, a bamboo cane, or something similar, must be secured to the wires at the required angle.

Set the first tree at about 1.8 m (6 ft) in from the end post, planting this and all subsequent trees at the desired angle. As the

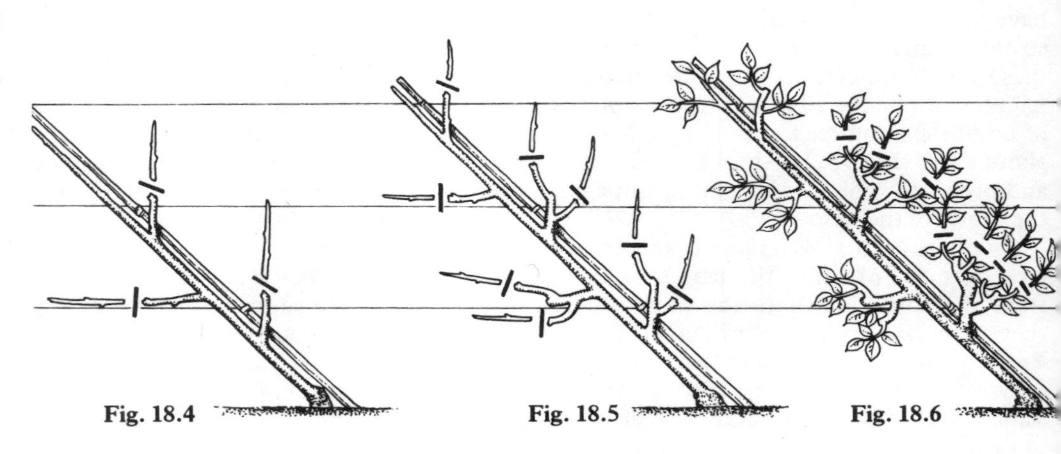

Fig. 18.4 Training a cordon. Plant an unpruned maiden, during early spring. Set it at 45° and support it against a cane and wires. Cut any sideshoots back to four buds, but do not shorten the leader.

Fig. 18.5 A year later the cordon will have developed a number of sub-laterals. These should be cut back to 2.5 cm (1 in) from their points of origin. Any new laterals should be cut back to four buds.

Fig. 18.6 The second summer after planting. From now on, summer pruning should be undertaken annually. Cut back the foliage-covered laterals to three leaves from the base. Also, cut back sub-laterals to 2.5 cm (1 in).

trees will invariably be grafted specimens, it is important when planting to ensure that the scion part of the graft union is in the uppermost position. This will prevent the union being forced open if the tree is inadvertently bent lower than it should be!

The training process will, again, take place over a three or four year period. Begin by planting an unpruned maiden tree in spring, and tie it in several places to the bamboo cane. Do not tip or prune the leading stem, but cut any sideshoots back to four buds (Fig. 18.4), the aim being to create a system of 'spurs', (short, stiff branchlets comprising flower buds). No further pruning is done until the following spring. In the meantime the reduced sideshoots will have produced sub-lateral branches, and some of the original growth on the plant will have transformed into flower buds.

Any new laterals should be pruned back to four buds a year after planting. At this time, the sub-laterals can be cut back to about 2.5 cm (1 in) from their points of origin (Fig.18.5). It is quite likely that a small number of fruits will be produced during the following summer. However, they will be sparse and small, and should not really be allowed to ripen. Patience, as always, is a virtue, and long-term welfare of the tree is preserved if these first apples are picked off as soon as they are seen. Indeed, the flowers in spring are best cut off as they appear, but take great care to leave the rosette of leaves intact.

From now on, a degree of pruning should take place each summer. Cut back the leafy lateral branches to three good leaves from the base, but not counting the basal rosette. At the same time, cut back the sub-laterals to about 2.5 cm (1 in) (Fig.18.6). The benefit of summer pruning is based partly on the fact that the removal of leaves during the growing season checks the root growth, and thus the shoot growth. It encourages the development of fruit buds, and by the cutting away of some surplus growth, air and light are allowed entrance to the centre of the tree framework, resulting in the buds, wood and eventual fruits being well-ripened.

Finally, when the leader stem has grown beyond the topmost horizontal wire, cut away the tip (which should be one-year-old wood), back to its junction with the previous year's growth.

Espalier training

This is a form of training, similar to that of the cordon, but where a system of branches are meticulously guided so that they grow in a pattern determined by the gardener. It was originally designed through a genuine need to grow fruit trees in limited spaces. However, apart from fruiting plants, there are many other trees and shrubs that respond well to the stringent restrictions of espalier training. *Acer palmatum* (Japanese maple), *Taxus baccata* (common yew) and some of the smaller magnolias will look marvellous if trained in this fashion.

The basic espalier form comprises a

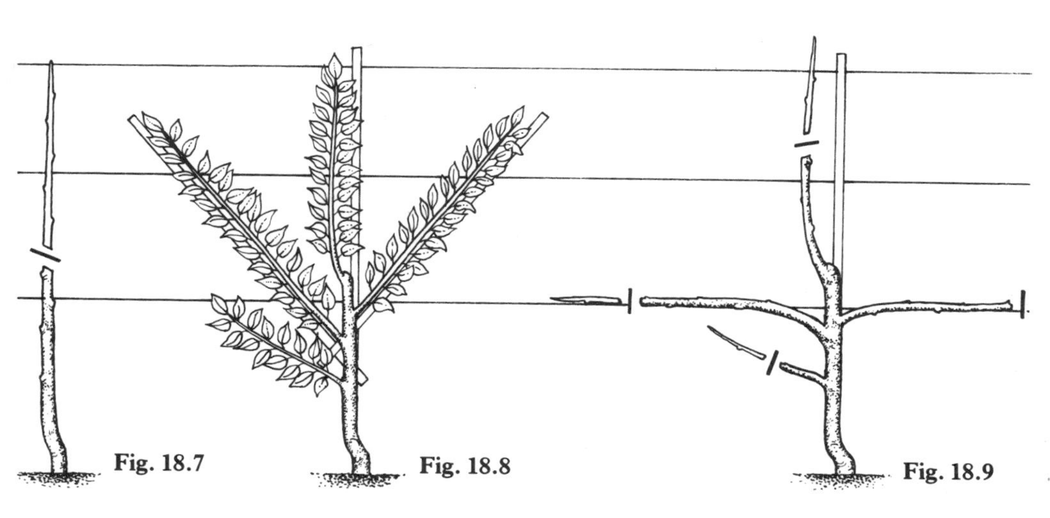

Fig. 18.7 Training an espalier. Plant a maiden, in early spring, against a system of supported wires. Immediately cut the stem back to around 37 cm (15 in) from the ground.

Fig. 18.8 Over the spring and summer months, the uppermost bud will shoot out and become a leader, which should be tied in, vertically. Side shoots should be tied in to the lowest wires.

Fig. 18.9 A year after planting, cut back the leader to within 45 cm (18 in) of the bottom branches, and reduce these by about one third.

single-stemmed tree, grown against a supporting structure (free-standing wires, or against a wall or fence), with horizontal branches coming out at intervals of about 45 cm (18 in). As with cordons, plant a maiden tree in early spring; it should be set vertically in the soil. Immediately cut the stem back to around 37 cm (15 in) from the ground (Fig. 18.7). The uppermost bud will shoot out and become the leader. As it grows it should be tied to a vertical supporting cane. By cutting the stem back, the lower buds will be forced to produce growth (Fig. 18.8). Two of the lowest will become the bottom horizontal arms, and these should be tied, as they grow, to the lowest wires (Fig. 18.9).

The following spring cut back the vertical leader to within 45 cm (18 in) of the bottom branches. This will encourage the buds immediately below the cut to shoot out, and the process of tying two branches to the wires is repeated. At the same time, as you cut back the leader, prune the horizontal branches by about one third. This will aid the formation of fruiting spurs.

This training technique is repeated until the tree has acquired the right number of horizontal branches, and the central leader has reached the top wire (Fig. 18.10).

Fig. 18.10 The training technique is repeated until a mature espalier is the result.

Training roses

An introduction to roses, and specifically their propagation by budding, was made in Chapter 16. The most popular group for decorating the summer garden has to be the large-flowered bush roses (known as 'hybrid teas'). The flowers are shapely and they sit on long, strong stems. There is a bewildering range of colours available, true blue and black being about the only absentees. The fragrance is usually medium or strong, and the blooms are perfect for exhibiting in competitions, or cutting for indoor decoration.

The aim when pruning hybrid tea roses is to encourage the development of strong basal growths, and to form an open-centred plant with an evenly spaced system of branches. When planting, make sure the union (where the bud was placed within the rind of the stem) is sited above soil level.

The first pruning cut should be made immediately after planting. Cut back each shoot at about 15 cm (6 in) from ground level. Always cut to an outward-facing bud. By early summer new shoots will have developed, and a few flowers may open. During late summer or early autumn the flowered stems should be tipped back, and any soft, unripe shoots cut out.

The annual routine pruning should be initiated late the following winter. Cut out any dead and diseased wood, followed by the removal of any inward growing or crossing stems. Then cut back each shoot to between 15–23 cm (6–9 in) from the ground level.

Second only to hybrid teas in popularity are the cluster-flowered bush roses ('floribundas'). These bear their flowers in trusses, and several blooms open at a time in each truss. Individually, the blooms are not so spectacular as the hybrid tea, but in a bedding display they are unrivalled. Indeed, with hybrid teas there are distinct flushes of bloom during the summer months; with floribundas there is more or less a continual flowering throughout summer and autumn.

They are more vigorous than hybrid teas, so generally they should be pruned less severely; it is a curious enigma that vigorous stems should be pruned lightly, and weak shoots need hard pruning! Undertake the same formative pruning as for hybrid tea roses, but remember that floribundas are larger plants, so they should be cut back to around 23–25 cm (9–10 in).

Floribundas were not widely grown until the 1950s, but even newer are the 'miniatures'. These can be used for growing in tubs and rock gardens, for edging flowerbeds, or as temporary indoor pot plants. Most varieties are 30–45 cm (12–18 in) high, and both flowers and leaves are small, in proportion with the rest of the plant. Miniatures are certainly worth growing, and each year sees a new crop of introductions by the larger rose nurseries.

The training of miniatures should follow the same lines, although one should not cut back newly planted subjects too drastically. Occasionally strong over-vigorous shoots are thrown up, which spoil the overall look of the plant. It is really better to remove these entirely when pruning takes place, in spring, so that the plant had a proper balanced framework throughout the growing season.

Standard roses are simply hybrid teas, floribundas or miniatures, often with a weeping habit, that are budded high up on long-stemmed rootstocks. Thus, formative training will be the same; the advantage to the gardener being that he or she will not have to bend down to do it!

Many gardeners are confused about the differences between climber and rambler roses. If you want to cover a wall or screen, it is usually better to choose a climber. These have larger flowers but smaller trusses and stiffer stems. Blooms are carried on a framework of mature wood. Planting, whether against a wall, fence, tree or free-standing pillar, should be carried out during early spring. Do not undertake pruning at this stage. The first summer flowers will appear on laterals from the old wood, and new growth will develop from the base of the plant. Cut off all flower trusses as they fade, to prevent the plant from putting its energy unnecessarily into the production of berries (also known as 'hips'). In late autumn or early winter cut back the flowered laterals, as well as some of the new shoots, by as much as a quarter of their length in some cases, to maintain the symmetry of the plant.

Weak and damaged wood should also be removed at this time, and finally all of the new shoots should be tied in to supporting wires or trellis.

Rambler roses, on the other hand, have long pliable stems, and bear huge trusses of smallish flowers. These are, perhaps, more colourful than climbers at certain times, but there is unfortunately only one flush of bloom. Growth is extremely vigorous, but the leaves need regular spraying to combat mildew, to which these roses are prone.

During autumn or spring, the rambler should be planted in the same way as the climber, but this time all of the shoots must be cut back, to around 23–28 cm (9–15 in) from the ground. During the spring and summer new shoots will develop, and these should be trained into place on horizontal support wires. Flowers are carried on new wood, that is, lateral shoots. These should be cut out during late summer, so that next year's flowering wood can be tied in.

Training shrubs

The final area that may cause concern to a few, is that of general garden shrubs, and whether or not they should undergo a period of training before they are put into a normal annual pruning routine. Some of our best-known shrubs do, indeed, benefit from pruning in the first growing season after propagation, and these include mainly deciduous shrubs, such as buddleia, cotoneaster, deutzia, forsythia, philadelphus, ribes and viburnum.

It is worth remembering that those plants which naturally throw out shoots from ground level – or below it – really need pruning. They will be better balanced, and make far better specimens. Some shrubs are grown specially for their colourful bark. Take, for example, Cornus (shrubby dogwood). It should be pruned back around 20 cm (8 in) from ground level at planting time. Over the spring and summer period it will throw out long stems covered with rich green foliage. When the leaves fall in autumn we are treated to magnificent red, orange or yellow stems, that often provide much needed colour during the winter months. At the end of winter, just before growth restarts, these stems should be reduced, practically back to their bases, so that the plant can throw out new shoots from the bottom again.

Generally it does no harm at all to reduce a deciduous shrub by about half its size, at planting time. You then give it every opportunity to develop branches in the most desirable places, and you will then, more than likely, end up with a superior, well-balanced specimen.

Evergreen species are, perhaps, a little less responsive to training, but it is possible with the majority to prune them to shape at a later time, preferably during late spring.

PART II

Plants and Their Propagation

HARDY ORNAMENTAL AND FRUITING PLANTS
TENDER ORNAMENTAL AND FRUITING PLANTS
HARDY AND TENDER VEGETABLES AND CULINARY HERBS

This extensive encyclopaedia covers the propagation of hardy ornamental and fruiting plants, tender ornamental and fruiting plants, and hardy and tender vegetables.

Hardy implies that the plants can be grown out of doors all the year round. Tender indicates that the plants are grown as greenhouse or house plants, or bedded out for the summer, in temperate climates, especially in areas subjected to frosts.

All commonly grown plants are included as well as many uncommon but nevertheless desirable kinds. All the genera are available.

Plants are listed under generic names. A common name follows if applicable. Plant type, or types, is indicated: for example, shrub, perennial, alpine, annual, bulb.

Then follows the method or methods of propagation, the time of year to carry it out, and the place and/or conditions. Additional tips and hints are given as considered necessary.

If the propagation of a species, or a particular group within the genera, differs from that of the genera as a whole, then the species or groups are discussed separately.

A few words about raising plants from seeds. When this method is indicated it mainly refers to collecting seeds from species (or buying) and not hybrid plants or cultivars. However, one can buy packets of hybrids or cultivars from seedsmen (especially of hardy and half-hardy annuals and perennials), but we are not implying that you should save your own seeds from these.

There are several methods of propagating many genera. We recommend you choose one which suits your purpose and the facilities at your disposal. Bear in mind that one method of propagation is not necessarily any better than another, and it is never our intention to imply this.

The methods of propagation and how to provide suitable conditions are fully described in Part I.

Budding – roses, trees, Chapter 16.
Bulbs, corms and tubers, vegetative propagation, Chapter 14.
Division of hardy and tender plants, Chapter 11.
Grafting – whip and tongue, saddle, spliced side, veneer, spliced side veneer, tomato, Chapter 15.
Layering – simple, serpentine, tip, mound, continuous (French), dropping, carnation, air, Chapter 12.
Leaf and leaf-bud cuttings, Chapter 10.
Plantlets, hardy and tender plants, Chapter 13.
Root cuttings, Chapter 9.
Seed storing and stratification, Chapter 4.
Seeds – sowing under cover, Chapter 6.
Seeds – sowing outdoors, Chapter 5.
Spores – ferns and other plants, Chapter 7.
Stem cuttings – softwood, basal, semi-ripe, hardwood, sections, eye, pipings, Chapter 8.

Bulbous plants come in many shapes and forms and often confusion exists over whether certain types come from bulbs at all! Perhaps they are from corms, tubers or rhizomes. Whichever the source, most bulbous plants may be propagated in a number of ways, such as from bulbils, bulblets, scales, division or scoring and scooping.

Cordons are single-stemmed fruiting trees – apples and pears being the most common. Generally purchased as grafted maidens, they are planted and trained at an angle of 45°, pointing northwards to gain maximum direct sunlight.

Seedlings of *Araucaria araucana* (monkey puzzle, Chile pine), a favourite conifer with the Victorians, are easily raised in warmth but very slow growing.

The mouse plant, *Arisarum probiscideum*, is an amusing perennial for partial shade and humus-rich soil, and is easily increased by spring division.

Cardiocrunum giganteum (giant lily) will take many years to reach flowering size from seeds or bulblets but the huge blooms are well worth waiting for.

The easiest method of increasing *Darlingtonia californica* (California pitcher plant, cobra lily) is to divide established clumps in mid-spring.

The dramatic border perennials, eremurus or foxtail lilies, can be propagated by division in early autumn, but take care to avoid damaging their fleshy roots.

Pumpkins are fast-growing half-hardy annuals cultivated for their huge fruits. Plants are raised in a heated greenhouse during spring and later planted outdoors.

CHAPTER 19

Hardy Ornamental and Fruiting Plants

KEY TO PLANT TYPES

A = Alpines
An = Annuals
An Bi G = Annual and biennial grasses
An G = Annual grasses
An H = Annual herbs
An P G = Annual and perennial grasses
Aq = Aquatics
B = Bulbs
Bi = Biennials
C = Conifers
Ca Aq = Carnivorous aquatics
Ca P = Carnivorous perennials
Cl = Climbers
Cl P = Climbing perennials
Cl S = Climbing shrubs
Cm = Corms
Cm P = Cormous perennials
F = Ferns
Fl Aq = Floating aquatics
M L P = Moisture-loving perennials
O = Orchids
P = Perennials
P G = Perennial grasses
R P = Rhizomatous perennials
S = Shrubs
S S = Sub-shrubs
S P = Shrubby perennials
Sub Aq = Submerged aquatics
T = Trees
Tu = Tubers
Tu P = Tuberous perennials
Tw S = Twining shrubs

ABELIA S
Semi-ripe cuttings in mid-summer, in heated propagating case, 18°C (65°F). Will also root in a garden frame.
Softwood cuttings in spring, heated propagating case or mist, 21°C (70°F).

ABELIOPHYLLUM S
Semi-ripe cuttings in mid-summer, in a heated propagating case, 18°C (65°F).

ABIES (Silver fir) C
Seeds sown outdoors in early spring. Some may not germinate till following year.
Veneer grafting early to mid-spring, greenhouse, in warmth, 21°C (70°F). Useful rootstock *Abies alba*.

ABUTILON S
Seeds early to mid-spring, heated propagating case, 18°C (65°F).
Semi-ripe cuttings mid- to late summer, heated propagating case, 18°C (65°F).

ACAENA (New Zealand burr) A
Seeds late winter/early spring, garden frame.
Division early to mid-spring, replanting immediately.

ACANTHOLIMON (Prickly heath) P
Seeds early to mid-spring, heated propagating case, 15.5°C (60°F).
Division early to mid-spring, potted and placed in garden frame.
Semi-ripe cuttings late summer, garden frame.

ACANTHOPANAX T/S
Seeds sown when ripe, or stratified, outdoors.
Root cuttings early/mid-winter, in heated propagating case, 18°C (65°F), or garden frame.

ACANTHUS (Bear's breeches) P
Seeds early to mid-spring, heated propagating case, 15.5°C (60°F).
Root cuttings early winter, garden frame.
Division early to mid-spring, replanting immediately.

Fig. 19.1 Seeds of acer (maple) are sown during early to mid-spring in an outdoor seed bed, where they generally germinate freely.

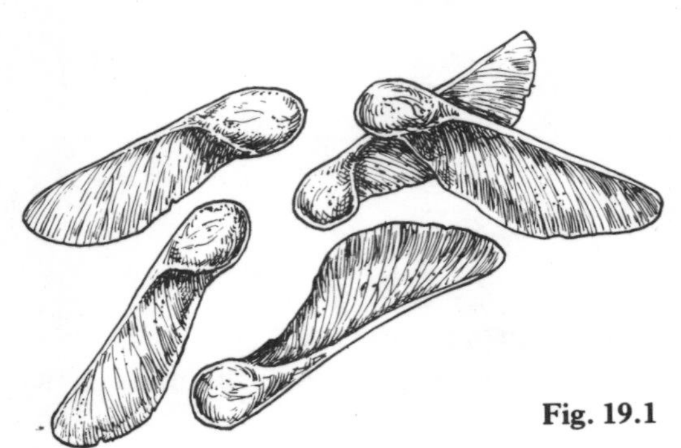

Fig. 19.1

ACER (Maple) T
Seeds early to mid-spring outdoors (Fig.19.1).
Whip and tongue grafting for cultivars, late winter/early spring, outdoors. Use species for rootstock.
Spliced side veneer grafting for *A. japonicum* and *A. palmatum* cultivars, greenhouse, with bottom heat, 21°C (70°F). Use species as rootstock.
Softwood cuttings for *A. japonicum* and *A.palmatum* cultivars, mid-spring to early summer, heated propagating case or mist unit, 21°C (70°F).

ACHILLEA (Yarrow) P/A
Soft basal cuttings mid- to late spring, heated propagating case, 15.5°C (60°F).
Division early to mid-spring. Alpine divisions can be potted and established in garden frame.
Seeds sown late spring/early summer, outdoors or garden frame.

ACONITUM (Monk's hood) P
Division early to mid-spring, replanting immediately.
Seeds sown when ripe, early autumn, garden frame.

ACORUS (Sweet flag) Aq
Division mid-spring to early summer, replanting immediately.

ACTAEA (Baneberry) P
Division early to mid-spring, replanting immediately.
Seeds sown late autumn or mid-spring, garden frame or cold greenhouse.

ACTINIDIA Cl
Seeds in early/mid-spring, heated propagating case, 15.5°C (60°F).
Semi-ripe cuttings mid- to late summer, heated propagating case or mist unit, 21°C (70°F).
Hardwood cuttings in early winter, greenhouse, provide bottom heat, 21°C (70°F).

ADIANTUM (Maidenhair fern) F
Division early to mid-spring, replanting immediately.
Spores as soon as ripe, in gentle heat, greenhouse.

ADONIS An/P
Seeds mid-spring or early autumn, outdoors, for annuals; when ripe, outdoors, for perennials.
Division for perennials, early to mid-spring, replanting immediately.

AEGOPODIUM (Ground elder) P
Division of the rhizomes in early to mid-spring, replanting immediately.
Root cuttings in early/mid-winter, garden frame.

AESCULUS (Horse chestnut) T/S
Seeds early to mid-autumn, outdoors.
Whip and tongue grafting late winter/early spring, outdoors. Use *A. hippocastanum* rootstocks.
Budding early summer to early autumn, stocks as above.
Root cuttings for *A. parviflora*, early to mid-winter, garden frame or outdoors.

AETHIONEMA (Stone cress) A
Seeds in winter, outdoors, in pots.
Softwood cuttings, early to mid-summer, heated propagating case, 15.5°C (60°F).

AGAPANTHUS (African lily) P
Division early to mid-spring, replanting immediately. Roots fleshy — take care.
Seeds early to mid-spring, heated propagating case, 15.5°C (60°F).

AGRIMONIA (Agrimony) P
Division in early to mid-spring, replanting immediately.

AGROPYRON P G
Division early to mid-spring, replanting immediately.

AGROSTEMMA (Corn cockle) An
Seeds early to mid-spring or early to mid-autumn, outdoors.

AILANTHUS (Tree of heaven) T
Seeds late winter/early spring, outdoors or in cool greenhouse.
Root cuttings early to mid-winter, garden frame.

AJUGA (Bugle) P
Division early to mid-spring, replanting immediately.
Plantlets pegged down until rooted, spring or summer.

AKEBIA Cl
Simple layering mid-spring to late summer, outdoors.
Semi-ripe cuttings late summer/early autumn, heated propagating case, 18°C (65°F).

ALBIZIA T/S
Seeds of hardy species sown in spring, heated propagating case, 21°C (70°F).

ALCHEMILLA (Lady's mantle) P
Division early to mid-spring, replanting immediately.
Seeds late spring/early summer, outdoors. They germinate very freely.

ALISMA (Water plantain) Aq
Division mid-spring to early summer, replanting immediately.
Seeds in spring, keeping compost permanently wet, heated propagating case, temperature 15.5°C (60°F).

ALLIUM (Ornamental onions) B
Seeds mid- to late winter, in pots, outdoors.
Bulblets removed when dormant, late summer/early autumn, planting outdoors.
Bulbils removed when developed, stored for winter and planted outdoors in mid-spring (Fig.19.2).

ALNUS (Alder) T/S
Seeds in autumn or early spring, outdoors.
Simple layering between mid-spring and late summer, outdoors.
Spliced side veneer grafting useful for cultivars, early spring, greenhouse, with bottom heat, 21°C (70°F). Use species as rootstocks.

ALOPECURUS (Foxtail) PG
Seeds early spring, heated propagating case, 15.5°C (60°F)

ALSTROEMERIA (Peruvian lily) P
Seeds mid-spring, heated propagating case, 18–21°C (65–70°F). Soak seeds in warm water for 12 hours.
Division mid-spring; try not to remove soil; replant immediately.

ALTHAEA (Hollyhock) An/P
Seeds late spring/early summer, outdoors. Annuals best sown late winter in heated propagating case, 15.5°C (60°F).

ALYSSUM (Madwort) An/P/A
Seeds in early spring, heated propagating case, 15.5°C (60°F), or outdoors early to mid-spring, for annuals.
Semi-ripe cuttings mid- to late summer, garden frame for perennials and alpines.

Fig. 19.2

Fig. 19.2 The bulbils which form on the flower stems of some alliums (ornamental onions) can be removed when developed, stored for the winter and planted outdoors in mid-spring.

AMARYLLIS B
Bulblets removed and replanted in summer when dormant.
Seeds mid-spring in heated propagating case, 15.5°C (60°F). Soak seeds for one hour. Seedlings can take eight years to flower.

AMELANCHIER (Snowy mespilus, June berry) S/T
Seeds early to mid-autumn, outdoors or garden frame. Fresh seeds recommended.
Simple layering mid- to late spring, outdoors.
Division by removing rooted suckers between late autumn and early spring, replanting immediately.

AMORPHA (Lead plant) S
Division between late autumn and early spring, replanting portions immediately.
Softwood cuttings mid-spring to early summer, rooted in a heated propagating case, 18°C (65°F).

AMPELOPSIS Cl
Hardwood or eye cuttings in early winter, under glass with bottom heat 15.5°C (60°F).
Softwood cuttings in early summer, rooted in heated propagating case or under mist, 18–21°C (65–70°F).

ANACYCLUS (Mount Atlas daisy) P/An
Seeds of perennials sown in autumn if fresh; stored seeds in late winter after a short cold spell. Sow annuals in spring. Heated propagating case, 10°C (50°F).
Soft basal cuttings mid-spring to early summer, heated propagating case, 15.5°C (60°F).

ANAGALLIS (Pimpernels) P/A/An
Seeds in early to mid-spring, heated propagating case, 15.5°C (60°F). Hardy annuals outdoors in spring.
Division in mid-spring for perennial species, replanting immediately.

ANAPHALIS (Pearly everlasting) P
Division in early to mid-spring, replanting immediately.
Basal cuttings mid- to late spring. They will root in a garden frame.
Seeds sown mid-spring and germinated in a garden frame.

ANCHUSA (Alkanet) P/An
Root cuttings for perennials, early to mid-winter, garden frame.
Seeds for perennials and annuals, early to late spring, outdoors.

ANDROMEDA (Bog rosemary) S
Division between late autumn and early spring, replanting immediately.
Seeds early to mid-spring, heated propagating case, 15.5°C (60°F).
Semi-ripe cuttings mid- to late summer, heated propagating case 15.5°C (60°F).

ANDROSACE (Rock jessamine) A
Division or offsets in early to mid-spring, replanting immediately. Alternatively offsets can be potted and established in a garden frame.
Seeds sown as soon as ripe, in pots, outdoors; bring into gentle heat in spring.

ANEMONE (Windflowers) P
Division early to mid-spring, replanting immediately. Divide spring-flowering kinds in late summer.
Seeds when ripe, early to mid-autumn, in pots outdoors.
Root cuttings of *A.* x *hybrida* (Japanese anemones), early to mid-winter, garden frame.

ANEMONOPSIS P
Division in early to mid-spring, replanting immediately.
Seeds in late spring, garden frame. Germination may be slow.

ANGELICA (Angelica) P
Seeds late summer/early autumn, or mid-spring, outdoors.

ANTENNARIA (Mountain everlasting) A
Division early to mid-spring, replanting immediately.
Seeds winter, in pots, outdoors.

ANTHEMIS (Chamomile) P/A
Division in early to mid-spring, replanting immediately.

Seeds late spring/early summer, outdoors.
Semi-ripe cuttings mid- to late summer, garden frame.
Basal cuttings of herbaceous sorts in mid- to late spring, garden frame.

ANTHERICUM (St Bernard's lily) P
Division in early or mid-spring, replanting immediately. Only divide congested clumps.

ANTHRISCUS (Chervil) An H
Annual herb
Seeds in early or mid-spring, outdoors. Can also sow in early autumn in a heated greenhouse – 7°C (45°F) – for winter use.

ANTHYLLIS S/P
Semi-ripe cuttings mid- to late summer, in a heated propagating case, 15.5°C (60°F).
Division early to mid-spring, replanting immediately.
Seeds early to mid-spring, heated propagating case, 15.5°C (60°F).

APONOGETON (Water hawthorn) Aq
Division between mid-spring and early summer, replanting immediately.

AQUILEGIA (Columbine) P/A
Seeds of perennials can be sown in late spring/early summer outdoors. Sow alpines in winter in pots and stand outdoors.

ARABIS (Rock cress) P
Division early to mid-spring, or early autumn, replanting immediately.
Semi-ripe cuttings, early to late summer, garden frame.
Seeds late spring/early summer, outdoors or garden frame.

ARALIA T/S
Root cuttings in early or mid-winter, garden frame or with bottom heat in greenhouse 15.5°C (60°F).
Seeds in mid-spring, outdoors.

ARAUCARIA (Monkey puzzle, Chile pine) C
Seeds early or mid-spring, in heated propagating case, 15.5°C (60°F), or garden frame. Growth of seedlings is very slow.

ARCTOSTAPHYLOS (Bearberry) S
Semi-ripe cuttings, ideally with a heel, taken in late summer and rooted in a greenhouse with gentle heat or in garden frame.
Seeds early autumn in garden frame, or mid-spring in a heated propagating case, 15.5°C (60°F).
Simple layering in spring, outdoors.

ARENARIA (Sandwort) A
Division early to mid-spring, replanting immediately.
Basal cuttings in summer and rooted in a garden frame.
Seeds in winter, in pots and placed outdoors.

ARGEMONE (Prickly poppy) An/P
Seeds in early spring, in heated propagating case, 18°C (65°F); or mid-spring outdoors.

ARISAEMA P
Offsets in early to mid-spring, when plants are starting into growth.
Seeds in spring in heated propagating case, 15.5°C (60°F); germination may be slow.

ARISARUM (Mouse plant) P
Division early to mid-spring, replanting portions immediately.

ARISTOLOCHIA (Dutchman's pipe) Cl
Seeds early to mid-spring, heated propagating case, 15.5°C (60°F).
Semi-ripe cuttings in mid- or late summer, rooted in heated propagating case, 18°C (65°F).

ARMERIA (Thrift) A
Division early to mid-spring, replanting immediately.
Semi-ripe basal cuttings mid- to late summer, rooted in a heated propagating case, 15.5°C (60°F), or in a garden frame.
Seeds sown in mid- to late winter in pots and placed outdoors.

ARNEBIA (Prophet flower) P
Seeds sown in mid-spring, heated propagating case, 15.5°C (60°F).
Root cuttings mid-winter, heated propagating case, 15.5°C (60°F).

ARONIA (Chokeberry) S
Semi-ripe cuttings mid-summer to mid-autumn, rooted in a heated propagating case 18°C (65°F).
Division between late autumn and early spring, replanting immediately.
Seeds early to mid-spring outdoors after winter stratification.

ARTEMISIA (Wormwood) S/P
Semi-ripe cuttings for shrubs, mid-to late summer, heated propagating case 15.5°C (60°F), or garden frame.
Division for perennials, early to mid-spring, replanting immediately.

ARUM Tu P
Division in early or mid-spring, replanting the portions immediately.

ARUNCUS (Goat's beard) P
Division in early or mid-spring, replanting immediately.
Seeds sown in spring outdoors or in garden frame.

ARUNDINARIA (Bamboo) P
Perennials
Division mid-spring, replanting the portions immediately.

ARUNDO PG
Perennial grasses
Division in early or mid-spring, replanting the portions immediately.

ASARUM (Wild ginger, asarabacca) P
Division early to mid-spring, replanting the portions immediately.
Seeds in spring, heated propagating case, 15.5°C (60°F).

ASCLEPIAS (Milkweed) P
Division early to mid-spring, replanting the portions immediately.
Seeds mid-spring, heated propagating case, up to 24°C (75°F), after freezing for two or three weeks. Germination may be slow.

ASPERULA (Woodruff) P/A/An
Division for perennial species and alpines in early to mid-spring, replanting the portions immediately.
Soft basal cuttings of alpine species in mid- to late spring, garden frame.
Seeds for annual species, early to mid-spring, out of doors where they are to flower.

ASPHODELINE (Asphodel) P
Division in early or mid-spring, replanting the divisions immediately.
Seeds in early spring, garden frame.

ASPHODELUS (Asphodel) P
Division in early or mid-spring, replanting the divisions immediately.
Seeds in early spring, garden frame.

ASPLENIUM (Spleenwort) F
Division in mid-spring, replanting portions immediately.
Spores sown as soon as ripe, heated propagating case, 18–21°C (65–70°F); keep moist and shaded, but light needed for germination.

ASTELIA P
Division of *A. nervosa* in early to mid-spring, replanting the portions immediately.

ASTER (includes Michaelmas daisies) P
Division early to mid-spring, replanting the portions immediately. Single rooted shoots can also be used.

ASTERANTHERA Cl
Semi-ripe cuttings in late summer, in heated propagating case, 18°C (65°F).

ASTILBE (False goat's beard) P
Division in early or mid-spring, replanting the portions immediately.
Seeds in spring, heated propagating case, 15.5°C (60°F). Keep compost moist.

ASTRAGALUS (Milk vetch) P/A
Seeds sown mid- to late winter in pots and placed out of doors.
Division where possible, early to mid-spring, replanting portions immediately.

ASTRANTIA (Masterwort) P
Division early to mid-spring, replanting the portions immediately.
Seeds in late spring or early summer

outdoors after chilling for four weeks. Alternatively sow fresh seed in trays in early autumn and germinate in a garden frame.

ATHROTAXIS C
Conifers
Seeds sown in early to mid-spring outdoors.
Semi-ripe cuttings taken in early to mid-autumn and rooted in a garden frame, or in a propagating case with gentle heat, about 15.5°C (60°F).

ATHYRIUM F
Division in mid-spring just as growth is about to start. Replant the portions immediately.
Spores sown as soon as ripe, heated propagating case, 18–21°C (65–70°F); keep moist and shaded, but remember that light is needed for the spores to germinate.

ATRIPLEX (Orach) S/An
Semi-ripe cuttings for propagating shrubby species. They can be taken in early to mid-autumn and rooted in a garden frame or under a low polythene tunnel.
Seeds of hardy annuals should be sown in early or mid-spring out of doors in their final positions.

AUBRIETA A
Seeds can be sown during mid- or late winter in pots and placed out of doors.
Softwood cuttings root easily during the period mid-spring to early summer, in gentle heat in a propagating case, or in a garden frame. Best material obtained from plants which have been cut back.
Hardwood cuttings can be taken in mid- or late autumn and rooted in a garden frame.
Division of established plants in early autumn, replanting the portions immediately in their permanent positions.

AUCUBA (Includes spotted laurel) S
Semi-ripe cuttings taken in early or mid-autumn root very easily in a garden frame, under a low polythene tunnel or even in a sheltered well-drained spot in the open ground.

AVENA (Oat) An Bi Orn G
Seeds sown in early or mid-spring out of doors in their final positions. Alternatively sow in the autumn if the soil is well drained.

AZARA S
Semi-ripe cuttings in late summer or early autumn, rooted in a heated propagating case, 18°C (65°F).
Hardwood cuttings in late autumn, inserted in the open ground or a garden frame.

AZOLLA (Fairy moss) Aq
Division mid-spring/early summer. These floating aquatics are easily separated and even split up themselves.

BACCHARIS S
Seeds sown in spring in a heated propagating case, 15.5°C (60°F).
Semi-ripe cuttings taken in summer root easily in a propagating case with gentle bottom heat, 15.5°C (60°F)

BALLOTA S/P
Softwood cuttings of shrubby species, mid-spring to early summer, easily rooted in a heated propagating case, 15.5°C (60°F).
Seeds in early or mid-spring for perennials, heated propagating case, 15.5°C (60°F).

BAPTISIA (False indigo) P
Division in early or mid-spring, replanting the portions immediately.
Seeds are sown in mid- or late spring and can be germinated in a garden frame or in an outdoor seed bed.

BARBAREA (Winter cress) P
Seeds sown late spring or early summer in outdoor seed bed.
Division in early or mid-spring, replanting immediately.

BELAMCANDA (Leopard flower) P
Seeds sown in spring in a heated propagating case, 21°C (70°F). A week's pre-chilling will assist germination.
Division of the stout rhizomes in early or mid-spring, replanting the portions immediately.

BELLIS (Daisy) P
Division of the tufts in early to mid-spring, replanting immediately.

Seeds, including those of the double-flowered daisies which are grown as biennials for spring bedding, are sown in late spring or early summer in an outdoor seed bed.

BELLIUM A

Division in mid-spring, replanting the divisions immediately on the rock garden.
Seeds sown in pots during late winter and placed outdoors or in a garden frame.

BERBERIDOPSIS Cl S

Softwood cuttings taken between mid-spring and early summer and rooted in a heated propagating case or under mist, 21°C (70°F).
Semi-ripe cuttings made from side shoots in mid-summer, rooted under mist or in a heated propagating case, 21°C (70°F).
Simple layering in mid- or late spring outdoors.
Seeds sown in early or mid-spring, heated propagating case, 15.5°C (60°F).

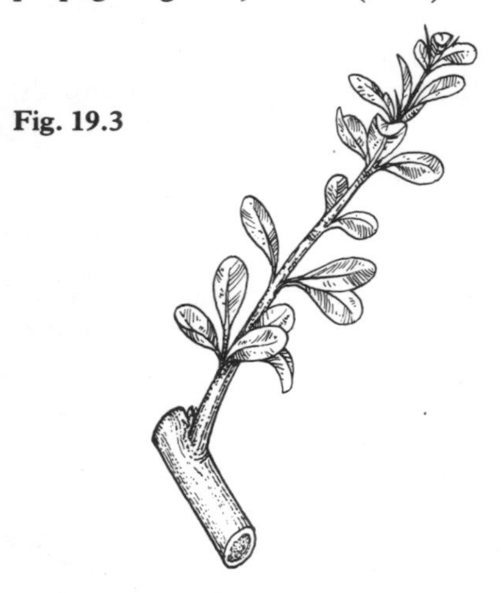

Fig. 19.3 Thin current year's semi-ripe side shoots of berberis (barberry) can be formed into mallet cuttings by removing them with a short portion of the main woody stem attached. This wood will ensure survival while they are forming roots.

BERBERIS (Barberry) S

Semi-ripe cuttings of thick-stemmed berberis during early to mid-autumn. If desired, an entire shoot can be removed and cut into 15 cm (6 in) sections, discarding the soft tip. Heated propagating case for fast rooting, 18°C (65°F), otherwise garden frame or low polythene tunnel.
Mallet cuttings for berberis with thin shoots, formed of current year's side shoots with a short portion of the main woody stem attached. The soft tips of the shoots are best removed. Heated propagating case for fast rooting, 18°C (65°F), otherwise garden frame or low polythene tunnel (Fig. 19.3).
Seeds sown in early or mid-spring in an outdoor seed bed, after winter stratification.

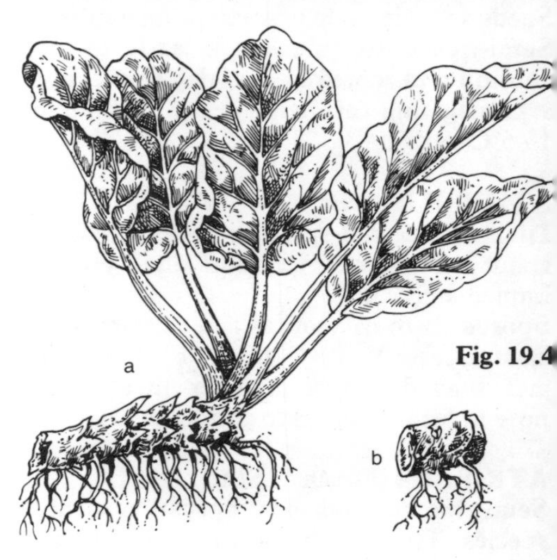

Fig. 19.4 Bergenias can be propagated in mid-winter by cutting some thick rhizomes (*a*) into 2.5–3.5 cm (1–1 ½ in) portions, each with at least one dormant growth bud (*b*). They are rooted in a heated propagating case.

BERGENIA P

Division of large established clumps after flowering in spring. Alternatively divide in early autumn provided the soil is very well drained.
Diced rhizomes in mid-winter. Remove some thick rhizomes from parent plants, wash thoroughly and then cut into 2.5–3.5 cm (1–1½ in) portions, ensuring each has at least one dormant growth bud. Press the sections, buds uppermost, into containers of perlite or vermiculite, to half their depth. Place in heated propagating case, 21°C (70°F) (Fig. 19.4).
Seeds early or mid-spring, in heated propagating case, 15.5–21°C (60–70°F). May take many months to germinate.

BESCHORNERIA P

Offsets, which look like suckers, can be removed in mid-spring and replanted outdoors.
Seeds in mid-spring, heated propagating case, 18°C (65°F).

BETULA (Birch) T

Seeds sown in autumn or early to mid-spring outdoors.

Spliced side veneer grafting in early spring. Use rootstocks of *Betula pendula* (silver birch). Greenhouse, with bottom heat, 21°C (70°F), and humidity.

BIGNONIA (DOXANTHA) Cl
Semi-ripe cuttings in summer, heated propagating case, 18–21°C (65–70°F).
Seeds in mid-spring, heated propagating case, 18–21°C (65–70°F).

BLECHNUM F
Spores when ripe, heated propagating case, 18–21°C (65–70°F); keep moist and shaded, but light needed for germination.
Division in mid-spring, replanting the portions immediately in their permanent positions.

BLETILLA O
Division by separating and replanting the pseudobulbs in early or mid-spring, or in late spring after flowering.

BOLAX A
Division of the cushions in early or late spring, replanting the portions immediately in permanent positions.
Cuttings taken in mid-summer and rooted in a garden frame.

BORAGO (Borage) P/An
Division for the perennial species, during early or mid-spring, replanting immediately.
Seeds for the annuals, sowing early or mid-spring outdoors where they are to flower.
Cuttings of young shoots in spring, garden frame.

BOYKINIA P
Seeds sown in late spring/early summer, outdoor seed bed.
Division in early or mid-spring, replanting the portions immediately.

BRACHYCOME (Swan River daisy) An/P
Seeds sown in early spring, heated propagating case, 18°C (65°F). Or outdoors in flowering sites during mid-spring.
Division in early or mid-spring for perennials, replanting immediately in flowering positions.

BRIZA (Quaking and pearl grasses)
Orn An P G
Seeds sown out of doors in permanent positions, either in early/mid-spring, or during early autumn.
Division for perennial species, early to mid-spring, replanting the portions immediately.

BRODIAEA Cm
Cormlets separated and replanted in early autumn, in outdoor nursery bed or in garden frame, to grow on to flowering size.
Seeds sown early spring and germinated in a heated propagating case, 15.5°C (60°F).

BROMUS (Brome grass) An P G
Seeds for annual species, early to mid-spring, or early to mid-autumn, outdoors in flowering positions.
Division for perennial species, early to mid-spring, replanting the small clumps immediately.

BROUSSONETIA (Paper mulberry) S/T
Seeds sown in spring in slight heat.
Suckers can be removed with some roots attached and replanted elsewhere.
Semi-ripe cuttings with a heel in late summer, rooted in heated propagating case, 15.5°C (60°F).
Simple layering in spring, outdoors.
Budding for cultivars in summer outdoors, using species as rootstocks.

BRUNNERA P
Division in early or mid-spring, replanting the portions immediately.
Root cuttings in mid- to late autumn, garden frame.
Seeds late spring or early summer, outdoor seed bed.

BUDDLEIA (Includes butterfly bush) S
Softwood cuttings as soon as available – mid-winter to early summer – heated propagating case or under mist, 21°C (70°F). Can also be rooted in a garden frame during early summer.
Hardwood cuttings late autumn to early winter, garden frame.
Seeds sown early to mid-spring in outdoor seed bed.

BULBINELLA P
Division early to mid-spring, taking care not to damage the fleshy roots, replanting the portions immediately.
Seeds sown mid- to late winter in pots and placed out of doors.

BULBOCODIUM Cm
Cormlets can be removed at planting time in early autumn and planted in an outdoor nursery bed.
Seeds best sown as soon as ripe, or in mid- to late winter, in pots and placed out of doors.

BUPHTHALMUM (Oxeye) P
Division of the clumps in early to mid-spring, replanting immediately.
Seeds sown in late spring or early summer in an outdoor seed bed.

BUPLEURUM (Hare's ear) S/P
Semi-ripe cuttings for shrubby species, early to late summer, heated propagating case, 15.5°C (60°F).
Division for perennials, early to mid-spring, replanting immediately.

BUTOMUS (Flowering rush) Aq
Division between mid-spring and early summer, replanting immediately.
Seeds sown as soon as ripe or in mid- to late spring, in gentle heat and keeping compost permanently wet.

BUXUS (Box) S
Semi-ripe cuttings early to mid-autumn, heated propagating case for fast rooting, 18°C (65°F), or garden frame/low polythene tunnel.
Seeds early to mid-spring, outdoor seed bed – a slow process.
Division for dwarf forms, simply pulling plants apart and replanting in spring.

CAESALPINIA S
Seeds sown in spring, after first soaking for several hours in warm water. Heated propagating case, 15.5°C (60°F).

CALAMINTHA (Calamint) P
Division in early or mid-spring, replanting the divisions immediately.
Seeds sown spring in outdoor seed bed.
Soft cuttings late spring/early summer, heated propagating case, 15.5°C (60°F).

CALCEOLARIA (Slipperwort) A/S
Seeds for alpine species sown as soon as ripe, or in mid- to late winter, in pots, outdoors or garden frame. Seeds of shrubby species in spring, heated propagating case, 18°C (65°F).
Division for alpine species in mid-spring, replanting portions immediately.
Semi-ripe cuttings of shrubby species in late summer/early autumn, heated propagating case, 15.5°C (60°F).

CALENDULA (Marigold) An
Seeds early to mid-spring, or early to mid-autumn if soil is very well drained, outdoors where the plants are to flower.

CALLA (Bog or water arum) Aq
Division between mid-spring and early summer, replanting the portions immediately.

CALLICARPA (Beautyberry) S
Seeds early or mid-spring, heated propagating case, 15.5°C (60°F).
Semi-ripe cuttings early to late summer, heated propagating case, 15.5–18°C (60–65°F).

CALLIRRHOE (Poppy mallow) An/P
Seeds sown in mid- or late spring outdoors, or in a garden frame, although plants dislike root disturbance.
Softwood cuttings can be used to propagate perennial species, late spring or early summer, garden frame.

CALLISTEMON (Bottle brush) S/T
Softwood or semi-ripe cuttings, ideally with heel of older wood, early to late summer, heated propagating case, or under mist, 18–21°C (65–70°F).
Seeds sown early spring, heated propagating case, 15.5–18°C (60–65°F).

CALLITRICHE (Water starwort) Aq
Soft cuttings between mid-spring and early summer, bunching several together with a lead weight and planting them in the pool.

CALLUNA (Heather, ling) S
Semi-ripe cuttings of short side shoots pulled off with a heel. Do not try to strip lower leaves, but pinch out tip. Early summer to early autumn, heated propagating case, mist unit, 18–21°C (65–70°F), or garden frame (Fig.19.5)
Layering (dropping) in mid-spring, outdoors. A plant is lifted and replanted to half its depth, spacing out the stems well. Each or most will then root.
Seeds sown as soon as ripe, in pots, overwintered in cold spot outdoors, transferred to heated greenhouse, 10°C (50°F), in spring to germinate.

CALOCEDRUS (Incense cedar) C
Semi-ripe cuttings, with at least 12 mm (½ in) of brown, well-ripened wood at the base. Early to mid-autumn, heated propagating case, mist unit, 21°C (70°F), or garden frame.
Seeds late winter or early spring, outdoor seed bed.

CALOCHORTUS (Mariposa lily) B
Seeds late winter/early spring, garden frame.
Bulbils planted in April in a garden frame.
Bulblets removed and planted in outdoor nursery bed when dormant in late summer or early autumn.

CALTHA (Marsh marigold) M L P
Division between mid-spring and early summer, once flowering is over, replanting the portions immediately.
Seeds sown when ripe, or in early spring, compost kept permanently wet, garden frame.

CALYCANTHUS (Sweet shrub) S
Simple layering between mid-spring and late summer, outdoors.
Seeds sown early to mid-spring, garden frame.

CAMASSIA (Quamash) B
Bulblets removed in early autumn, when dormant, and replanted in nursery bed.
Seeds best sown as soon as ripe, in summer, alternatively late winter/early spring, in pots, garden frame or outdoors.

CAMELLIA S
Semi-ripe cuttings, wounded at base by removing a sliver of bark to expose the wood, late summer, heated propagating case or mist unit, 21°C (70°F).
Leaf-bud cuttings, again wounded at base, late summer, heated propagating case or mist unit, 21°C (70°F).
Spliced side veneer grafting in early spring, greenhouse, with bottom heat, 21°C (70°F), and humidity. Use seedlings of *C.japonica* as rootstocks.
Simple layering between mid-spring and late summer, outdoors.
Seeds best sown as soon as ripe in the autumn, heated propagating case, 15.5–18°C (60–65°F).

CAMPANULA (Bellflower) P A Bi
Seeds of alpines in mid- or late winter, in pots, placed outdoors. Border perennials and biennials can be sown in late spring/early summer in an outdoor seed bed.
Division for border perennials and alpines, early to mid-spring, replanting portions immediately.
Soft basal cuttings for border perennials and alpines, mid-spring to early summer, heated propagating case, 15.5°C (60°F), or garden frame.

CAMPSIS (Trumpet creeper, trumpet vine) Cl
Semi-ripe cuttings mid- to late summer, heated propagating case or mist unit, 15.5–21°C (60–70°F).
Hardwood cuttings mid- to late autumn, garden frame.
Root cuttings early to mid-winter, heated propagating case, 15.5°C (60°F), or garden frame.
Seeds early to mid-spring, heated propagating case, 15.5°C (60°F).

CANTUA S/T
Semi-ripe cuttings with slight heel, early to mid-summer, heated propagating case, 15.5°C (60°F).

CARAGANA (Pea tree) S/T
Seeds sown in autumn, or in spring after being soaked in warm water, outdoor seed bed.

Fig. 19.5 Heathers may be propagated from semi-ripe cuttings of short side shoots pulled off with a heel. The lower leaves of calluna cuttings (*a*) should not be stripped off, but this is recommended for erica cuttings (*b*).

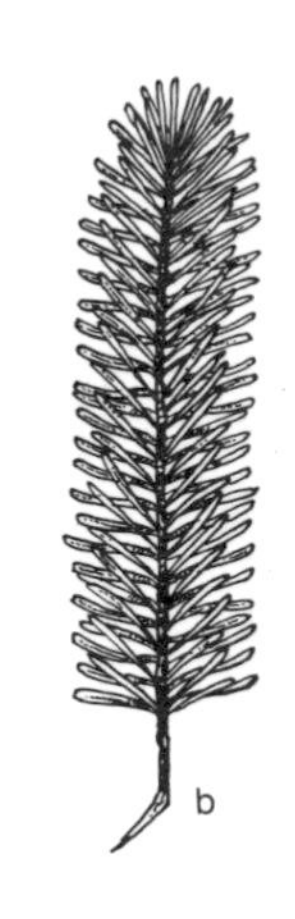

Semi-ripe cuttings mid- to late summer, heated propagating case, 18°C (65°F).
Simple layering between mid-spring and late summer, outdoors.

CARDAMINE P
Division in early or mid-spring, replanting the portions in their permanent positions.
Seeds sown late spring/early summer in outdoor seed bed.

CARDIOCRINUM (Includes the giant lily) B
Seeds sown as soon as ripe, autumn, in pots or pans and placed in a garden frame to germinate. Seedlings can take up to seven years to come into flower.
Bulblets removed when dormant, in early autumn, and replanted immediately in nursery bed or final positions.

CARDUNCELLUS P
Seeds sown in late spring in outdoor seed bed.
Division where possible in early to mid-spring, replanting immediately in flowering position.
Root cuttings in early or mid-winter, garden frame or cool greenhouse.

CAREX (Sedge) P
Division in early or mid-spring, replanting immediately in permanent positions.

CARLINA (Carline thistle)
Perennials
Seeds sown late spring in outdoor seed bed, or in a garden frame.

CARPENTERIA (Tree anemone) S
Seeds early to mid-spring, heated propagating case, 15.5–18°C (60–65°F).
Simple layering mid-spring to late summer, outdoors.
Soft or semi-ripe cuttings early to late summer, heated propagating case or mist unit, 18–21°C (65–70°F).
Hardwood cuttings 8 cm (3 in) long taken in mid-spring and rooted in a heated propagating case, 18°C (65°F).

CARPINUS (Hornbeam) T
Seeds ideally sown as soon as ripe in autumn, or early to mid-spring, in outdoor seed bed.
Spliced side veneer grafting in early spring, for cultivars, greenhouse, with bottom heat, 21°C (70°F). Use *C. betulus* as rootstock.
Semi-ripe cuttings in late summer for cultivars, wounding the base by removing a sliver of wood and rooting under mist, 21°C.

CARTHAMUS (False saffron, safflower) An
Seeds sown in early to mid-spring, heated propagating case, 15.5°C (60°F).

CARYA (Hickory) T
Seeds sown as soon as ripe in pots and placed in a garden frame.

CARYOPTERIS (Bluebeard) S
Softwood basal cuttings mid-spring to early summer, heated propagating case or mist unit, 18°C (65°F). Can also be rooted under low polythene tunnel or in garden frame.
Semi-ripe cuttings mid- to late summer, heated propagating case, 18°C (65°F).

CASSINIA S
Semi-ripe cuttings with a heel, early to mid-autumn, garden frame or low polythene tunnel.

CASSIOPE S
Seeds sown as soon as ripe, or in early spring, garden frame.
Semi-ripe cuttings late summer, heated propagating case, 15.5°C (60°F).
Simple layering between mid-spring and late summer, outdoors.
Layering (dropping) in mid-spring, outdoors. A plant is lifted and replanted to half its depth, spacing out the stems well. Each or most will then root.

CASTANEA (Sweet chestnut) T
Seeds early to mid-autumn, outdoor seed bed.
Whip and tongue grafting for cultivars, early spring, outdoors. Use *C. sativa* as rootstocks.

CATALPA (Indian bean tree) T
Seeds early to mid-spring, heated propagat-

ng case, 15.5°C (60°F), or garden frame. Alternatively late spring in outdoor seed bed.
Semi-ripe cuttings late summer in heated propagating case, 18°C (65°F).
Root cuttings early to mid-winter, garden frame.

CATANANCHE (Cupid's dart) P
Seeds late winter to early spring, heated propagating case, 15.5°C (60°F). Or sow outdoors in late spring/early summer.
Root cuttings early to mid-winter, garden frame.

CAUTLEYA P
Division of established clumps in early or mid-spring, replanting the portions immediately.

CEANOTHUS S
Semi-ripe cuttings late summer/early autumn, heated propagating case or mist unit, 18–21°C (65–70°F).
Root cuttings late autumn to early winter, heated propagating case, 18°C (65°F).
Seeds early to mid-spring, heated propagating case, 15.5°C (60°F).

CEDRUS (cedar) C
Seeds early spring, outdoor seed bed.
Veneer grafting for cultivars and forms, early spring, greenhouse, with bottom heat 21°C (70°F), and high humidity. Use *C. deodara* (deodar) as a rootstock.

CELASTRUS (Staff vine, bittersweet) Cl S
Hardwood cuttings late autumn to early winter, heated propagating case, 15.5°C (60°F).
Simple layering between mid-spring and late summer, outdoors.
Root cuttings in late winter, garden frame.
Seeds early to mid-spring, after winter stratification, outdoor seed bed. Sex of seedlings will not be known until they flower.

CELMISIA S/P
Soft basal cuttings mid-spring to early summer, heated propagating case, 15.5°C (60°F).
Division early to mid-spring, replanting immediately.
Seeds sown as soon as ripe, as they have a short viability period, in pots, garden frame or outdoors.

CELTIS (Nettle tree, hackberry) T
Seeds sown late winter or early spring, heated propagating case, 15.5°C (60°F).
Simple layering where practicable, between mid-spring and late summer, outdoors.
Hardwood cuttings in late autumn, garden frame.

CENTAUREA (Cornflower) An/P/S
Seeds of annuals are sown in early or mid-spring outdoors where they are to flower. Alternatively sow in early autumn if soil is well drained.
Division for perennials in early or mid-spring, replanting immediately.
Semi-ripe cuttings for shrubby species in late summer, heated propagating case 15.5°C (60°F).

CENTRANTHUS (Valerian) P
Seeds late spring or early summer in outdoor seed bed.
Softwood cuttings between mid-spring and early summer, heated propagating case, 15.5°C (60°F).

CEPHALARIA (Includes giant scabious) P
Seeds sown late spring/early summer in outdoor seed bed.
Division in early or mid-spring, replanting the portions immediately.
Root cuttings early to mid-winter, garden frame.

CEPHALOTAXUS (Plum yew) C
Seeds early to mid-spring, outdoor seed bed.
Semi-ripe cuttings, ensuring the base of each has some well-ripened brown wood, early autumn, heated propagating case or mist unit for fast rooting, 21°C (70°F), alternatively garden frame.

CERASTIUM (Snow in summer) A/P
Seeds mid- to late winter, in pots, placed outdoors.

Division in early or mid-spring, replanting the portions immediately in their permanent positions.
Soft of semi-ripe cuttings between mid-spring and late summer, heating propagating case, 15.5°C (60°F), or garden frame.

CERATOSTIGMA (Leadwort) S/P
Semi-ripe cuttings for shrubby kinds, late summer, heated propagating case, 18°C (65°F), or garden frame.
Division for shrubs and perennials, early to mid-spring replanting immediately.
Suckers with roots attached can be removed in spring, potted and established in a garden frame. Applies to shrubby species.

CERCIDIPHYLLUM T
Seeds sown in early or mid-spring and germinated in garden frame.
Simple layering between mid-spring and late summer, outdoors.
Softwood cuttings in late spring, heated propagating case, 18°C (65°F).

CERCIS (Judas tree) T
Seeds sown in early or mid-spring and germinated in a heated propagating case, 15.5°C (60°F), or in a garden frame.

CHAENOMELES (Ornamental quince) S
Simple layering between mid-spring and late summer, outdoors. Layers can take two years to form a good root system.
Seeds sown early to mid-spring in outdoor seed bed, after winter stratification.
Semi-ripe cuttings of side shoots, with heel, mid- or late summer, heated propagating case, 15.5°C (60°F).
Root cuttings mid-winter, obtained from near crown of plant, garden frame.

CHAEROPHYLLUM P
Seeds sown in spring, outdoor seed bed.

CHAMAECYPARIS (False cypress) C
Semi-ripe cuttings, with at least 12 mm (½ in) of brown, well-ripened wood at the base. Early to mid-autumn, heated propagating case or mist unit, 18–21°C (65–70°F). Alternatively, insert in a garden frame.
Veneer grafting for cultivars which are difficult from cuttings or slow growing Early spring, greenhouse, with bottom heat 21°C (70°F), and high humidity. Use *C lawsoniana* (Lawson cypress) as rootstocks.
Seeds sown early to mid-spring in outdoor seed bed.

CHAMAEDAPHNE S
Simple layering between mid-spring and late summer, outdoors.
Semi-ripe cuttings mid-summer to early autumn, heated propagating case, 15.5°C (60°F).
Seeds, covered only very lightly, early to mid-spring, garden frame.

CHAMAEMELUM (Chamomile) P
Division in early or mid-spring, replanting the portions immediately in permanent positions.
Softwood cuttings between mid-spring and early summer, heated propagating case, 15.5°C (60°F).
Seeds late spring/early summer, outdoor seed bed.

CHEIRANTHUS (Wallflower) P
Seeds sown late-spring/early summer in outdoor seed bed. Wallflowers are generally grown as biennials.
Softwood cuttings of side shoots, with a heel, taken between mid-spring and early summer and rooted in a heated propagating case, 15.5°C (60°F), or garden frame.

CHELONE (Turtlehead) P
Division in early or mid-spring, replanting immediately.
Seeds sown in early spring, heated propagating case, 15.5°C (60°F), or during mid-spring in a garden frame.
Softwood basal cuttings between mid-spring and early summer, using garden frame.

CHIASTOPHYLLUM A
Division early to mid-spring, replanting immediately.
Softwood cuttings between mid-spring and early summer, in heated propagating case, 15.5°C (60°F), or garden frame.
Seeds sown in spring and germinated in garden frame.

CHIMONANTHUS (Winter sweet) S
Simple layering between mid-spring and late summer, outdoors.
Seeds (imported) sown in early or mid-spring, heated propagating case or garden frame. A slow process to reach flowering size.
Softwood cuttings late spring/early summer, heated propagating case or mist unit, 18–21°C (65–70°F), but may experience low percentage take.

CHIONANTHUS (Fringe tree) S/T
Simple layering between mid-spring and late summer, outdoors.
Whip and tongue grafting in early spring, outdoors, using *Fraxinus excelsior* (common ash) for rootstocks.
Seeds can be sown in autumn, or in spring after winter stratification, in outdoor seed beds. Alternatively sow late winter/early spring in heated propagating case, 15.5°C (60°F), or in garden frame.

CHIONODOXA (Glory of the snow) B
Bulblets when dormant, in early autumn, replanting in nursery bed.
Seeds sown as soon as ripe in an outdoor seed bed and the seedlings left for several years to develop.

CHOISYA (Mexican orange blossom) S
Semi-ripe cuttings late summer to mid-autumn, heated propagating case, 18°C (65°F), garden frame or low polythene tunnel.

CHORDOSPARTIUM T
Seeds sown in late winter, heated propagating case, 15.5°C (60°F).

CHRYSANTHEMUM An/P/A
Seeds for annual species, early to mid-spring outdoors where they are to flower.
Division for early-flowering perennial border types, other perennial species like *C.maximum* (shasta daisy) and alpine species, early to mid-spring, replanting the portions immediately in their flowering positions.
Soft basal cuttings for early-flowering perennial border types, mid-winter to mid-spring, heated propagating case, 15.5°C (60°F). Also for other perennials and alpines, but early to mid-spring, rooting them in a garden frame (Fig. 19.6).
Cuttings of alpine species in early to mid-summer, rooting them in a garden frame.

Fig. 19.6

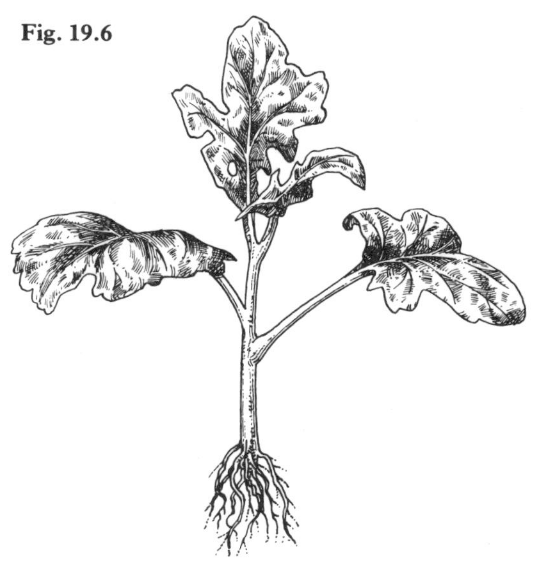

Fig. 19.6 Soft basal cuttings taken in spring offer a means of propagating chrysanthemums. They are cut off as close as possible to the crown of the plant but occasionally can be removed with a few roots attached as shown here, when they are known as Irishman's cuttings.

CHRYSOGONUM P
Division in early to mid-spring, replanting immediately.

CHUSQUEA (Bamboo) P
Division in early or mid-spring, replanting the portions immediately.

CIMICIFUGA (Bugbane) P
Division in early to mid-spring, replanting the divisions immediately in their permanent positions.
Seeds sown when ripe and germinated in a garden frame.

CIRSIUM (Thistle) P
Division where possible in early or mid-spring, replanting immediately.
Seeds sown in late spring or early summer in an outdoor seed bed.

CISTUS (Rock rose) S
Semi-ripe cuttings late summer to mid-autumn, garden frame/low polythene tunnel, or heated propagating case, 18°C.
Seeds early to mid-spring, heated propagating case, 15.5°C (60°F), or garden frame.

CLADANTHUS An
Seeds sown early to mid-spring outdoors where the plants are to flower.

CLADRASTIS (Yellow wood) T
Seeds sown early to mid-spring, outdoor seed bed or garden frame, after soaking in hot water.
Root cuttings early to mid-winter, garden frame.

CLARKIA An
Seeds sown in early to mid-spring, or early autumn if the soil is very well drained and the site sheltered; outdoors where the plants are to flower.

CLEMATIS (Virgin's bower) Cl/P
Leaf-bud cuttings for climbers, mid-spring to early summer, heated propagating case 18°C (65°F).
Serpentine layering for climbers, between mid-spring and late summer, outdoors.
Division for border perennials, early to mid-spring, replanting the portions immediately.
Seeds for climbers and perennials, when ripe in mid-autumn, garden frame. Seeds to be sown in early spring should be stratified over winter or pre-chilled in moist peat for several weeks in a refrigerator.
Soft basal cuttings for border perennials, mid-to late spring, garden frame.

CLERODENDRUM S/T/Cl
Semi-ripe cuttings mid- to late summer, heated propagating case, 18°C (65°F).
Root cuttings early to mid-winter, garden frame.
Suckers removed with roots in early to mid-spring and replanted outdoors.

CLETHRA (Includes sweet pepper bush) S
Semi-ripe cuttings late summer or early autumn, heated propagating case, 18°C (65°F).
Division early spring, replanting immediately in permanent positions.
Seeds early to mid-spring, heated propagating case, 15.5°C (60°F).

CLEYERA S/Sm T
Semi-ripe cuttings of side shoots with a slight heel, mid- to late summer, heated propagating case, 18°C (65°F).

CLINTONIA P
Division in early or mid-spring, replanting the portions immediately.
Seeds sown in late spring, outdoor seed bed.

CODONOPSIS (Bonnet bellflower) P
Seeds sown early spring and germinated in a garden frame.
Soft basal cuttings mid- to late spring, garden frame or heated propagating case, 15.5°C (60°F).

COIX (Job's tears) An G
Seeds sown late winter/early spring in heated propagating case, 15.5°C (60°F), or in late spring outdoors where plants are to flower.

COLCHICUM (Autumn crocus) Cm
Seeds sown as soon as ripe, mid-summer, and germinated in garden frame. They may take up to a year and a half to germinate.
Cormlets removed when dormant in mid-summer and replanted in outdoor nursery bed or flowering positions.

COLLETIA S
Semi-ripe cuttings mid-summer to early autumn, heated propagating case, 15.5°C (60°F).
Seeds sown early to mid-spring and germinated in a heated propagating case, 15.5°C (60°F), or garden frame.

COLLINSIA An
Seeds sown early to mid-spring, or early autumn if the soil is very well drained, outdoors where the plants are to flower. Several sowings at intervals in the spring will ensure a succession of flowers.

COLUTEA (Bladder senna) S
Semi-ripe cuttings late summer/early autumn, heated propagating case, 15.5°C (60°F).
Seeds early to mid-spring, heated propagating case, 18–21°C (65–70°F). Before sowing

Fig. 19.7

Fig. 19.7 The seeds of colutea (bladder senna), which are produced in decorative inflated pods, are sown during spring and germinated in a heated propagating case, after chipping them and soaking in water.

chip the seeds and soak them in water as they have hard coats (Fig. 19.7).

CONVALLARIA (Lily of the valley) P
Division when flowering is over, spring, replanting the portions immediately.
Seeds sown as soon as ripe, garden frame.

CONVOLVULUS An/S
Annuals and shrubs
Seeds for annual species, sown in early or mid-spring, or in early autumn if the soil is well drained, outdoors where the plants are to flower. Alternatively sow in early spring in heated propagating case, 15.5°C (60°F). The latter also for shrubby species.
Semi-ripe cuttings with a heel for shrubby species, late summer/early autumn, heated propagating case, 15.5°C (60°F).
Root cuttings early to mid-winter for shrubby species, heated propagating case, 15.5°C (60°F).

COPROSMA S/T
Semi-ripe cuttings mid- to late summer, heated propagating case, 18°C (65°F).

CORDYLINE (Cabbage palm) T
Cuttings (stem sections) of *C. australis* and *C.indivisa* late spring/early summer, heated propagating case, 18–21°C (65–70°F).
Seeds mid-spring, heated propagating case, 15.5–18°C (60–65°F) after soaking for short period in warm water to soften seed coats. Germination may take several months.

COREOPSIS (Tick seed) P/A
Seeds for annuals, early to late spring, outdoors where they are to flower. Make several sowings at intervals to procure a succession of flowers. Can also sow in early autumn if the soil is very well drained. Perennials sown late spring in outdoor seed bed.
Division for perennial species in early to mid-spring, replanting the divisions (which should have several shoots) immediately.

CORIARIA S
Seeds sown when ripe, or early to mid-spring, heated propagating case, 15.5°C (60°F).
Semi-ripe cuttings late summer/early autumn, heated propagating case, 15.5°C (60°F).
Suckers removed with roots when dormant and immediately replanted.
Simple layering between mid-spring and late summer, outdoors.

CORNUS (Cornel, dogwood) S/T
Hardwood cuttings of shrubby species like *C. alba* in late autumn or early winter, rooted in a garden frame.
Softwood cuttings for *C. florida* and *C. kousa*, early to mid-summer, mist propagation unit, 18–21°C (65–70°F).
Seeds sown in early to mid-spring, after winter stratification, garden frame or heated propagating case, 15.5°C (60°F). Germination may be slow.
Suckers with roots attached can be removed from some shrubby species, early to mid-spring, replanting immediately.
Simple layering between mid-spring and late summer, outdoors.
Division for *C. canadensis* (*Chamaepericlymenum canadense*), which is a perennial, early to mid-spring, replanting the portions immediately.

COROKIA (Includes wire-netting bush) S
Seeds sown early to mid-spring, heated propagating case, 15.5°C (60°F), or garden frame.

Semi-ripe cuttings late summer to early autumn, heated propagating case, 18°C.

CORONILLA (Crown vetch) S
Semi-ripe cuttings late summer/early autumn, heated propagating case, 15.5°C (60°F). Alternatively insert in garden frame during mid-autumn.
Seeds early to mid-spring, heated propagating case, 15.5°C (60°F).

CORTADERIA (Pampas grass) PG
Division in early or mid-spring, replanting the portions immediately.

CORTUSA P
Seeds, sown as soon as ripe in a garden frame, germinate freely.
Division in early or mid-spring, replanting the portions immediately.

CORYDALIS P/A
Division in early or mid-spring, or immediately after flowering for spring blooming species.
Seeds sown as soon as ripe and germinated in a garden frame.

CORYLOPSIS (Winter hazel) S
Simple layering between mid-spring and late summer, outdoors.
Softwood cuttings mid-spring to early summer, heated propagating case, 18°C (65°F). A difficult method with low percentage rooting.

CORYLUS (Hazel, filbert) S/T
Simple layering mid-spring to late summer, outdoors.
Seeds sown in autumn when ripe, or early to mid-spring after a period of chilling, outdoor seed bed.

COTINUS (Smoke tree) S/T
Simple layering between mid-spring and late summer, outdoors.
Semi-ripe cuttings late summer/early autumn, heated propagating case, 18–21°C (65–70°F).
Softwood cuttings early summer, mist unit or heated propagating case, 18–21°C (65–70°F).

Fig. 19.8

Fig. 19.8 Cotoneasters form their seeds in berries. They need to be stratified over winter and then in spring sown in an outdoor seed bed. Seeds of some species may not germinate until the second year.

COTONEASTER S
Semi-ripe cuttings mid- to late summer, heated propagating case or mist unit, 18–21°C (65–70°F).
Simple layering between mid-spring and late summer, outdoors.
Seeds early to mid-spring, after winter stratification, outdoor seed bed. Some species germinate in first year, others in second year (Fig. 19.8).

COTULA A
Division early to mid-spring, replanting the portions immediately.
Seeds sown in mid- to late winter, in pots, and placed out of doors.

CRAMBE (Includes colewort and seakale) P
Root cuttings early to mid-winter, garden frame.
Division in early or mid-spring, replanting immediately.

CRASSULA A
Division in early or mid-spring, replanting immediately, or pot and establish in garden frame.
Cuttings in summer, heated propagating case, 15.5°C (60°F).

\+ **CRATAEGOMESPILUS** (*Crataegus* × *Mespilus*) T
Simple or air layering between mid-spring and late summer, outdoors.
Whip and tongue grafting early spring, outdoors, using *Crataegus monogyna* (hawthorn) as rootstock.
Budding mid- to late summer, outdoors, rootstock as above.

CRATAEGUS (Thorn) T/S
Seeds sown early to mid-spring, after two winters' stratification, outdoor seed bed.
Whip and tongue grafting for cultivars, early spring, outdoors, using *C. monogyna* (hawthorn) as rootstock.
Budding for cultivars early to late summer, outdoors, rootstock as above.
Simple or air layering between mid-spring and late summer, outdoors.

CREPIS (Hawkweed) P/A/An
Division early or mid-spring for perennials and alpines, replanting the portions immediately.
Seeds of perennials or alpines sown in late spring/early summer in a garden frame or outdoor seed bed. Annuals are sown in early to mid-spring, or early autumn if the soil is well drained, outdoors where they are to flower.

CRINODENDRON (Lantern tree) T
Semi-ripe cuttings with a heel mid-to late summer, heated propagating case, 15.5°C (60°F), or mist unit.

CRINUM B
Division of the bulb clusters in mid-to late spring, replanting immediately.
Bulblets removed mid-to late spring and replanted immediately in nursery bed. They may bloom in three years.
Seeds sown as soon as ripe, heated propagating case, 21°C (70°F). The large seeds can be sown singly in small pots. Seedlings may take five years to reach flowering size.

CROCOSMIA *(Montbretia)* Cm
Division when the clumps become congested, early to mid-spring, replanting immediately in flowering site.
Seeds sown as soon as ripe and germinated in garden frame or cool greenhouse. Seeds which have been stored over winter should be pre-chilled for three weeks before sowing in heated propagating case, 15.5°C (60°F).

CROCUS Cm
Cormlets removed when dormant and planted in outdoor nursery bed to grow on to flowering size.
Seeds sown in summer as soon as ripe, in pots and placed in a garden frame or outdoors. Stored seeds can be sown in mid- to late winter.

CRYPTOGRAMMA (Mountain parsley fern) F
Division early to mid-spring, replanting the portions immediately.
Spores sown as soon as ripe, heated propagating case, 18–21°C (65–70°F); keep moist and shaded, but light needed for germination.

CRYPTOMERIA (Japanese cedar) C
Seeds early to mid-spring, heated propagating case, 15.5°C (60°F), or outdoor seed bed.
Semi-ripe cuttings early autumn, rooted in garden frame.
Veneer grafting for cultivars, early spring, greenhouse, with bottom heat 21°C (70°F) and high humidity. Use *C. japonica* as rootstock. Not a very popular method today.

CUNNINGHAMIA C
Seeds sown in early to mid-spring in heated propagating case, 15.5°C (60°F).
Semi-ripe cuttings ideally taken with a heel, late summer, heated propagating case, 18–21°C (65–70°F).

× **CUPRESSOCYPARIS** *(Chamaecyparis × Cupressus)* (Leyland cypress) C
Semi-ripe cuttings early to mid-autumn, heated propagating case, mist unit, 18–21°C (65–70°F) or garden frame. Cuttings should have at least 12 mm (½ in) of brown, well-ripened wood at the base.

CUPRESSUS (Cypress) C
Semi-ripe cuttings early to mid-autumn, ensuring each has at least 12 mm (½ in) of brown well-ripened wood at the base. Heated propagating case, mist unit, 18–21°C (65–70°F) or garden frame.
Veneer grafting for cultivars, especially where difficulty is experienced rooting cuttings. Early spring, greenhouse, with bottom heat, 21°C (70°F) and high humidity. Use *C. macrocarpa* as rootstock.
Seeds sown early to mid-spring in outdoor seed bed.

CURTONUS Cm
Division in mid-spring, replanting the clumps immediately. Each clump must consist of some corms, with roots and leaves.

CYANANTHUS A
Soft basal cuttings taken between mid-spring and early summer and rooted in heated propagating case, 15.5°C (60°F), or garden frame.
Seeds sown in mid-or late winter, in pots, and placed out of doors.

CYCLAMEN (Sowbread) Tu
Seeds sown as soon as ripe, or mid- to late winter, in pots and placed outdoors.

CYDONIA (Quince) T
Seeds sown early to mid-spring, in outdoor seed bed.
Simple or air layering between mid-spring and late summer, outdoors.
Whip and tongue grafting for cultivars, early spring, outdoors, using *C. oblonga* as rootstocks.
Hardwood cuttings late autumn, nursery bed outdoors.

CYMBALARIA A
Division in early or mid-spring, replanting the portions immediately.
Softwood cuttings mid-spring to early summer, heated propagating case, 15.5°C (60°F).
Seeds sown mid-to late winter, in pots and placed outdoors.

CYNOGLOSSUM (Hound's tongue) P/Bi
Seeds sown late spring/early summer in outdoor seed bed.
Division for perennials in early to mid-spring, replanting immediately.

CYPRIPEDIUM (Lady's slipper) O
Division in early to mid-spring, replanting the portions immediately

CYSTOPTERIS (Bladder fern) F
Division in mid-spring, provided plants have formed several crowns, replanting the portions immediately.
Bulbils for the species *C. bulbifera*. These form on the fronds and when they fall to the ground they take root and develop into new plants.
Spores sown in autumn, as soon as ripe. They will germinate in a garden frame.

CYTISUS (Broom) S
Seeds sown in early to mid-spring, after soaking in warm water to soften the hard coats, heated propagating case, 18–21°C (65–70°F), or garden frame.
Semi-ripe cuttings with a heel mid-to late summer, heated propagating case, mist unit 18-21°C (65-70°F), or garden frame.

DABOECIA (St Dabeoc's heath) S
Semi-ripe cuttings of short side shoots pulled off with a heel. Early summer to early autumn, heated propagating case, mist unit, 18–21°C (65–70°F) or garden frame.
Layering (dropping) in mid-spring, outdoors. A plant is lifted and replanted to half its depth, spacing out the stems well. Each or most will then root.
Seeds sown as soon as ripe, in pots, over-wintered in cold spot outdoors, transferred to heated greenhouse, 10°C (50°F), in spring to germinate.

DACTYLIS (Cocksfoot) P G
Division of clumps early to mid-spring, replanting immediately.
Seeds sown late spring/early summer in outdoor seed bed.

DANAE (Alexandrian laurel) S
Division in early or mid-spring, replanting in permanent positions.
Seeds sown early to mid-spring, outdoor seed bed.

DAPHNE S
Simple layering between mid-spring and late summer, outdoors.
Semi-ripe cuttings early to mid-summer, heated propagating case or mist unit, 21°C (70°F). This is not an easy method and cuttings may take several months to root.
Seeds sown as soon as ripe, or even in the green condition, in containers and placed out of doors. Only a small percentage will germinate in the first year. More will follow

in the second spring. This is the best method for *D. mezereum* but also suitable for other species.

DARLINGTONIA (California pitcher plant, cobra lily) P
Division mid-spring, replanting immediately.
Seeds sown in late winter in cool greenhouse, after soaking in water for a day or two. Cover seeds lightly with fine peat. Place pot in tray of water. Best results with fresh seeds.

DAVIDIA (Handkerchief tree) T
Air layering between mid-spring and late summer, outdoors.
Cuttings of current year's shoots in early summer, wounded at base and rooted in mist unit, 21°C (70°F).
Semi-ripe cuttings mid-to late summer, wounded at base, heated propagating case, 21°C (70°F).
Seeds sown in autumn after being stratified in moist peat for a year. Outdoor seed bed.

DECAISNEA S
Seeds sown early to mid-spring and germinated in a garden frame. Better would be seeds sown as soon as ripe during autumn in pots and placed outdoors. Transfer to greenhouse at onset of germination.

DECUMARIA Cl
Softwood cuttings in spring, rooting them in heated propagating case, 15.5°C (60°F).

DELPHINIUM P/An
Basal cuttings for perennials as soon as ready in spring (ensure bases are not hollow), heated propagating case, 15.5°C (60°F).
Division for perennials, early to mid-spring, replanting the portions immediately in flowering positions.
Seeds for perennials, spring, garden frame or greenhouse. Maximum temperature 15.5°C (60°F). Sow annuals (popularly known as larkspur) outdoors in flowering positions during early/mid-spring or early autumn.

DENDROMECON (Tree poppy) S
Seeds sown in early to mid-spring, heated propagating case, 15.5°C (60°F). Germination is very slow.
Root cuttings early to mid-winter, heated propagating case, 15.5°C (60°F).
Semi-ripe cuttings mid-to late summer, heated propagating case, 15.5°C (60°F).

DENTARIA (Toothwort) P
Division in spring or after flowering, replanting portions immediately.
Seeds sown in spring in a garden frame.

DESCHAMPSIA (Hair grass) P G
Seeds sown late spring/early summer in an outdoor seed bed. Plants self-sow freely.
Division of the tufted clumps in early to mid-spring, replanting immediately.

DESFONTAINEA S
Seeds sown early to mid-spring in heated propagating case, 18°C (65°F).
Simple layering between mid-spring and late summer, outdoors.
Semi-ripe cuttings early to mid-autumn, heated propagating case, 18°C (65°F). Alternatively insert in garden frame during mid-autumn.

DEUTZIA S
Semi-ripe cuttings between early and late summer, heated propagating case, 18°C (65°F), or garden frame.
Hardwood cuttings late autumn/early winter, garden frame.
Softwood cuttings late spring, rooted in a mist unit, 21°C (70°F).

DIANTHUS (Carnations, pinks) P/A/An/Bi
Semi-ripe cuttings, also known as pipings, for perennials and alpines, taken in mid-or late summer and rooted in a garden frame. They will root in a matter of weeks.
Layering of young shoots in mid-to late summer outdoors for perennial border carnations and pinks.
Seeds for perennial and annual carnations, late winter to early spring, heated propagating case, 15.5°C (60°F). Alpines can be sown during winter in pots and placed outdoors. Sow the biennial

Dianthus barbatus (sweet william) during late spring or early summer in an outdoor seed bed.

DICENTRA (Includes bleeding heart) P
Division during early/mid-spring, or immediately after flowering for spring-flowering species, replanting portions immediately. Alternatively cut up dormant crowns in autumn, each piece consisting of growth buds and roots, replanting immediately, or box up and overwinter in garden frame.
Root cuttings early to mid-winter, garden frame.
Seeds early spring, heated propagating case, 15.5°C (60°F). Look out also for self-sown seedlings.

DICTAMNUS (Burning bush) P
Seeds sown as soon as ripe in an outdoor seed bed. Spring-sown seed can take a year to germinate.
Division in spring, but portions are slow to re-establish.

DIERAMA (Wand flower) Cm P
Division of clumps during early to mid-spring, replanting the portions immediately. Alternatively lift when dormant, detach small offsets and plant in nursery bed to grow on to flowering size.
Seeds sown early to mid-spring, garden frame or heated propagating case, 15.5°C (60°F), requiring light to germinate.

DIERVILLA (Bush honeysuckle) S
Semi-ripe cuttings between early and late summer, heated propagating case, 15.5°C (60°F) or garden frame.
Hardwood cuttings late autumn or early winter, garden frame.

DIGITALIS (Foxglove) Bi/P
Seeds sown in late spring/early summer, outdoor seed bed. Seeds are very tiny so handle carefully. Alternatively sow in a seed tray. Some species self-sow very freely.

DIONYSIA A
Offsets can be carefully removed in mid-summer and rooted in pots in a garden frame.
Seeds sown as soon as ripe in pots, placed out of doors or in a cool shaded garden frame.

DIOSPYROS (Persimmon) T
Seeds sown late winter to early spring, following stratification over winter, heated propagating case.
Whip and tongue grafting late winter, outdoors, using rootstocks of *D. virginiana* or *D. kaki* seedlings.
Shield budding in summer, outdoors, rootstocks as for grafting.

DIPELTA S
Semi-ripe cuttings late summer/early autumn, heated propagating case, 15.5°C (60°F), or garden frame.
Hardwood cuttings late autumn, garden frame.
Seeds sown in outdoor seed bed in spring.

DIPSACUS (Teasel) Bi
Seeds sown in late spring or early summer in an outdoor seed bed.

DIPTERONIA T
Trees
Layering, simple or air, between mid-spring and late summer, outdoors.
Semi-ripe cuttings with a heel, mid-to late summer, heated propagating case, 18°C (65°F). Not a reliable method.

DISANTHUS S
Simple layering between mid-spring and late summer, outdoors.
Semi-ripe cuttings late summer/early autumn, heated propagating case, 18°C (65°F).

DISPORUM (Fairy bells) P
Seeds sown early to mid-spring, heated propagating case, 15.5°C (60°F).
Division early to mid-spring, replanting the portions immediately.

DISTYLIUM S/T
Semi-ripe cuttings late summer/early autumn, heated propagating case or mist-propagation unit, 21°C (70°F).
Simple layering between mid-spring and late summer, outdoors.

DODECATHEON (Shooting stars) P
Division early to mid-spring, replanting the portions immediately.
Seeds sown as soon as ripe in a garden frame.

DORONICUM (Leopard's bane) P
Division after flowering, replanting the portions immediately in their permanent positions.
Seeds sown in late spring/early summer in an outdoor seed bed.

DORYCNIUM P/S S
Seeds sown early to mid-spring, heated propagating case or garden frame.
Softwood cuttings for shrubby species in early summer, heated propagating case 15.5°C (60°F).
Division in early spring for clump-forming species, replanting the portions immediately.

DOUGLASIA A
Seeds sown when ripe or in mid-to late winter, in pots and placed in garden frame or outdoors.
Cuttings can be taken in early summer and rooted in a garden frame.

DRABA (Whitlow grass) A
Seeds sown late winter/early spring in pots and placed in a garden frame or outdoors.
Cuttings formed or single rosettes, taken mid-summer and rooted in a garden frame.
Division where possible, immediately after flowering, replanting the portions straight away in permanent positions.

DRACOCEPHALUM (Dragon's head) P/An
Division for perennials, early to mid-spring, replanting the portions immediately.
Seeds for perennials, late spring/early summer, outdoor seed bed. Annuals can be sown in mid-spring outdoors where they are to flower.
Soft cuttings for perennials between mid-spring and early summer, heated propagating case, 15.5°C (60°F).

DRACUNCULUS (Dragon plant) Tu P
Division of clumps early to mid-spring, replanting the portions immediately in flowering positions.

DRIMYS S/T
Shrubs and trees
Simple layering mid-spring to late summer, outdoors.
Semi-ripe cuttings late summer or early autumn, wounded at base, rooting them in a garden frame, heated propagating case or mist unit, 18–21°C (65–70°F).
Softwood cuttings for *D.winteri*, mid-spring, wounded at base and placed in a mist unit.

DROSERA (Sundew) P
Division early or mid-spring, replanting the portions immediately in permanent positions.
Seeds sown early to mid-spring, unheated propagating case. Sow on surface of finely chopped live sphagnum moss or peat. Keep moist and humid.

DRYAS (Includes mountain avens) A
Division early to mid-spring, cutting off rooted stems and replanting immediately.
Semi-ripe cuttings with a heel late summer, rooting them in a garden frame.
Seeds sown as soon as ripe, or mid-to late winter, in pots, outdoors or garden frame.

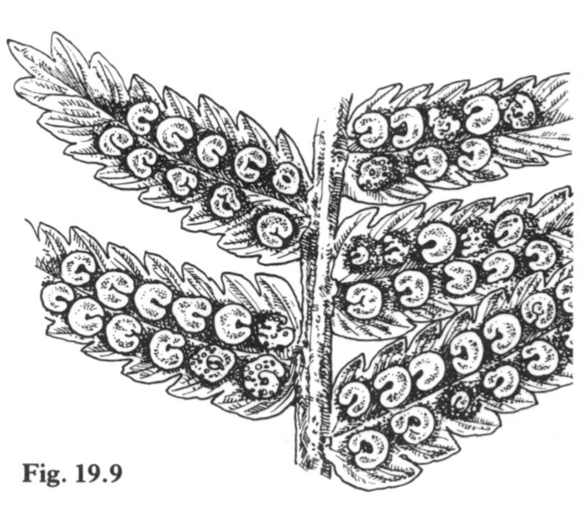

Fig. 19.9

Fig. 19.9 Dryopteris (buckler fern) species may be propagated from spores as soon as ripe. Here spore cases on the undersides of the fronds are shown.

DRYOPTERIS (Buckler fern) F
Division early to mid-spring, replanting immediately.
Spores sown as soon as ripe, heated prop-

agating case, 18–21°C (65–70°F); keep moist and shaded, but light needed for germination (Fig. 19.9).
Plantlets or 'bulbils' produced on leaves. Pick off when well developed and plant in seedtray, garden frame.

ECHINACEA (Purple cone flower) P
Division early to mid-spring, replanting portions immediately.
Root cuttings early to mid-winter, garden frame.
Seeds mid-spring, sowing in an outdoor seed bed.

ECHINOPS (Globe thistle) P
Root cuttings early to mid-winter, garden frame.
Seeds sown late spring/early summer in an outdoor seed bed.
Division early to mid-spring, replanting portions immediately. Not the best method as plants do not relish being disturbed.

Fig. 19.10
Elaeagnus are propagated from semi-ripe cuttings during autumn, rooting them in heat. First wound the base by removing a thin slice of bark as this encourages rooting.

ECHIUM (Viper's bugloss) An/Bi
Seeds sown in early spring or early autumn, outdoors in flowering positions. Can also sow in late winter/early spring in heated propagating case, 15.5°C (60°F).

EDGEWORTHIA S
Cuttings can be taken in the spring and rooted in a garden frame.
Seeds sown in late winter and germinated in heated propagating case, 18°C (65°F).

EDRAIANTHUS (Grassy bells) A
Seeds sown mid-to late winter in pots and germinated outdoors or in a garden frame.
Softwood cuttings mid-to late spring, garden frame.

ELAEAGNUS S
Semi-ripe cuttings, wounded at the base by removing a thin slice of bark to encourage rooting, early to mid-autumn, heated propagating case or mist-propagation unit, 21°C (70°F) (Fig. 19.10).
Seeds sown when ripe in autumn, outdoor seed bed. Seed to be sown in spring should be stratified over winter.

ELODEA (Includes Canadian pondweed) Sub Aq
Cuttings mid-spring to early summer, bunching them together with a lead weight and dropping them in the water.
Division mid-spring to early summer, replacing the portions immediately in the water.

ELSHOLTZIA S S
Soft or semi-ripe cuttings early to mid-summer, with or without a heel, heated propagating case, 15.5°C (60°F) or garden frame.

ELYMUS (Lyme grass) P G
Division early to mid-spring, replanting the tufts immediately.

EMBOTHRIUM (Chilean fire bush) S
Root cuttings early or mid-winter, heated propagating case, 21°C (70°F).
Seeds sown early to mid-spring, heated propagating case, 15.5°C (60°F).
Suckers can be removed in spring with some roots attached and immediately replanted elsewhere.
Cuttings can be taken in early summer and rooted in a mist-propagation unit, 21°C (70°F). Not an easy method for amateurs.

EMPETRUM (Crow berry) S A
Layering (dropping) can be carried out in mid-spring, outdoors.
Semi-ripe cuttings between early and late summer, rooting them in a garden frame.

ENDYMION (Bluebell) B
Bulbs
Bulblets can be removed when dormant in early or mid-autumn (planting time) and replanted in flowering positions or in nursery bed.
Seeds sown in autumn or mid- to late winter, in pots, and placed outdoors or in a cool garden frame. Alternatively sow in an outdoor seed bed during autumn.

ENKIANTHUS S
Simple layering between mid-spring and late summer, outdoors.
Semi-ripe cuttings late summer or early autumn, heated propagating case, 18–21°C.

Seeds sown early to mid-spring, barely covered with compost, heated propagating case, 18–21°C (65–70°F). Best to use fresh seeds.

EOMECON (Snow poppy) P
Division early to mid-spring, replanting the portions immediately.
Root cuttings early to mid-winter, garden frame.

EPILOBIUM (Willow herb) P/A
Division in early or mid-spring, replanting the portions immediately.
Root cuttings make a suitable method of propagation for some species, early to mid-winter, garden frame.
Seeds sown in spring and germinated in garden frame.
Basal cuttings taken in mid-or late spring and rooted in a garden frame.

EPIMEDIUM (Barrenwort) P
Division as soon as flowering is over, replanting the portions immediately.
Seeds sown as soon as ripe, summer, garden frame.

ERANTHIS (Winter aconite) Tu P
Division when dormant, late summer or early autumn, replanting the portions immediately.
Seeds sown as soon as ripe, usually sometime in spring, in pots and germinated in a garden frame.

EREMURUS (Foxtail lily, giant asphodel) P
Division early autumn, replanting the portions immediately and taking care not to damage the fleshy roots.
Seeds sown as soon as ripe, heated propagating case, 15.5°C (60°F). Older seeds take longer to germinate. Seedlings take at least five years to reach flowering size.

ERICA (Heath, heather) S
Semi-ripe cuttings of short side shoots pulled off with a heel. Tips can be pinched out. Early summer to early autumn, heated propagating case, mist unit, 18–21°C (65–70°F) or garden frame.
Layering (dropping) in mid-spring, outdoors. A plant is lifted and replanted to half its depth, spacing out the stems well. Each or most will then root.
Seeds sown as soon as ripe, in pots, overwintered in cold spot outdoors, transferred to heated greenhouse, 10°C (50°F), in spring to germinate.

ERIGERON (Flea bane) P
Division early to mid-spring, replanting the portions in flowering positions immediately.
Seeds sown mid- to late spring in pots and germinated in a garden frame, or in outdoor seed bed.

ERINACEA (Hedgehog broom) S
Semi-ripe cuttings mid-to late summer, heated propagating case, 15.5°C (60°F), or garden frame where rooting will be slower.
Seeds sown early to mid-spring and germinated in a heated propagating case or garden frame.

ERINUS (Fairy foxglove) A
Seeds germinate very freely when sown in spring, either in pots and germinated in a garden frame, or in an outdoor seed bed.
Division in early to mid-spring, replanting the portions immediately.

ERIOBOTRYA (Loquat) S/T
Seeds sown early to mid-spring and germinated in a heated propagating case, 15.5°C (60°F).
Semi-ripe cuttings late summer to early autumn, heated propagating case, 18–21°C (65–70°F).

ERIOGONUM (Umbrella plant, wild buckwheat) P/S
Seeds sown in spring, heated propagating case, 15.5°C (60°F).
Division for perennials, early to mid-spring, replanting immediately.

ERIOPHYLLUM P/S S
Seeds sown during mid-spring in an outdoor seed bed.
Division for perennial kinds during early or mid-spring, replanting the portions immediately in their permanent positions.

ERITRICHIUM (Alpine forget-me-not) A
Seeds sown in mid-to late winter in pots and placed out of doors in a cold aspect to expose them to frost, which assists germination.
Division can be carried out in mid-spring, replanting the portions immediately in their flowering positions. The difficult *E. nanum* is best left undisturbed, though.

ERODIUM (Stork's bill) A
Semi-ripe cuttings can be taken in early or mid-summer, using the lower shoots, and rooted in a heated propagating case, 15.5°C (60°F), or garden frame. Not suitable for mat-forming species, such as *E. chamaedryoides*.
Division during early or mid-spring, replanting the portions immediately.
Seeds sown as soon as ripe during summer, in pots, and placed outdoors to germinate.
Root cuttings for some species, such as the mat-forming kinds, early spring, rooting them in a garden frame.

ERYNGIUM (Sea holly) P/Bi
Seeds for perennials and biennials, sown in late spring or early summer in an outdoor seed bed. Alternatively germinate in a garden frame.
Root cuttings early to mid-winter, inserting them in a garden frame.
Division for species which form clumps, during early or mid-spring, replanting the portions immediately into their flowering positions.

ERYSIMUM An/Bi/P
Seeds sown during late spring or early summer in an outdoor seed bed, where they will germinate freely.
Cuttings of side shoots, with a hell, can be taken from perennial kinds during mid-or late summer and rooted in a garden frame, where young plants should be overwintered.

ERYTHRONIUM (Dog's tooth violet, trout lily) Cm P
Seeds sown as soon as ripe, during summer, in pots, and placed in a shaded garden frame. If seeds are saved until spring they may take up to a year to germinate. Compost must not be allowed to dry out. Seedlings can take at least five years to reach flowering size.
Offsets can be removed when the plants are dormant in summer, replanting them immediately in a nursery bed to grow on to flowering size, which may take at least three years.

ESCALLONIA S
Semi-ripe cuttings mid-summer to mid-autumn, rooting them in a garden frame, under a low polythene tunnel or in a heated propagating case, 15.5°C (60°F).
Hardwood cuttings taken mid- to late autumn and inserted in the open ground. Recommended only for mild areas.
Seeds sown in early or mid-spring in a heated propagating case, 15.5°C (60°F).

ESCHSCHOLZIA (Californian poppy) A
Seeds sown in early or mid-spring, or in early autumn in mild climate and if the soil is well drained, outdoors where the plants are to flower.

EUCALYPTUS (Gum tree) T
Seeds provide the only means of raising eucalyptus, ideally collected locally. When buying imported seeds try to ensure they originate from trees at high altitudes. This will ensure the subsequent trees are hardy. The seeds of some species, particularly very hardy alpine kinds, should be given cold stratification for two months prior to sowing. Sow late winter or early spring, covering the small seeds with 3 mm (⅛ in) of fine compost. Germinate the seeds in a heated propagating case, 15.5°C (60°F).

EUCOMIS (Pineapple flower) B
Bulblets can be removed in early or mid-autumn and replanted immediately, either in a nursery bed to grow on, or in permanent positions.
Seeds can be sown in mid-spring in a heated propagating case, 15.5°C (60°F).

EUCRYPHIA T/S
Layering, either simple or air layering, between mid-spring and early autumn, outdoors.
Semi-ripe cuttings late summer, using side shoots which are not flowering, ideally with

a heel of older wood, or wounded at the base by removing a thin slice of wood. Root in a heated propagating case, 15.5–18°C (60–65°F). Alternatively insert cuttings under mist in early spring, first wounding them at the base. Not an easy method.
Seeds sown early to mid-spring in a heated propagating case, 15.5°C (60°F).

EUONYMUS (Includes spindle trees) S/T
Simple layering between mid-spring and late summer, outdoors.
Semi-ripe cuttings can be taken in mid- to late summer, ideally with a heel, and rooted in a garden frame. Alternatively take cuttings of evergreen species in early to mid-autumn and insert them either in a garden frame or under a low polythene tunnel.
Division for *E. radicans* and other types with a similar habit of growth, early to mid-spring, replanting the portions immediately in their permanent positions.
Seeds early to mid-spring in an outdoor seed bed, after being stratified over the winter. Bear in mind the seeds may not germinate during the first season.

EUPATORIUM (Hemp agrimony) P
Division for perennial species, early to mid-spring, replanting the portions immediately in flowering positions.

EUPHORBIA (Spurge) P/An/S
Division for perennial species, early to mid-spring, or in early autumn for spring-flowering kinds, replanting the portions immediately.
Seeds also for perennial species, sowing in late spring or early summer, in a garden frame or outdoor seed bed. Seeds for the annual species *E. marginata* (snow on the mountain), sowing in early to mid-spring outdoors where the plants are to flower.
Softwood cuttings for shrubby species, mid-spring to early summer, rooting them in a heated propagating case, 15.5°C (60°F).

EURYA S
Semi-ripe cuttings with a slight heel of older wood, mid- to late summer, heated propagating case, 15.5–18°C (60–65°F).

EURYOPS S
Soft cuttings in spring, heated propagating case, 15.5°C (60°F).
Semi-ripe cuttings mid-summer, rooting them in a garden frame.

EXOCHORDA (Pearlbush) S
Simple layering between mid-spring and late summer, outdoors.
Softwood cuttings late spring to early summer, rooting them in a heated propagating case, 18–21°C (65–70°F). Alternatively insert them in a mist-propagation unit.
Seeds sown late winter and germinated in a garden frame.

FABIANA S
Semi-ripe cuttings taken in late summer and rooted in a heated propagating case, 18–21°C (65–70°F). Or take cuttings in mid-autumn and insert in a garden frame.
Softwood cuttings taken in mid-spring and rooted in a heated propagating case or under mist, 18–21°C (65–70°F).

FAGUS (Beech) T
Spliced side veneer grafting for cultivars during early spring, greenhouse, bottom heat and high humidity, 21°C (70°F). The rootstock is *F. sylvatica* (common beech).
Whip and tongue grafting for cultivars, early spring, outdoors. Rootstock as above.
Seeds for species, sown as soon as collected in autumn, or early to mid-spring (store in moist peat), in an outdoor seed bed.

× **FATSHEDERA** (Tree ivy, aralia ivy) S
Semi-ripe cuttings prepared from lateral shoots, or the tips of main shoots, mid-summer to early autumn, rooting them in a heated propagating case, 15.5°C (60°F). Cuttings can also be rooted in a garden frame.

FATSIA (Japanese aralia) S
Semi-ripe cuttings late summer, rooted in a heated propagating case, 15.5°C (60°F).
Seeds sown early to mid-spring and germinated in a heated propagating case, 13°C (55°F).
Suckers removed in early spring, potted and established in a heated propagating case.

Air layering can be carried out in spring, outdoors.

FEIJOA S
Semi-ripe cuttings with a slight heel, mid- to late summer, rooted in a heated propagating case, 15.5–18°C (60–65°F).
Simple layering between mid-spring and late summer, outdoors.
Seeds sown early to mid-spring, heated propagating case, 15.5°C (60°F).

FERULA (Includes giant fennels) P
Seeds sown as soon as ripe, or in spring, where the plants are to flower. The plants dislike disturbance.

FESTUCA (Fescue) P G
Division early to mid-spring, replanting the clumps immediately in their permanent positions.
Seeds sown in late spring or early summer in an outdoor seed bed with light, well-drained soil.

FICUS (Includes figs) T
Hardwood cuttings for *Ficus carica* (fig), early to mid-winter, inserting them in deep pots and rooting in a heated propagating case, 18–21°C (65–70°F).
Simple layering if practicable, summer, outdoors.

FILIPENDULA (Meadowsweet) P
Division early to mid-spring, replanting the portions immediately.
Seeds sown as soon as ripe, or in early spring, and germinated in a garden frame, or in a heated propagating case with gentle heat, 10–13°C (50–55°F).

FOENICULUM (Fennel) C
Division early to mid-spring, replanting the portions immediately.
Seeds sown early to mid-spring in an outdoor seed bed.

FONTANESIA S
Semi-ripe cuttings early to mid-autumn, garden frame.
Simple layering between mid-spring and late summer, outdoors.

FORSYTHIA (Golden bells) S
Simple layering carried out between mid-spring and late summer.
Semi-ripe cuttings taken early to mid-summer and rooted in a heated propagating case, 15.5°C (60°F), or garden frame.
Hardwood cuttings taken late autumn or early winter and rooted in a garden frame. In mild areas with very well-drained soil the cuttings can be rooted in the open ground.

FOTHERGILLA S
Simple layering between mid-spring and late summer, outdoors.
Semi-ripe cuttings mid-summer under mist or in a heated propagating case, 21°C (70°F). Wound base of cutting by removing a thin slice of wood. Not an easy method of propagation.
Seeds sown early to mid-spring in a heated propagating case, 15.5°C (60°F). Use a peaty compost.

FRAGARIA (Strawberry, including ornamental) P
Layering runners in early or mid-summer outdoors.
Division in early or mid-spring, replanting the portions immediately in their flowering positions.
Seeds sown during early or mid-spring, heated propagating case, 15.5°C (60°F).

FRANCOA P
Division during early or mid-spring, replanting the portions immediately.
Seeds sown in early or mid-spring in a garden frame or cool greenhouse.

FRANKENIA (Sea heath) P/S S
Softwood cuttings taken between mid-spring and early summer and rooted in a heated propagating case, 15.5°C (60°F).
Division in early or mid-spring, replanting the portions immediately.

FRANKLINIA (Franklin tree) S/Sm T
Softwood cuttings taken in spring and rooted in a heated propagating case, 18°C.
Hardwood cuttings taken in late autumn or early winter and rooted in a garden frame.
Seeds sown in spring in an outdoor seed bed.

Simple layering between mid-spring and late summer, outdoors.

FRAXINUS (Ash) T
Seeds sown in the autumn ideally, or during early/mid-spring, in an outdoor seed bed.
Whip and tongue grafting for the cultivars, late winter to early spring, outdoors. Use *F. excelsior* (common ash) as a rootstock.
Chip budding for cultivars during late summer, using the same rootstock as for grafting.

FREMONTODENDRON (FREMONTIA) S
Softwood cuttings late spring or early summer, rooted in a heated propagating case, 18°C (65°F).
Seeds sown late winter in a heated propagating case, 15.5°C (60°F), after soaking in tepid water for two days.

FRITILLARIA (Fritillary) B
Bulblets removed when dormant in early to mid-autumn (planting time) and planted in a garden frame or outdoor nursery bed to grow on.
Scooping and scoring for species with large bulbs, late summer or early autumn, 21°C (70°F).
Seeds sown as soon as they are ripe in summer, in pots, germinating them in a garden frame or slightly heated greenhouse. Purchased seeds, sown in spring, may be very slow to germinate.

FUCHSIA S
Softwood cuttings in late spring as soon as available, rooting them in a heated propagating case, 15.5°C (60°F).
Semi-ripe cuttings late summer, garden frame, or heated propagating case, 15.5°C.

GAGEA B
Seeds sown in autumn or winter, in pots and germinated outdoors or in a garden frame.
Bulblets removed at planting time in early autumn and planted in an outdoor nursery bed to grow on to flowering size.

GAILLARDIA (Blanket flower) P/A
Division for perennials, early to mid-spring, replanting the portions immediately.
Root cuttings for perennials, early to mid-winter, inserting them in a garden frame.
Seeds for perennials, sown in late spring or early summer in an outdoor seed bed. Alternatively sow in late winter or early spring in a heated propagating case, 15.5°C (60°F). Annuals can be sown in early to mid-spring outdoors where they are to flower. Alternatively under glass as for perennials.

Fig. 19.11

Fig. 19.11 Galanthus (snowdrops) may be propagated by division immediately after flowering. Clumps should be separated out to individual bulbs.

GALANTHUS (Snowdrop) B
Division of congested clumps immediately after flowering, replanting the bulbs immediately in their permanent positions. The clumps should be separated out to individual bulbs (Fig. 19.11).
Seeds sown as soon as ripe, in pots, and germinated out of doors or in a shaded garden frame. Seedlings can take five years to reach flowering size. Snowdrops also self-sow.

GALEGA (Goat's rue) P
Basal cuttings taken as soon as ready in the spring, rooting them in a heated propagating case with gentle heat.
Division during early or mid-spring, replanting the portions immediately. It will be found that the clumps are rather tough.
Seeds sown late spring or early summer in an outdoor seed bed.

GALEOBDOLON (LAMIASTRUM) (Yellow archangel) P
Division during early or mid-spring,

replanting the portions immediately.
Rooted stolons can be removed in early or mid-spring and immediately replanted in permanent positions. The plant spreads rapidly by means of these stolons.

GALIUM (Bedstraw) P
Division during early or mid-spring, replanting the portions immediately in flowering positions.
Seeds sown in mid-spring in an outdoor seed bed.

GALTONIA (Summer hyacinth) B
Bulblets removed when dormant in early or mid-spring (planting time) and immediately replanted in a nursery bed. They are not freely produced.
Seeds sown late winter in pots and germinated outdoors or in a garden frame. Seedlings take four or five years to attain flowering size.

GARRYA (Silk-tassel bush) S
Semi-ripe cuttings taken in late summer and rooted in a heated propagating case, 18–21°C (65–70°F). Alternatively take them in mid-autumn and root in a garden frame.
Hardwood cuttings prepared from shoots removed at their base can be taken in late winter and rooted in a mist-propagation unit, 21°C (70°F).
Simple layering between mid-spring and late summer, outdoors. Layers can take up to two years to form good root systems.

GAULTHERIA S
Semi-ripe cuttings taken in late summer or early autumn and rooted in a heated propagating case, 15.5°C (60°F), or garden frame.
Layering (dropping) during mid-spring outdoors. A plant is lifted and replanted to half its depth, spacing out the stems well.
Division early to mid-spring, replanting the portions immediately in their permanent positions.
Seeds should be surface sown on fine peat during autumn and germinated in a garden frame. Best sown in pots or pans.
Suckers can be removed from some species, provided they are well rooted, during autumn and immediately replanted.

× **GAULNETTYA** S
Semi-ripe cuttings late summer to early autumn, rooted in a heated propagating case, 15.5°C (60°F) or garden frame.
Division in early or mid-spring, replanting the portions immediately in their permanent positions.

GAURA P/An
Seeds can be sown in late winter or early spring in a heated propagating case, 15.5°C (60°F), or during early to mid-spring outdoors in flowering positions.

GENISTA (Broom and gorse) S
Seeds sown in late winter or early spring and germinated in a heated propagating case, 15.5°C (60°F), after chipping and soaking in tepid water to soften hard seed coats.
Semi-ripe cuttings early to mid-autumn, rooted in a heated propagating case, 15.5°C (60°F), or garden frame. Cuttings of *G. hispanica* can be taken in early summer and rooted under a low polythene tunnel. Worth trying for other genistas.

GENTIANA (Gentian) P/A
Division in early or mid-spring, replanting the portions immediately.
Soft basal cuttings during mid- or late spring, rooting them in a heated propagating case, 15.5°C (60°F).
Seeds sown as soon as they are ripe, in pots, placing them outdoors or in a garden frame. Subjected to frost, they generally germinate freely at the onset of warmer weather in spring, or they can be taken into a heated greenhouse to germinate.

GERANIUM (Crane's bill) P/A
Division during early or mid-spring, replanting the portions immediately.
Seeds sown in early autumn or early spring, in pots, and germinated in a garden frame. Some species set copious amounts of seeds and will self-sow freely.
Root cuttings taken early to mid-winter and inserted in a garden frame.

GEUM (Avens) P
Division during early or mid-spring, replanting the portions immediately.

Seeds sown as soon as ripe during summer, alternatively during spring, in pots and placed in a garden frame.
Plantlets produced on runners by some species like *G. reptans* can be removed during summer, potted and overwintered in a well-ventilated garden frame. Or peg down *in situ* until rooted.

GILIA An/Bi/P
Seeds sown during early to mid-spring, outdoors where the plants are to flower. Alternatively sow in early autumn in mild areas provided the soil is very well drained.

GILLENIA P
Division during early or mid-spring, replanting the portions immediately.
Basal cuttings as soon as available in spring, rooting them in a heated propagating case, 15.5°C (60°F).
Seeds sown during late spring or early summer in an outdoor seed bed.

GINKGO (Maidenhair tree) T
Seeds (imported) sown during early or mid-spring and germinated in a heated propagating case, 18°C (65°F). Alternatively sow in a garden frame or outdoor seed bed.
Air layering between mid-spring and late summer, outdoors.
Simple layering if feasible, between mid-spring and late summer, outdoors.
Spliced side veneer grafting for cultivars, early spring, in greenhouse, with bottom heat and high humidity, 21°C (70°F). The rootstock is *G. biloba*.

GLAUCIDIUM P
Division during early or mid-spring, replanting the portions immediately. Take care not to damage the thick rhizomes.
Seeds sown during spring in a greenhouse or garden frame, in a temperature of 10–13°C (50–55°F.)

GLAUCIUM (Horned poppy) An/Bi/P
Seeds sown in early or mid-spring, outdoors where the plants are to flower. Horned poppies do not transplant well.

GLECHOMA (Ground ivy) P
Soft cuttings taken any time between mid-spring and late summer and rooted in a garden frame or slightly heated propagating case.
Division in mid-spring, replanting the portions immediately.
Serpentine layering in spring or summer outdoors, by pinning the trailing stems to the ground, when they will quickly take root.

GLEDITSIA (Locust) T
Seeds sown in late winter in a heated propagating case, 15.5°C (60°F), after soaking in hot water for an hour to soften the hard seed coats.
Whip and tongue grafting for cultivars, in early spring outdoors, the rootstock being *G. triacanthos*.
Spliced side veneer grafting for cultivars, early spring, in greenhouse with bottom heat and high humidity, 21°C (70°F). Rootstock as above.

GLOBULARIA (Globe daisy) P/S S
Seeds sown early to mid-spring in a garden frame.
Division if feasible in early or mid-spring, replanting the portions immediately.
Semi-ripe cuttings for shrubby species in mid- or late summer/early autumn, rooting them in a garden frame.

GLYCERIA P G
Division of the clumps during early or mid-spring, replanting the portions immediately in their permanent positions.

GLYCYRRHIZA (Liquorice) P
Division during early to mid-spring, replanting the portions immediately in their flowering positions.

GNAPHALIUM (Cudweed) P/An/S S
Seeds sown in spring in a heated propagating case, 15.5°C (60°F). The seeds need light to germinate.
Division of perennial species, early to mid-spring, replanting the portions immediately.

GODETIA An
Seeds sown early to mid-spring, or early autumn in mild areas provided the soil is very well drained, outdoors where the plants

are to flower. Cover only very lightly with fine soil.

GREVILLEA S
Shrubs
Semi-ripe cuttings of side shoots taken in early or mid-summer and rooted in a heated propagating case or mist-propagation unit, 18–21°C (65–70°F). Alternatively cuttings can be taken in the spring and rooted as above.
Seeds sown in early or mid-spring in a heated propagating case, 15.5°C (60°F), after soaking in water for 48 hours.

GRISELINIA S
Semi-ripe cuttings taken between early and mid-autumn and rooted in a garden frame or under a low polythene tunnel. In mild areas with well-drained soil it is also possible to root cuttings in the open ground. These are very easy shrubs to propagate from cuttings.
Seeds sown in early or mid-spring and rooted in a heated propagating case, 15.5°C (60°F).

GUNNERA P
Division during mid-spring, replanting the portions immediately in their permanent positions. With the large *G. manicata* and *G. chilensis* aim to remove the smaller lateral crowns with some roots attached and pot them into suitable-sized pots. Overwinter the young plants in a frost-free but cool greenhouse.
Seeds sown in early or mid-spring and germinated in a heated propagating case, 15.5°C (60°F). Overwinter seedlings in a frost-free but cool greenhouse.

GYMNOCLADUS (Kentucky coffee tree) T
Root cuttings taken early to mid-winter and rooted in a heated propagating case, 15.5°C (60°F).
Seeds sown late winter or early spring in a heated propagating case, 15.5°C (60°F). First file away part of the hard outer seed-coat to improve germination.

GYPSOPHILA (Chalk plant) P/A/An
Seeds for all types. Seeds of perennials can be sown late spring or early summer in an outdoor seed bed. Those of alpine species are best sown in pots during mid- to late winter and germinated outdoors or in a garden frame. Seeds or annuals can be sown during early or mid-spring, outdoors in their flowering positions.
Soft basal cuttings for perennials, as soon as available in mid-or late spring, rooting them in a heated propagating case, 15.5°C (60°F).
Soft cuttings of alpine species in spring, rooting them in a garden frame.
Division for mat-forming alpine species in early spring, replanting the portions immediately in permanent positions.

HABERLEA P (grown as A)
Leaf cuttings taken between early and late summer. Use entire mature leaves and root them in a heated propagating case, 15.5°C (60°F), or in a shaded garden frame.
Seeds sown as soon as ripe and germinated in a shaded garden frame.

HABRANTHUS B
Bulblets can be removed and replanted when the parent bulbs are dormant in the autumn (planting time). Best to leave established groups undisturbed for as long as possible.
Seeds sown as soon as they are ripe, in pots, and placed in a garden frame.

HACQUETIA P
Division during early or mid-spring, replanting the portions immediately.
Seeds sown as soon as ripe, in pots, germinating them in a garden frame.

HAKEA (Pincushion tree) S/Sm T
Seeds sown in spring and germinated in a heated propagating case, 15.5°C (60°F). The seedlings should spend their first winter under glass.
Semi-ripe cuttings with a heel taken in early or mid-summer and rooted in a heated propagating case, 15.5°C (60°F). Overwinter young plants under glass. Not an easy method of propagation. Better results from seeds.

HAKONECHLOA P G
Division of the clumps in early or mid-

The aquatic *Hyrdocharis morsus-ranae* (frogbit) may be propagated by removing rooted runners in the period mid-spring to early summer which are simply dropped into water.

Lewisias can be propagated by division, seeds and offsets. Also, they are among the few hardy plants which can be propagated from leaf cuttings.

The bizarre moisture-loving perennial *Lysichiton americanum* (skunk cabbage) should be divided after flowering or in autumn, taking care to avoid root damage.

Nertera granadensis (N. depressa) forms creeping stems which root at the nodes so a single plant can, if desired, be split into many small pieces.

Phyteuma comosum (horned rampion) is a choice alpine plant which can be increased by division during early to mid-spring, or from seeds in early spring.

The striking ornamental thistle *Silybum marianum* (milk thistle), grown for its foliage, is a biennial easily raised from seeds sown outdoors *in situ*.

Any offsets produced by the succulent *Agave americana* (century plant) can be removed during spring and established in pots in a heated propagating case.

Raise *Chamaedorea elegans* (parlour palm, dwarf mountain palm) from suckers if produced, or from seeds germinated in a high temperature, both in spring.

spring, replanting the portions immediately in their permanent positions.

HALESIA (Snowdrop tree) T/S
Simple layering if possible, between mid-spring and late summer, outdoors. Layers can take up to two years to produce a good root system.
Softwood cuttings in late spring or early summer and rooted in a heated propagating case or mist-propagation unit, 18–21°C (65–70°F). Cuttings may be slow to root.
Seeds sown as soon as ripe, in autumn, placing them in a temperature of 15.5°–21°C (60–70°F) for up to three months, then outdoors in a temperature of 0.5–4.5°C (33–40°F) for another three months. Return to greenhouse at onset of germination. Without this warm/cold treatment seeds may not germinate the first year.
Root cuttings taken in winter and rooted in a garden frame.

× HALIMIOCISTUS (HALIMIUM × CISTUS) S
Semi-ripe cuttings taken in early or mid-summer and rooted in a heated propagating case, 18°C (65°F). Good results also from cuttings taken during mid-or late summer and inserted in a garden frame.

HALIMIUM S
Semi-ripe cuttings taken in early or mid-summer and rooted in a heated propagating case, 18°C (65°F). Good results also from cuttings taken during mid-or late summer and inserted in a garden frame.

HALIMODENDRON (Salt tree) S
Simple layering between mid-spring and late summer, outdoors.
Whip grafting (without the 'tongue') during late winter in greenhouse, with bottom heat 18°C (65°F). Rootstocks are two-year-old seedlings of *Caragana arborescens*. Graft close to their base.

HAMAMELIS (Witch hazel) S/T
Simple layering between mid-spring and late summer outdoors. Layers take two years to produce a good root system.
Air layering between mid-spring and late summer, outdoors.
Spliced side veneer grafting during early spring, in greenhouse with bottom heat, 21°C (70°F), and high humidity. Use seedlings of *H. virginiana* as a rootstock, or young plants of *Distylium racemosum*.
Seeds sown when ripe in the autumn and germinated in a garden frame. Germination can take at least 18 months, sometimes more.
Softwood cuttings removed close to the older wood in late spring and rooted in a mist-propagation unit, 21°C (70°F). Overwinter young plants in a cold greenhouse, keeping the compost on the dry side.

HAPLOPAPPUS S/P
Soft or semi-ripe cuttings for shrubby species, between late spring and mid-summer, rooting them in a heated propagating case, 15.5°C (60°F).
Division for perennials in early or mid-spring, replanting the portions immediately in their flowering positions.

HEBE (Shrubby veronica) S
Softwood cuttings taken between mid-spring and early summer, rooting them in a heated propagating case, 15.5°C (60°F). A good method for tender species and cultivars, young plants of which should be overwintered in cool but frost-free conditions.
Semi-ripe cuttings taken between mid-summer and mid-autumn, rooting them in a garden frame or under a low polythene tunnel. Both types of cuttings root very easily.

HEDERA (Ivy) Cl
Leaf-bud cuttings taken from juvenile growth in early or mid-summer and rooted in a heated propagating case, 15.5°C (60°F), mist unit or garden frame.
Semi-ripe cuttings of shoot tips during mid- or late summer, heated propagating case, 15.5°C (60°F), mist unit or garden frame. Cuttings taken from adult growth will produce bushy plants, so take cuttings from juvenile growth if climbers are required.
Serpentine layering between mid-spring and late summer, outdoors. Layers root comparatively quickly.

HEDYSARUM S/P
Seeds of perennials sown in an outdoor nursery bed during late spring or early summer. Seeds of shrubby species can be sown in spring and germinated in a garden frame.
Simple layering between mid-spring and late summer for shrubby species. The layers take about a year to produce a good root system.
Semi-ripe cuttings for shrubby species, late summer, garden frame.

HELENIUM (Sneezeweed) P
Division in early or mid-spring, replanting the portions immediately.
Seeds sown late spring or early summer in an outdoor seed bed.

HELIANTHEMUM (Rock rose) S
Semi-ripe cuttings taken with a heel in early or mid-summer and rooted in a heated propagating case, 15.5°C (60°F). Mid-summer cuttings can also be rooted in a garden frame.
Seeds sown in early or mid-spring in a heated propagating case.

HELIANTHUS (Sunflower) P/An
Division for perennials during early or mid-spring, replanting the portions immediately.
Seeds for annuals, sowing during early or mid-spring outdoors in their flowering positions. Perennials can also be raised from seeds, sowing in late spring in an outdoor seed bed.

HELICHRYSUM (Straw flower) P/A/S
Seeds for perennials, sowing during late spring in an outdoor seed bed.
Division for perennials and alpines, early to mid-spring, replanting the portions immediately.
Semi-ripe cuttings prepared from lateral shoots, between early and late summer, for perennials, alpines and shrubby species, rooting them in a heated propagating case with gentle warmth, or in a garden frame.

HELICTOTRICHON P G
Division of the clumps during early or mid-spring, replanting the tufts immediately in their permanent positions.
Seeds can be sown in late spring or early summer in an outdoor seed bed.

HELIOPSIS (Oxeye) P
Division of the clumps during early or mid-spring, replanting the portions immediately.
Basal cuttings taken as soon as ready in the spring and rooted in a heated propagating case, 15.5°C (60°F).
Seeds sown during late spring or early summer in an outdoor seed bed.

HELIPTERUM (ACROCLINIUM, RHODANTHE) (Everlasting, strawflower) An
Seeds sown during early to mid-spring outdoors where the plants are to flower.

HELLEBORUS (Hellebore, Lenten rose, Christmas rose) P
Division immediately after flowering, replanting the portions immediately. Best to lift and divide only large well-established plants as hellebores resent disturbance.
Seeds sown as soon as they are ripe in containers and germinated in a shaded garden frame. Some species may self-sow quite freely, especially *H. orientalis*, so watch out for seedlings around the parent plants.

HEMEROCALLIS (Day lily) P
Division of the clumps during early or mid-spring, replanting the tufts immediately in their flowering positions. If desired, hemerocallis are easily split into quite small divisions, each consisting of a single shoot with roots attached.
Seeds sown late spring or early summer in an outdoor seed bed or in a garden frame.

HEPATICA (Liverleaf) P
Division as soon as flowering is over, replanting the portions immediately in their permanent positions, in soil which is well enriched with peat or leafmould.
Seeds sown in mid-spring, very shallowly in a moisture-retentive, shaded, outdoor seed bed.

HERMODACTYLUS (Snake's head iris) Tu P
Division of the clumps after flowering replanting the portions immediately.

HERNIARIA (Herniary, rupturewort) A/P
Division for perennial species in early or mid-spring, replanting the portions immediately.
Seeds sown shallowly in the spring outdoors.

HESPERIS (Sweet rocket) P/Bi
Division of perennials in early or mid-spring, replanting the portions immediately in their flowering positions.
Seeds sown in mid-spring in a sunny outdoor seed bed. The species *H. matronalis* self-sows freely so watch out for seedlings around the parent plants. These can be transplanted in the spring.

HEUCHERA (Alum root, coral flower) P
Division during early or mid-spring, replanting the portions immediately.
Seeds sown during early or mid-spring in containers and germinated in a garden frame; or sow late spring or early summer in an outdoor seed bed.

× HEUCHERELLA (HEUCHERA × TIARELLA) P
Division during early to mid-spring, replanting the portions immediately. Seeds are not produced so plants cannot be raised by this method.

HIBISCUS (Shrubby mallow) S/An
Spliced side veneer grafting for cultivars of the hardy shrub, *H. syriacus,* early spring in a greenhouse with bottom heat, 21°C (70°F), and high humidity. The rootstocks are seedlings of *H. syriacus.*
Semi-ripe cuttings for the hardy shrub, prepared from side shoots, ideally with a heel, late summer or early autumn, heated propagating case, 18°C (65°F). Cuttings are slow to root and become established.
Simple layering for the hardy shrub, between mid-spring and late summer, outdoors.
Seeds for the hardy shrub, sowing during early or mid-spring in an outdoor seed bed, where they germinate readily. Seeds of hardy annuals can be sown at the same time, but direct into their flowering positions.

HIERACEUM (Hawkweed) P
Division early to mid-spring, replanting the portions immediately in their permanent positions. Hawkweeds often spread by means of stolons.
Seeds sown in late spring or early summer in an outdoor seed bed.

HIPPOPHAE (Sea buckthorn) S
Seeds sown during early or mid-spring in an outdoor seed bed, after being stratified over winter. Alternatively sow direct in seed bed in autumn.
Suckers complete with roots can be lifted and replanted elsewhere in early spring.
Simple layering between mid-spring and late summer, outdoors.
Root cuttings taken during early or mid-winter and rooted in a garden frame.

HOHERIA (Ribbon wood) S/Sm T
Semi-ripe cuttings taken during late summer or early autumn and rooted in a heated propagating case, 18°C (65°F). Alternatively they can be inserted in a garden frame.
Simple layering between mid-spring and late summer, outdoors.
Air layering if simple layering is not possible, between mid-spring and late summer, outdoors.

HOLBOELLIA Cl
Softwood cuttings taken in spring as soon as available and rooted in a heated propagating case, 15.5°C (60°F).
Serpentine layering between mid-spring and late summer, outdoors.
Seeds sown during spring in a heated propagating case, 15.5°C (60°F).

HOLCUS P G
Division in early or mid-spring, replanting the tufts immediately in their permanent positions.
Seeds if available, sown late spring in an outdoor seed bed.

HOLODISCUS S/Sm T
Semi-ripe cuttings taken late summer or early autumn and rooted in a heated propagating case, 15.5°C (60°F), or garden frame.

Hardwood cuttings taken in late autumn and inserted in a garden frame.
Simple layering between mid-spring and late summer, outdoors.
Seeds sown during spring in a heated propagating case, 21°C (70°F), after winter stratification at 5°C (41°F).

HORDEUM (Foxtail barley, squirrel tail grass) An/P G
Seeds sown early to mid-spring in their flowering positions.

HORMINUM (Dragon-mouth) A
Division early or mid-spring, replanting portions immediately in permanent positions.
Seeds sown in early or mid-spring, very shallowly in an outdoor seed bed.

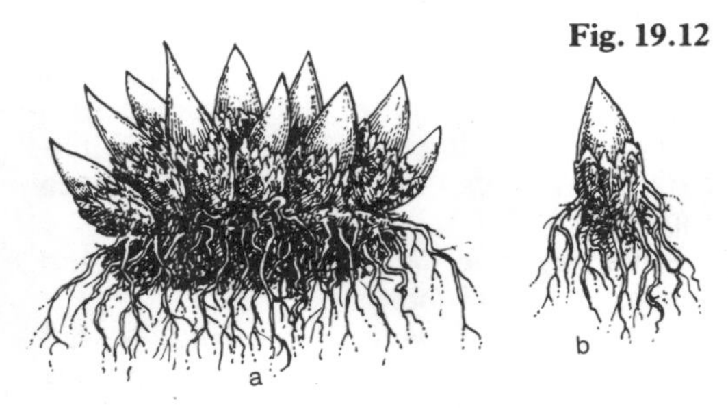

Fig. 19.12 Clumps of hosta (plantain lily) (*a*) can be split down to single buds with roots attached (*b*) in spring if many new plants are required. Pot individually and establish in a shaded garden frame.

HOSTA (Plantain lily) P
Division early to mid-spring, replanting the portions immediately. If many new plants are required clumps can be split down to single buds with roots attached. Careful use of a knife is called for here. These divisions can be potted and established in a shaded garden frame before planting in permanent positions (Fig. 19.12).
Seeds sown in mid-spring in a shaded garden frame.

HOTTONIA (Water violet) Aq
Division between mid-spring and early summer, replacing the portions immediately in the water.

HOUSTONIA (Bluets) P/A
Division early to mid-spring, replanting the portions immediately in their permanent positions.

HOUTTUYNIA P
Division early to mid-spring, replanting the portions immediately in their flowering positions.
Seeds sown in spring in a garden frame, keeping the compost moist.

HUMULUS (Hop) Cl P
Climbing herbaceous perennials
Division during early or mid-spring, replanting the portions immediately.
Seeds sown during mid-spring in a garden frame.

HUTCHINSIA A
Division during early or mid-spring, replanting the portions immediately in flowering positions.
Seeds sown mid- to late winter in pots and germinated outdoors or in a well-ventilated garden frame.

HYACINTHUS (Hyacinth) B
Bulblets removed early to mid-autumn (planting time) and replanting in a nursery bed to grow on to flowering size.
Scooping and scoring to produce bulbils, carried out in late summer or early autumn in a temperature of 21°C (70°F).
Seeds sown as soon as ripe in pots and germinated outdoors or in a garden frame.

HYDRANGEA S (inc Cl S)
Semi-ripe cuttings mid- to late summer, rooted in a heated propagating case, 15.5°C (60°F) or garden frame. Alternatively, cuttings taken in late summer or early autumn can be rooted in the open ground.
Softwood cuttings spring or early summer, rooted in a heated propagating case, 15.5°C (60°F) or garden frame.
Serpentine layering for the climbing species, between mid-spring and late summer, outdoors.
Seeds sown during early or mid-spring in a heated propagating case.

HYDROCHARIS (Frogbit) Aq
Rooted runners are detached in the period mid-spring to early summer and dropped into the water.
Winter resting buds sink to the bottom of the pool in autumn and arise again in spring to develop into new plants.

HYPERICUM (St John's wort) S/P/A
Semi-ripe cuttings for shrubs, early to mid-autumn, rooted in a garden frame or under a low polythene tunnel.
Softwood cuttings for *H. calycinum* in mid-spring, rooting them in a heated propagating case, 15.5°C (60°F).
Soft or semi-ripe cuttings for perennials and alpines, early to mid-summer, heated propagating case with gentle warmth, or garden frame.
Division for perennials and alpines, some shrubs such as *H. calycinum,* early to mid-spring, replanting the portions immediately in their permanent positions.
Seeds sown early to mid-spring, germinating them in a heated propagating case, 15.5°C (60°F), or garden frame.

HYPOXIS (Star grass) Cm or R P
Division of the clumps when they are dormant, in autumn, replanting the portions immediately in their permanent positions.

HYSSOPUS (Hyssop) P
Soft basal cuttings early to late spring, rooted in a heated propagating case with gentle warmth, or in a garden frame.
Division during early or mid-spring, replanting the portions immediately.
Layering (dropping) in mid-to late spring, outdoors. A plant is lifted and replanted to half its depth, spacing out the stems well. Each or most will then root.
Seeds sown in early spring and germinated in a garden frame. Alternatively sow during mid-spring in an outdoor seed bed.

IBERIS (Candytuft) An/P/A
Seeds of annuals are sown at intervals during early to late spring outdoors where the plants are to flower. May also be sown in early autumn in mild areas, provided the soil is very well drained.
Soft or semi-ripe cuttings for perennials and alpine species, early to mid-summer, rooted in a heated propagating case with gentle warmth, or in a garden frame.

IDESIA T
Seeds sown early to mid-spring in a heated propagating case, 18°C (65°F).
Air or simple layering between mid-spring and late summer, outdoors.
Semi-ripe cuttings in early autumn, heated propagating case, 18–21°C (65–70°F).

ILEX (Holly) T/S
Semi-ripe cuttings between mid-autumn and early winter, heated propagating case, mist propagation unit, 18–21°C (65–70°F), or garden frame. To encourage rooting, the bases of the cuttings should be wounded by removing a thin slice of wood.
Simple layering between mid-spring and late summer, outdoors. Layers take two years to form a good root system.
Seeds sown during early and mid-spring in an outdoor seed bed, after stratification for two winters. Even so the seeds may be slow to germinate, sometimes taking up to two or three years.
Budding can be used for cultivars, mid-to late spring or late summer to early autumn, outdoors. Rootstocks are three-year-old seedlings of *Ilex aquifolium.* 1.2 m (4 ft) high rootstocks can be used to form standards, inserting three buds at the top.

ILLICIUM (Anise tree) S/Sm T
Semi-ripe cuttings during late summer or early autumn, heated propagating case, 18°C (65°F).
Simple layering in spring, outdoors. The layers take about 18 months to form a good root system.

INCARVILLEA P
Division during early or mid-spring, replanting the portions immediately. The clumps are quite difficult to divide as they are tough.
Seeds sown during early or mid-spring in a heated propagating case, 18°C (65°F). They can take eight weeks to germinate.

INDIGOFERA (Indigo) S/P
Semi-ripe cuttings for shrubby species in mid- or late summer, rooting them in a heated propagating case, 18°C (65°F). Alternatively they can be rooted in a garden frame in mid-summer. Overwinter young plants in a cold greenhouse.
Seeds sown during early or mid-spring in a heated propagating case, 15.5°C (60°F).

INULA (Includes elecampane) P
Division during early or mid-spring, replanting the portions immediately in their flowering positions.
Seeds sown in late spring or early summer in an outdoor seed bed.

IONOPSIDIUM (Violet cress) An
Seeds sown in succession from mid-spring to mid-summer, outdoors where the plants are to flower. Seeds can also be sown in early autumn in mild areas, for early flowering the following year. Cover seedlings with cloches over winter.

IPHEION B
Bulblets, which are produced in abundance, can be removed in late summer when the parent bulbs are dormant, and replanted immediately in permanent positions. Bulblets can be kept in a cool place until autumn, but ensure they do not become dry. Clumps of ipheion can be lifted and divided at two- or three-yearly intervals.
Seeds sown in mid- or late winter in pots and germinated outdoors.

IRIS P/B
Division for species which form rhizomes, such as the bearded, beardless and water irises. This should be carried out immediately after flowering, ensuring each division has a rhizome with roots and a fan of leaves. Replant immediately in permanent positions.
Bulblets for species which grow from bulbs, such as dwarf alpine types, and the Dutch, Spanish and English irises. Remove bulblets in early or mid-autumn (planting time) and replant immediately in a nursery bed to grow on to flowering size.
Seeds for all types, sown as soon as ripe in pots and germinated outdoors.

ISATIS (Woad) B
Seeds sown late spring or early summer in an outdoor seed bed.

ITEA (Includes Virginia willow) S
Semi-ripe cuttings taken during late summer or early autumn and rooted in a heated propagating case, 18°C (65°F), or garden frame. To encourage rooting, wound the base of each cutting by removing a thin slice of bark.
Simple layering between mid-spring and late summer, outdoors.

IXIA (Corn lily) Cm
Cormlets can be removed in mid- or late autumn (planting time) and immediately replanted in a nursery bed to grow on to flowering size.
Seeds sown during early spring in a heated propagating case, 15.5°C (60°F). Seedlings take up to three years to start producing flowers.

IXIOLIRION (Ixia lily) B
Bulbets removed and immediately replanted at planting time – mid-autumn, or early spring in cold areas.

JASIONE (Sheep's-bit scabious) P
Division during early or mid-spring, replanting the portions immediately in their flowering positions.
Seeds sown late spring or early summer in an outdoor seed bed.

JASMINUM (Jasmine) Cl/S
Serpentine layering for climbing species, mid-spring to late summer, outdoors. Layers should be well rooted within a year.
Semi-ripe cuttings mid- or late summer, heated propagating case, mist unit, 18–21°C (65–70°F), garden frame or low polythene tunnel.
Hardwood cuttings taken late autumn and inserted in a garden frame.
Seeds sown in autumn and placed in a garden frame to germinate.

JEFFERSONIA A
Division during early or mid-spring, replanting the portions immediately in their permanent positions.
Seeds sown as soon as they are ripe and germinated in a garden frame. One needs patience for this method as seeds can take a year and a half to germinate, and young plants may not flower until three to five years old.

JUGLANS (Walnut) T
Seeds sown during early or mid-spring in an

outdoor seed bed. Store over winter in slightly damp peat, in polythene bags, ensuring cool conditions. Alternatively sow in autumn, but guard against rodents. Seedlings should be protected from spring frosts.
Whip and tongue grafting for cultivars during early spring, outdoors. The rootstock is *J. regia*.

JUNCUS (Rush, bog rush) P
Division between mid-spring and early summer, replanting the portions immediately in their permanent positions. The clumps are tough and may need to be separated with the help of a knife.

JUNIPERUS (Juniper) C
Semi-ripe cuttings taken between early autumn and mid-winter and rooted in a heated propagating case, 18–21°C (65–70°F), or garden frame. Rooting is slow in the latter. Cuttings can also be taken in mid-spring and rooted in a garden frame or, for quicker results, under mist.
Veneer grafting for cultivars, early spring, greenhouse, with bottom heat, 21°C (70°F), and high humidity. A suitable rootstock is the parent species: e.g. *J. chinensis* for *J. chinensis* cultivars. Do not graft dwarf or slow-growing cultivars.
Simple layering between mid-spring and late summer, outdoors. Layers take a year to form a good root system.
Seeds sown during autumn, or early/mid-spring, in an outdoor seed bed.

KALMIA (Includes calico bush and sheep laurel) S
Simple layering between mid-spring and late summer, outdoors. Layers take 18 to 24 months to form a good root system.
Semi-ripe cuttings taken late summer or early autumn and rooted in a heated propagating case, 18°C (65°F). Not an easy method.
Seeds sown early or mid-spring in a heated propagating case, 15.5°C (60°F). They are very fine so should hardly be covered with compost.

KALMIOPSIS S
Semi-ripe cuttings taken in mid- or late summer and rooted in a heated propagating case, 18°C (65°F).
Simple layering between mid-spring and late summer, outdoors. Layers can take up to two years to form a good root system.

KERRIA (Jew's mallow) S
Semi-ripe cuttings early to mid-summer, rooted in a garden frame.
Hardwood cuttings mid- to late autumn, rooted in a garden frame or in the open ground.
Simple layering between mid-spring and late summer, outdoors. Layers should be well rooted within a year.
Division of established plants during early spring, replanting the portions immediately.

KIRENGESHOMA P
Division during early or mid-spring, replanting the portions immediately.
Seeds sown during early or mid-spring in a heated propagating case, 15.5°C (60°F). Only lightly cover the seeds. Germination may be slow.

KNAUTIA P
Division in early or mid-spring, replanting the portions immediately.
Seeds sown in late spring or early summer in outdoor seed bed.

KNIPHOFIA (Red-hot poker, torch lily) P
Division during early or mid-spring, replanting the tufts immediately.
Seeds sown during mid- to late spring in an outdoor seed bed.

KOELREUTERIA (Golden-rain tree) T
Root cuttings taken during early or mid-winter and rooted in a heated propagating case, 15.5°C (60°F). Not a particularly successful method.
Seeds sown during late winter in a heated propagating case, 15.5°C (60°F), after winter stratification. Should germinate in three weeks.

KOLKWITZIA (Beauty bush) S
Softwood cuttings mid- to late spring, rooted in a mist-propagation unit, 18–21°C (65–70°F).

Semi-ripe cuttings mid- to late summer, heated propagating case, 18°C (65°F), or garden frame.

× LABURNOCYTISUS (CYTISUS × LABURNUM) T
Whip and tongue grafting during late winter or early spring, outdoors. The rootstock is *Laburnum anagyroides*.
Budding in mid-summer, outdoors. Rootstock as above.

LABURNUM (Golden-rain tree, golden chain) T
Hardwood cuttings taken late autumn or early winter and rooted in a heated propagating case, 18°C (65°F), or outdoors.
Whip and tongue grafting for cultivars, early spring, outdoors. The rootstock is *L. anagyroides* (common laburnum).
Budding for cultivars, mid-summer, outdoors. Rootstock as above.
Seeds sown early or mid-spring in an outdoor seed bed. Germination usually very good.

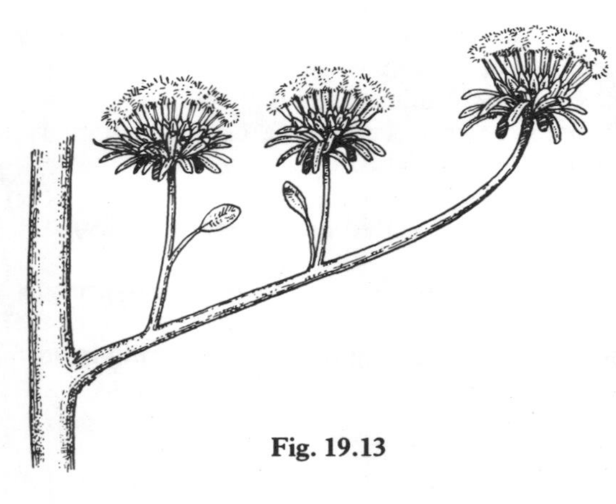

Fig. 19.13

Fig. 19.13 Plants in the *Compositae* family, such as lactuca (flowering lettuce) often have tiny 'parachutes' attached to their seeds to aid distribution. Seeds of lactuca are sown in spring or early summer in an outdoor seed bed.

LACTUCA (Flowering lettuce) P
Seeds sown late spring or early summer in an outdoor seed bed (Fig. 19.13).
Division early or mid-spring, replanting the portions immediately in their permanent positions.

LAGURUS (Hare's-tail grass) An G
Seeds sown during early or mid-spring outdoors in their flowering positions. May also be sown during late summer or early autumn in a garden frame, the seedlings overwintered in this (Fig. 19.14).

Fig. 19.14

Fig. 19.14 Seed heads of lagurus (hare's-tail grass) are very ornamental. They contain copious seeds, which may be sown outdoors in flowering positions during spring, or late summer/early autumn in a garden frame.

LAMARCKIA (Goldentop) An G
Seeds sown during mid- or late spring outdoors in their flowering positions. May also be sown during late summer or early autumn in a garden frame, the seedlings overwintered in this.

LAMIUM (Dead nettle) P
Division early or mid-spring, replanting the portions immediately.
Seeds sown during late spring or early summer in an outdoor seed bed.

LARIX (Larch) C
Seeds sown during early or mid-spring in an outdoor seed bed.
Veneer grafting for cultivars, early spring, greenhouse, with bottom heat, 21°C (70°F), and high humidity. The rootstock is *L. decidua* (common or European larch).

LATHYRUS (Includes sweet pea)
An/P, some Cl
Seeds for the annual sweet pea, sown during early to mid-autumn in a garden frame, the seedlings overwintered in this. Alternatively sow during early or mid-spring outdoors where the plants are to flower. Perennials can also be raised from seeds, sown during early or mid-spring in a garden frame or greenhouse. Before sowing seeds of lathyrus they should be soaked in water for 12 hours, or nicked with a knife,

to ensure quick germination, as they have hard coats.
Division for some perennial species, early or mid-spring, replanting the portions immediately.

LAURUS (Sweet bay, bay laurel) T
Semi-ripe cuttings taken early or mid-autumn and rooted in a garden frame or under a low polythene tunnel.
Seeds sown during early or mid-spring in an outdoor seed bed.

LAVANDULA (Lavender) S
Semi-ripe cuttings taken early or mid-autumn and rooted in a garden frame or under a low polythene tunnel.
Layering (dropping) during mid-spring, outdoors. A plant is lifted and replanted to half its depth, spacing out the stems well. Each or most will then root.
Seeds sown during spring in a heated propagating case, 15.5°C (60°F). Germination may be slow.

LAVATERA (Mallow) P/An/Bi
Seeds for annuals sown during early or mid-spring, or early autumn in mild areas, outdoors where the plants are to flower. Biennials can be sown during late spring or early summer in an outdoor seed bed. Perennials are sown during spring in a heated propagating case, 21°C (70°F).
Semi-ripe cuttings for perennials species, mid-to late summer in a heated propagating case, 15.5°C (60°F).

LAYIA (Tidy tips) An
Seeds sown during early or mid-spring outdoors where the plants are to flower. In mild areas with well-drained soil a sowing can be made in early autumn.

LEDUM S
Division early or mid-spring, replanting the portions immediately.
Simple layering between mid-spring and late summer, outdoors. Layers take a year to form a good root system.
Semi-ripe cuttings taken with a heel during mid- or late summer and rooted in a heated propagating case, 15.5°C (60°F).
Seeds sown during early or mid-spring in a heated propagating case, 15.5°C (60°F), with shade. Sow on the surface of fine sphagnum peat. This is a slow process.

LEIOPHYLLUM (Sand myrtle) S
Simple layering between mid-spring and late summer, outdoors. Layers take a year to form a good root system.
Semi-ripe cuttings, with a slight heel, taken mid-to late summer and rooted in a heated propagating case, 15.5°C (60°F).
Seeds sown during spring in a garden frame. Use a peaty compost.

LEMNA (Duckweed) Fl Aq
Division between mid-spring and early summer, replacing the portions immediately in the water. In practice, division simply consists of scooping up a handful of the tiny plants.

LEONTOPODIUM (Edelweiss) P, inc A
Seeds sown in mid- or late winter, in pots, and germinated outdoors or in a well-ventilated garden frame.
Division of established plants during early or mid-spring, replanting the portions immediately.

LEPTOSPERMUM S/T
Semi-ripe cuttings taken during mid- to late summer and rooted in a heated propagating case or mist unit, 15.5°C (60°F). Wound them at the base by removing a thin slice of bark to expose the wood. Alternatively cuttings can be taken during mid-autumn and rooted in a garden frame. This is a slower process.
Seeds sown during early or mid-spring in a heated propagating case, 18°C (65°F). May be slow to germinate.

LESPEDESA (Bush clover) P/S
Basal cuttings taken as soon as available in spring and rooted in a heated propagating case, 15.5°C (60°F).
Division during early or mid-spring, replanting the portions immediately.
Seeds sown early spring and germinated in a heated propagating case, 15.5°C (60°F).

LEUCANTHEMUM A
Cuttings taken early or mid-summer, nodal

or with a heel, and rooted in a garden frame. Overwinter young plants in this.

LEUCOJUM (Snowflake) B
Bulblets removed in early autumn (planting time), or summer for autumn-flowering species, and immediately replanted in nursery bed to grow on to flowering size.
Seeds sown as soon as ripe, or mid-to late winter, in pots, and germinated outdoors. Seedlings can take four to six years to reach flowering size.

LEUCOTHOE S
Semi-ripe cuttings taken during late summer and rooted in a heated propagating case, mist-propagation unit, 18–21°C (65–70°F) or garden frame.
Simple layering between mid-spring and late summer, outdoors. Layers take a year to form a good root system.
Seeds sown during early spring in a heated propagating case, 15.5°C (60°F). Sow on pure sphagnum peat and keep shaded.

LEVISTICUM (Lovage) P
Division during early or mid-spring, replanting the portions immediately.
Seeds sown during late spring or early summer in an outdoor seed bed.

LEWISIA A
Division during early or mid-spring, replanting the rosettes immediately in their flowering positions.
Seeds sown as soon as ripe, in pots, and germinated outdoors. Storing in a refrigerator for four weeks prior to sowing may help to speed and improve germination.
Leaf cuttings, using whole leaves, early summer, rooting them in a garden frame or greenhouse.
Offsets removed early summer, potted and rooted in a garden frame.

LEYCESTERIA (Himalaya honeysuckle) S
Semi-ripe cuttings late summer or early autumn, rooted in a garden frame.
Hardwood cuttings late autumn or early winter, rooted in a garden frame.
Seeds sown in autumn, or spring after being stored in moist peat, and germinated in a garden frame.

LIATRIS (Gayfeather, blazing star, button snakeroot) P
Division early or mid-spring, replanting the portions immediately in permanent positions.
Seeds sown early spring in a garden frame.

LIBERTIA P
Division during early or mid-spring, replanting the portions immediately in permanent positions.
Seeds sown as soon as ripe, or in early spring, and germinated in a garden frame or greenhouse.

LIGULARIA P
Division early or mid-spring, replanting the portions immediately in flowering positions.
Seeds sown late spring or early summer in an outdoor seed bed.

LIGUSTRUM (Privet) S
Softwood cuttings early summer, rooted in a garden frame.
Semi-ripe cuttings mid-to late summer, heated propagating case, 18°C (65°F), or garden frame.
Hardwood cuttings, especially for *L. ovalifolium* and *L. vulgare*, late autumn or early winter, outdoors.
Simple layering between mid-spring and late summer, outdoors. Layers should be well rooted within a year.
Seeds sown in autumn, or during early or mid-spring after winter stratification, in an outdoor seed bed.

LILIUM (Lily) B
Scales autumn or spring, rooted in a heated propagating case, 21°C (70°F).
Bulblets removed autumn (planting time) and immediately replanted in a nursery bed to grow on.
Bulbils can be 'sown' in pots after collection (as soon as mature) and placed in a garden frame to grow on for a year.
Seeds sown as soon as ripe and germinated in a garden frame. Some species do not produce top growth for many months after sowing, first forming small bulbs under the

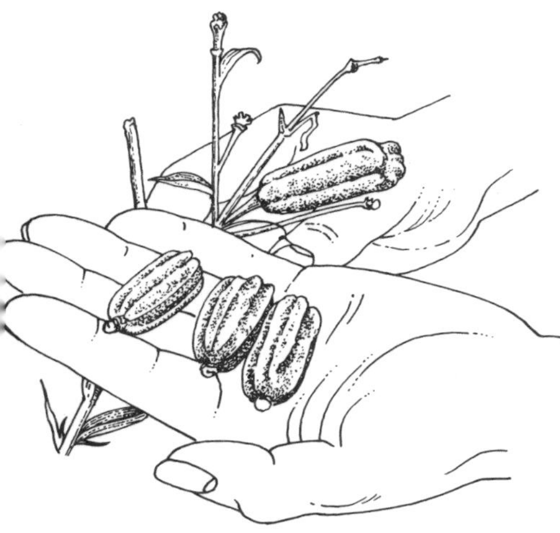

Fig. 19.15

Fig. 19.15 Collect seeds of lilium (lily) as soon as ripe and germinate in a garden frame. The seeds are contained in fat pods, as shown here, and need to be removed from these.

compost. Non-fresh lily seed can take 18-24 months to germinate. Some lilies take up to three years to reach flowering size (Fig. 19.15).

LIMNANTHES (Poached-egg flower) An
Seeds sown early or mid-spring outdoors in flowering positions. Seeds can also be sown in early autumn in mild areas with well-drained soil.

LIMONIUM (Sea lavender, statice) An/P
Seeds of annual species can be sown during mid-spring outdoors where the plants are to flower but plants will be late coming into bloom. Seeds of perennial species are sown during early spring in a heated propagating case, 15.5°C (60°F).
Division for perennial species, early to mid-spring, planting the portions immediately.
Root cuttings for perennial species, early to mid-winter, garden frame.

LINANTHUS An
Seeds sown early or mid-spring outdoors where the plants are to flower. Seeds can alos be sown during early autumn in mild areas, provided the soil is well drained.

LINARIA (Toadflax) An/P
Seeds for annuals, sown early or mid-spring, outdoors where the plants are to flower. Seeds can also be sown during early autumn in mild areas, provided the soil is well drained. Seeds of perennial species can be sown during early or mid-spring and germinated in a garden frame.
Division for perennial species, early to mid-spring, replanting the portions immediately.
Soft cuttings for perennial species, mid-spring to early summer, rooting them in a heated propagating case, 15.5°C (60°F).

LINDERA (Includes spicebush) S/T
Simple layering between mid-spring and late summer, outdoors.
Semi-ripe cuttings mid-autumn, garden frame.

LINUM (Flax) An/P/A/S S
Seeds for annual species, sowing early or mid-spring outdoors where the plants are to flower. Seeds of perennials, alpines and sub-shrubs can be sown during early or mid-spring in a garden frame.
Soft basal cuttings for perennials and alpines, late spring or early summer, heated propagating case, 15.5°C (60°F).
Softwood cuttings for sub-shrubs, late spring or early summer, heated propagating case, 15.5°C (60°F).
Semi-ripe cuttings for perennials and alpines, mid-to late summer, heated propagating case, 15.5°C (60°F) or garden frame.

LIPPIA (Lemon-scented verbena) S
Softwood cuttings as soon as available between mid-spring and early summer, rooting them in a heated propagating case, 18°C (65°F) or under mist.

LIQUIDAMBAR (Sweet gum) T
Seeds sown during autumn soon after collection in an outdoor seed bed. Fresh seeds germinate well in the first spring. Older, imported seeds should be stratified over the first winter then sown in spring. They may not germinate until the next spring. Seedlings should be protected from frosts in the spring.
Air layering between mid-spring and late summer, outdoors.

LIRIODENDRON (Tulip tree) T
Seeds sown in autumn, or stratified over winter and sown in spring. Sow in trays and as soon as germination occurs transfer to a

heated greenhouse. Alternatively sow in outdoor seed bed. Seedlings should be pricked off into pots.
Spliced side veneer grafting for cultivars, early spring, greenhouse, bottom heat, 21°C (70°F), and high humidity. The rootstock is *L. tulipifera.*
Air layering between mid-spring and late summer, outdoors

LIRIOPE (Lilyturf) P
Division early to mid-spring, replanting the tufts immediately in their permanent positions.
Seeds sown as soon as ripe and germinated in a garden frame.

LITHOCARPUS T/S
Seeds sown as soon as ripe in an outdoor seed bed, or in a greenhouse in cold areas.
Air or simple layering between mid-spring and late summer, outdoors.

LITHOSPERMUM (Gromwell) S S/P
Semi-ripe cuttings of side shoots with a heel taken in mid- to late summer and rooted in a garden frame.
Seeds sown during early or mid-spring in a garden frame.

LOBELIA P
Division during early or mid-spring, replanting the portions immediately in their flowering positions.
Seeds sown during early spring and germinated in a heated propagating case, 15.5°C.

LOBULARIA (Sweet alyssum) An
Seeds sown during late winter or early spring and germinated in a heated propagating case, 10–13°C (50–55°F). Alternatively sow between mid- and late spring outdoors where the plants are to flower.

LOMATIA S/T
Semi-ripe cuttings with a heel taken during late summer and rooted in a heated propagating case, 18–21°C (65–70°F).
Seeds sown early or mid-spring and germinated in a heated propagating case, 15.5°C (60°F).

LONAS (African daisy) An
Seeds sown during early or mid-spring outdoors where the plants are to flower.

LONICERA (Honeysuckle) S/Cl
Semi-ripe cuttings for climbers, early or mid-summer, rooted in a heated propagating case, 18°C (65°F). Alternatively take cuttings during mid- to late autumn and root in a garden frame. Shrubby species can also be propagated from semi-ripe cuttings, taken during mid- to late summer and rooted in a heated propagating case or garden frame.
Hardwood cuttings for shrubby species, taken late autumn or early winter and rooted in a garden frame, or outdoors in mild areas.
Serpentine layering for climbers, between mid-spring and late summer, outdoors. Layers should be well rooted within a year.
Simple layering for shrubby species, between mid-spring and late summer, outdoors. Layers should be well rooted within a year.
Seeds for shrubs and climbers, sown as soon as ripe in an outdoor seed bed. Alternatively stratify over winter and sow during early or mid-spring.

LOROPETALUM S
Simple layering between mid-spring and late summer, outdoors.
Semi-ripe cuttings with a heel taken mid-to late summer and rooted in a heated propagating case or under mist, 18–21°C.
Seeds sown during early or mid-spring and germinated in a heated propagating case, 18°C (65°F).
Spliced side veneer grafting during early spring, in greenhouse with bottom heat, 21°C (70°F), and high humidity. Use seedlings of *Hamamelis virginiana* as a rootstock.

LOTUS (Includes bird's-foot trefoil) P
Seeds sown during early or mid-spring in a heated propagating case, 15.5°C (60°F).

LUNARIA (Honesty) Bi/P
Seeds sown late spring or early summer in an outdoor seed bed.
Division for perennial species, early to mid-spring, replanting the portions immediately in flowering positions.

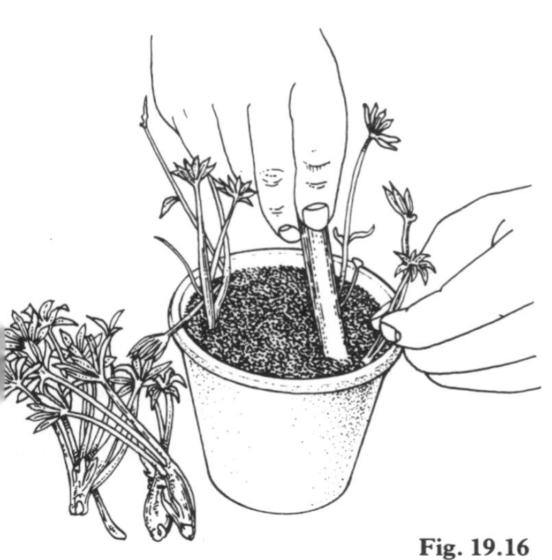

Fig. 19.16

Fig. 19.16 Perennial lupins (lupinus) may be propagated from soft basal cuttings as soon as available in spring, rooting them in a garden frame.

LUPINUS (Lupin) An/P/S S
Seeds for annuals, early or mid-spring outdoors where the plants are to flower; or early autumn if the soil is well drained. Perennials and sub-shrubs can also be raised from seeds, sown during early spring and germinated in a garden frame. Seeds may be soaked in tepid water for a day prior to sowing, to soften the hard seed coats.
Soft basal cuttings for perennials, as soon as available during early to late spring, rooting them in a garden frame (Fig. 19.16).
Semi-ripe cuttings for sub-shrubs, mid- to late summer, rooting them in a garden frame.

LUZULA (Woodrush) P
Division during early or mid-spring, replanting the tufts immediately in their permanent positions.

LYCHNIS (Campion) P
Division early or mid-spring, replanting the portions immediately in their flowering positions.
Seeds sown during late spring or early summer in an outdoor seed bed.
Soft basal cuttings taken during mid- to late spring and rooted in a garden frame.

LYCIUM (Box thorn) S
Suckers with roots attached removed during early or mid-spring and immediately replanted elsewhere.
Simple layering between mid-spring and late summer, outdoors. Layers take a year to form a good root system.
Semi-ripe cuttings with a heel mid- to late summer, rooting them in a heated propagating case, 18°C (65°F).
Seeds sown during early or mid-spring and germinated in a heated propagating case, 15.5°C (60°F).

LYSICHITON (Skunk cabbage) P
Division when flowering is over, or in autumn, replanting the portions immediately in permanent positions. Take care to avoid damaging the roots.
Seeds sown as soon as ripe and germinated in a garden frame. Keep the compost wet by standing the container in water. Seeds may also be sown in a moist seed bed outdoors.

LYSIMACHIA (Loosestrife) P
Division early or mid-spring, replanting the portions immediately.
Basal cuttings as soon as available during mid-or late spring, rooting them in a garden frame.
Seeds sown during late spring or early summer in an outdoor seed bed.

LYTHRUM (Includes purple loosestrife) P
Division early or mid-spring, replanting the portions immediately in flowering positions.
Basal cuttings as soon as available during mid-to late spring, rooting them in a garden frame.
Seeds sown during late spring or early summer in an outdoor seed bed.

MACLURA (Osage orange) T
Root cuttings taken during early winter, potted individually and rooted in a heated propagating case, 15.5°C (60°F).
Seeds sown during autumn or early spring in a heated propagating case, 15.5°C (60°F).

MAGNOLIA T/S
Simple layering between mid-spring and late summer, outdoors. Layers take one or two years to form a good root system.
Air layering between mid-spring and early summer, outdoors.
Softwood cuttings taken between mid-

spring and early summer, slightly wounding them at the base, and rooted in a heated propagating case or mist unit, 21°C (70°F).
Semi-ripe cuttings for the evergreen *M. grandiflora*, taken during late summer, slightly wounded at the base, and rooted in a heated propagating case or mist unit, 21°C (70°F).
Seeds sown preferably during autumn, or spring (in which case store in slightly damp sand), and germinated in a garden frame or greenhouse.

× MAHOBERBERIS (BERBERIS × MAHONIA) S
Semi-ripe cuttings with a slight heel early to mid-autumn, rooted in a garden frame. They take a year to form a good root system.

Fig. 19.17

Fig. 19.17 Leaf-bud cuttings taken during mid-autumn or mid-spring and rooted in heat provide an excellent means of propagating mahonia (including Oregon grape).

MAHONIA (Includes Oregon grape) S
Leaf-bud cuttings taken during mid-autumn or mid-spring and rooted in a heated propagating case or mist unit, 21°C (70°F) (Fig. 19.17).
Seeds sown when ripe during late summer, or early to mid-spring, germinating them in a garden frame or greenhouse.
Division for *M. aquifolium* (Oregon grape), early to mid-spring, replanting the portions immediately in permanent positions.

MALCOLMIA (Virginian stock) An
Seeds sown between early spring and mid-summer, outdoors where the plants are to flower. Sow batches of seeds in succession to ensure continuity of flowers. Seeds may also be sown during early autumn if the soil is well drained.

MALOPE An
Seeds sown during early or mid-spring outdoors where the plants are to flower. Plants may self-sow.

MALUS (Apples, crab apples) T
Whip and tongue grafting for cultivars during early spring, outdoors. Apple cultivars are grafted onto Malling Merton and Malling apple rootstocks (not usually available to amateur gardeners). Cultivars of ornamental crab apples can be grafted onto the same rootstocks, or *M. sylvestris* (common crab apple.)
Budding for cultivars early to mid-summer, outdoors. Rootstocks as above.
Seeds sown during early spring in an outdoor seed bed, after winter stratification.

MALVA (Mallow) P
Soft basal cuttings during mid-spring, as soon as available, rooting them in a garden frame or heated propagating case, 15.5°C (60°F).
Seeds sown early to mid-spring and germinated in a garden frame or heated propagating case, 15.5°C (60°F).

MARGYRICARPUS (Pearl fruit) S
Semi-ripe cuttings with a slight heel from mid- to late summer and rooted in a heated propagating case, 15.5°C (60°F).
Simple layering between mid-spring and late summer, outdoors.
Seeds sown during early and mid-spring and germinated in a garden frame.

MARRUBIUM (Horehound) P
Division during early to mid-spring, replanting the portions immediately in their flowering positions.
Seeds sown early to late spring in an outdoor seed bed.
Cuttings rooted in an outdoor nursery bed during mid-spring.

MARSDENIA Tw S/S S
Soft cuttings taken in mid- or late spring and rooted in a heated propagating case, 15.5°C (60°F).

MATTEUCCIA (Ostrich fern) F
Division during early or mid-spring,

replanting the portions immediately in their permanent positions.
Spores sown as soon as ripe, heated propagating case, 18–21°C (65–70°F); keep moist and shaded, but light needed for germination.

MATTHIOLA (Stock) An/Bi
Seeds sown according to type. Night-scented stock (annual) sown early to mid-spring outdoors where plants are to flower. Brompton stock (biennial) sown early to mid-summer and germinated in a garden frame. East Lothian stock (biennial) and Beauty of Nice stock (annual) sown late winter/early spring and germinated in a heated propagating case, 15.5°C (60°F); alternatively sow mid- to late summer and germinate in a garden frame. Ten-week stock (annual) sown late winter/early spring and germinated in a heated propagating case, 15.5°C (60°F).

MAZUS A
Division early to mid-spring, replanting the portions immediately in flowering positions.
Seeds sown as soon as they are ripe and germinated in a garden frame.

MECONOPSIS (Includes blue and Welsh poppies) P/An/Bi
Division for perennials, early to mid-spring, replanting the portions immediately in their flowering positions.
Seeds sown as soon as they are ripe, as they have a short viability period, germinating them in a garden frame.

MELIOSMA T/S
Seeds sown late winter or early spring and germinated in a heated propagating case, 15.5°C (60°F).
Semi-ripe cuttings taken in mid-summer and rooted in a heated propagating case, 15.5°C (60°F).
Simple layering mid-spring to late summer, outdoors.

MELISSA (Balm) P
Division during early or mid-spring, replanting the portions immediately in their permanent positions.
Seeds sown mid- to late spring in permanent positions.

MELITTIS (Bastard balm) P
Division early or mid-spring, replanting the portions immediately in their flowering positions.
Seeds sown mid- to late spring in permanent positions.

MENTHA (Mint) P/A
Division during early to mid-spring, replanting the portions immediately in their permanent positions.
Soft basal cuttings between mid-spring and early summer, rooting them in a garden frame, or heated propagating case with a temperature of 15.5°C (60°F).

MENTZELIA (BARTONIA) An/P/S
Seeds for annual species, sown during early or mid-spring in flowering positions.

MENYANTHES (Bog bean) Aq
Division between mid-spring and early summer, replanting the portions immediately in their permanent positions.

MENZIESIA S
Semi-ripe cuttings with a heel taken mid- to late summer and rooted in a heated propagating case, 15.5°C (60°F).
Simple layering between mid-spring and late summer, outdoors.
Seeds sown during early or mid-spring and germinated in a heated propagating case, lime-free compost, 13°C (55°F).

MERTENSIA (Smooth lungwort) P
Seeds sown as soon as ripe and germinated in a garden frame.
Division during early spring, replanting the portions immediately in their flowering positions. Not easy to divide and needs to be done carefully.

MESPILUS (Medlar) T
Seeds sown during early or mid-spring in an outdoor seed bed, after stratification for 15 months.
Simple or air layering between mid-spring and late summer, outdoors.
Whip and tongue grafting early spring,

outdoors. Use *Pyrus communis* (pear) or *Cydonia oblonga* (quince) as a rootstock.

METASEQUOIA (Dawn redwood) C
Hardwood cuttings taken late autumn or early winter and rooted in a heated propagating case or under mist, 21°C (70°F). May also be rooted in a garden frame, but a slower method.
Seeds sown during early or mid-spring and germinated in a heated propagating case, 15.5°C (60°F), or garden frame.

MICHELIA T/S
Trees or shrubs
Semi-ripe cuttings taken with a heel in late summer or early autumn and rooted in a heated propagating case, 18°C (65°F).
Simple or air layering between mid-spring and late summer, outdoors.

MILIUM (Millet) P G
Division early or mid-spring, replanting the tufts immediately in final positions.
Seeds sown during late spring or early summer in an outdoor seed bed.

MIMULUS (Monkey flower) P, rarely S S
Seeds sown late winter or early spring in a heated propagating case, 15.5°C (60°F).
Soft cuttings between mid-spring and early summer, rooting them in a heated propagating case, 15.5°C (60°F).
Division early or mid-spring, replanting the portions immediately in flowering positions.

MISCANTHUS P G
Division during early or mid-spring, replanting the clumps immediately in their permanent positions.

MITCHELLA (Partridge berry) Tr P
Division early or mid-spring, replanting the portions immediately in their permanent positions.
Simple layering between mid-spring and late summer, outdoors.
Semi-ripe cuttings mid- to late summer, heated propagating case, 15.5°C (60°F).
Seeds sown in autumn in a garden frame.

MITELLA (Mitrewort) P
Division during early or mid-spring, replanting the portions immediately in their permanent positions.
Seeds sown early or mid-spring in a heated propagating case, 15.5°C (60°F), or garden frame.

MITRARIA Cl S
Division in early or mid-spring, replanting the portions immediately in their flowering positions.
Cuttings taken in spring or summer and rooted in a garden frame.

MOLINIA P G
Division early to mid-spring, replanting the tufts immediately in their permanent positions.
Seeds sown late spring/early summer in an outdoor seed bed.

MOLTKIA P/S S
Semi-ripe cuttings, mid-summer, heated propagating case, 15.5°C (60°F).
Seeds early to mid-spring, garden frame.
Simple layering in early autumn, outdoors.

MOLUCCELLA (Bells of Ireland) An
Seeds sown during late winter or early spring in a heated propagating case, 15.5°C (60°F); alternatively sow outdoors during mid-spring where the plants are to flower.

MONARDA (Bergamot, horsemint) P/A
Division for perennials, early to mid-spring, replanting the portions immediately where they are to flower.
Seeds for annuals, sown during spring in heated propagating case, 15.5°C (60°F), or in a garden frame.

MONTIA P/A
Seeds sown as soon as ripe in garden frame.
Division for perennials, early or mid-spring, replanting the portions immediately in their permanent positions.

MORINA P
Division during early or mid-spring, replanting the portions immediately in their flowering positions. Not always possible as plants have thick tap root.
Seeds sown during early or mid-spring in a

arden frame, or late spring in an outdoor seed bed.

MORISIA P (for rock garden)
Root cuttings early spring, rooted in a garden frame or greenhouse.
Seeds sown spring and germinated in a garden frame.

MORUS (Mulberry) T
Simple layering if possible between mid-spring and late summer, outdoors.
Hardwood cuttings taken in late autumn or early winter and rooted in an outdoor nursery bed.

MUEHLENBECKIA (Wire plant)
Cl or Cr S
Semi-ripe cuttings mid- to late summer rooted in a garden frame. Alternatively mid- to late autumn, garden frame or low polythene tunnel.

MUSCARI (Grape hyacinth) B
Bulblets separated in early autumn (planting time) and replanted elsewhere.
Seeds sown as soon as ripe, or in mid- to late winter, in pots, and germinated outdoors or in a garden frame. Many species self-sow freely. Bulbs take two to four years to reach flowering size.

MUTISIA Cl S
Semi-ripe cuttings taken mid- or late summer and rooted in a heated propagating case or under mist, 18–21°C (65–70°F).
Seeds sown during early or mid-spring and germinated in a heated propagating case.

MYOSOTIDIUM P
Seeds sown during early or mid-spring, individually in small pots, and germinated in a greenhouse or garden frame.

MYOSOTIS (Forget-me-not) An/Bi/P
Seeds sown during mid- to late spring in an outdoor seed bed, or in trays and germinated in a garden frame.
Division for perennials if possible during early or mid-spring, replanting the portions immediately in their flowering positions.
Soft cuttings for perennials, early to late summer, rooting them in a garden frame or greenhouse.

MYRICA (Wax and bog myrtles) S
Semi-ripe cuttings taken with a heel during mid- to late summer and rooted in a heated propagating case, 18°C (65°F).
Seeds sown as soon as ripe and placed in a garden frame. Transfer to heated greenhouse in late winter.
Division during early or mid-spring, replanting the portions immediately in their permanent positions.

MYRICARIA (False tamarisk) S
Hardwood cuttings taken during mid- to late autumn and rooted in an outdoor nursery bed.

Fig. 19.18

Fig. 19.18 Propagate the aquatic myriophyllum (water milfoil) from soft cuttings taken in spring, bunching them together with a strip of lead to act as a weight and dropping them into a pool.

MYRIOPHYLLUM (Water milfoil) Aq
Soft cuttings taken between mid-spring and early summer, bunched together with strip of lead to act as a weight and dropped into pool (Fig. 19. 18).
Division of congested colonies between mid-spring and early summer, replacing the portions immediately in the pool.

MYRRHIS (Myrrh) P
Seeds sown as soon as ripe, preferably in flowering position.
Division during early or mid-spring, replanting the portions immediately in their permanent positions.

MYRTUS (Myrtle) S
Semi-ripe cuttings with a heel mid- to late summer, rooted in a heated propagating case or under mist, 18–21°C (65–70°F). Alternatively, take cuttings in late autumn and root in a garden frame.
Seeds sown early or mid-spring and germinated in a heated propagating case.

NANDINA (Heavenly bamboo, sacred bamboo) S
Semi-ripe cuttings with a heel mid- to late summer, garden frame or heated propagating case, 15.5°C (60°F). Alternatively insert cuttings, heeled or nodal, in a garden frame during late autumn.
Seeds sown as soon as ripe and placed in a garden frame.

NARCISSUS (Daffodil) B
Bulblets removed and replanted in nursery bed during early autumn (planting time).
Seeds sown as soon as ripe or during mid- to late winter, in pots, and germinated outdoors.
Scooping or scoring to produce bulbils, carried out in late summer or early autumn in a temperature of 21°C (70°F).

NEILLIA S
Semi-ripe cuttings taken mid- to late summer and rooted in a heated propagating case, 18°C (65°F).
Hardwood cuttings taken in late autumn and rooted in a garden frame.

NEMOPHILA (Baby blue eyes) A
Seeds sown during early or mid-spring where the plants are to flower. A sowing can also be made in early autumn if the soil is well drained.

NEPETA (Catmint) P
Division between early and mid-spring, replanting the portions immediately in their permanent positions.
Soft basal cuttings taken between mid-spring and early summer, rooting them in a heated propagating case, 15.5°C (60°F).
Seeds sown during early or mid-spring in a heated propagating case, 15.5°C (60°F).

NERINE (Includes Guernsey lily) B
Bulblets removed and replanted in nursery bed during late summer (planting time).
Seeds sown as soon as ripe and germinated in a greenhouse or garden frame.

NERTERA P
Division during early or mid-spring, replanting the portions immediately in their permanent positions. The creeping stems root at the nodes so plants can if desired be split down into small pieces.
Seeds sown during early or mid-spring and germinated in a heated propagating case, 15.5°C (60°F).

NICANDRA (Shoo-fly, apple of Peru) An
Seeds sown during late winter or early spring in a heated propagating case, 15.5°C (60°F). Alternatively sow during mid- to late spring outdoors.

NIEREMBERGIA (Cupflower) P/S S
Division for perennials, early or mid-spring, replanting the portions immediately in their flowering positions. The creeping *N. repens* is very easily divided as it roots at the nodes.
Semi-ripe cuttings for shrubby species, mid- to late summer, rooted in a heated propagating case, 15.5° (60°F).

NIGELLA (Love-in-a-mist) An
Seeds sown during early or mid-spring outdoors where the plants are to flower. A sowing may also be made during early autumn in mild areas, provided the soil is well drained, for earlier flowers the following year. Ideally cover seedlings with cloches over winter.

NOMOCHARIS B
Seeds sown mid-autumn or early spring and germinated in a heated propagating case, 10°C (50°F).
Scales spring or summer, rooted in a heated propagating case, 21°C (70°F).

NOTHOFAGUS (Southern beech) T
Seeds sown during late autumn or early winter in an outdoor seed bed. Alternatively sow in spring after artificial chilling in a fridge for six weeks (mix seeds with moist peat).

Semi-ripe cuttings taken during mid- or late summer and rooted in a heated propagating case or under mist, 18–21°C (60–65°F). Rooting percentage may be low.
Simple or air layering between mid-spring and late summer outdoors.

NOTHOLIRION B
Bulblets removed and immediately replanted outdoors after flowering. Parent bulbs die when flowering is over.
Seeds sown as soon as ripe and germinated in a garden frame.

NOTOSPARTIUM (Southern broom) S
Seeds sown during early or mid-spring and germinated in a heated propagating case, 15.5°C (60°F), or garden frame.

NUPHAR (Yellow pond lily, spatterdock) Aq
Division of the tuberous roots during mid-spring and early summer, replanting the portions immediately in their permanent positions.
Eyes of growing points removed from the roots during early summer and potted to grow on in a garden frame, in containers of water.
Seeds sown as soon as ripe in pots which are just submerged in containers of water. Germinate in a heated greenhouse, temperature 15.5°C (60°F).

NYMPHAEA (Water lily) Aq
Division of the tuberous roots during mid-spring and early summer, replanting the portions immediately in their permanent positions (Fig. 19. 19).
Eyes or growing points removed from the roots during early summer and potted to grow on in a garden frame, in containers of water.
Seeds sown as soon as ripe in pots which are just submerged in containers of water. Germinate in a heated greenhouse, temperature 15.5°C (60°F).

NYMPHOIDES (Fringed water lily) Aq
Division during mid-spring to early summer, replanting the portions immediately in their permanent positions.

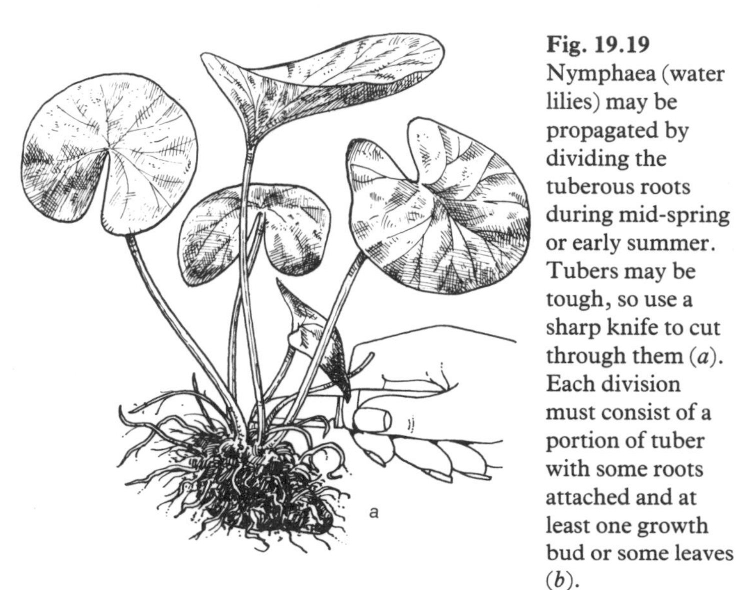

Fig. 19.19

Fig. 19.19
Nymphaea (water lilies) may be propagated by dividing the tuberous roots during mid-spring or early summer. Tubers may be tough, so use a sharp knife to cut through them (*a*). Each division must consist of a portion of tuber with some roots attached and at least one growth bud or some leaves (*b*).

Cuttings of runners taken between mid-spring and early summer and rooted in pots in a pool.

NYSSA (Tupelo tree) T
Seeds sown in early spring, after cold winter stratification, in an outdoor seed bed.
Simple layering between mid-spring and late summer, outdoors. Root into pots, as nyssa dislikes root disturbance. Rooting takes one to two years.
Air layering between mid-spring and late summer, outdoors. Rooting takes one to two years.

OENOTHERA (Evening primrose, sundrops) An/Bi/P
Seeds for all types. Sow annuals and perennials during early or mid-spring in a heated propagating case, 15.5°C (60°F), or in a garden frame. Biennials can be sown in midsummer and germinated in a garden frame.
Division for perennials, early to mid-spring, replanting the portions immediately in their flowering positions.
Soft cuttings for perennials, between mid-spring and early summer, rooting them in a garden frame.

OLEARIA (Daisy bush) S
Semi-ripe cuttings taken during late summer or early autumn and rooted in a garden frame or under a low polythene tunnel.

OMPHALODES (Navelwort) P/A/An
Division during early or mid-spring, replanting the portions immediately in their flowering positions.
Seeds sown during early or mid-spring and germinated in a garden frame. Hardy annuals can be sown during spring in their flowering positions.

ONOCLEA (Sensitive fern) F
Division of the rhizomes during early or mid-spring, replanting the portions immediately in their permanent positions.
Spores sown as soon as ripe, heated propagating case, 18–21°C (65–70°F); keep moist and shaded, but light needed for germination.

ONONIS (Restharrow) P/S S/An
Seeds sown during early or mid-spring and germinated in a heated propagating case, 15.5°C (60°F).
Soft cuttings for perennials, late spring or early summer, rooting them in a heated propagating case, 15.5°C (60°F).
Semi-ripe cuttings for shrubby species, mid- to late summer, heated propagating case, 15.5°C (60°F), or garden frame.

ONOPORDUM (Includes Scotch thistle) Bi
Seeds sown during late spring in an outdoor nursery bed. Alternatively sow direct in flowering positions. Seeds can also be sown during early spring and germinated in a heated propagating case, 15.5°C (60°F).

ONOSMA P/S S
Semi-ripe cuttings early to late summer, rooting them in a heated propagating case, 15.5°C (60°F), or garden frame.
Seeds sown during early or mid-spring in a heated propagating case, 15.5°C (60°F). Germination is slow.

OPHIOPOGON (Lilyturf, mondo grass) P
Division during early or mid-spring replanting the tufts immediately in their permanent positions.

ORIGANUM (Marjoram) P/S S
Seeds sown late winter or early spring and germinated in a heated propagating case 15.5°C (60°F).
Soft basal cuttings taken late spring or early summer and rooted in a heated propagating case, 15.5°C (60°F).
Division where feasible, early to mid-spring, replanting the portions immediately in their permanent positions.

ORNITHOGALUM B
Bulblets removed during early autumn (planting time) and replanted immediately. They take one or two years to reach flowering size.
Seeds sown as soon as ripe, or during mid or late winter, in pots, and germinated outdoors or in a garden frame.

ORONTIUM (Golden club) Aq
Division during mid-spring to early summer, replanting the portions immediately in their permanent positions.
Seeds sown as soon as ripe in pans submerged in shallow water and germinated in a heated greenhouse with a temperature of about 15.5°C (60°F).

OSMANTHUS S
Semi-ripe cuttings taken early or mid autumn and rooted in a heated propagating case or under mist, 18°C (65°F).
Simple layering between mid-spring and late summer, outdoors. Layers take one to two years to form a good root system.

× OSMAREA S
Semi-ripe cuttings taken early to mid-autumn and rooted in a heated propagating case, 18°C (65°F) or garden frame.

OSMARONIA (OEMLERIA) (Indian plum, oso berry) S
Suckers removed with roots between late autumn and early spring and replanted immediately elsewhere.
Simple layering between mid-spring and late summer, outdoors.
Seeds sown in early spring, after cold winter stratification, in an outdoor seed bed.

OSMUNDA (Includes royal fern) F
Division between early and mid-spring, replanting the portions immediately in their permanent positions. Densely packed crowns may not divide successfully.
Spores sown as soon as ripe, heated propagating case, 18–21°C (65–70°F); keep moist and shaded, but light needed for germination.

OSTEOMELES S
Semi-ripe cuttings taken mid- to late summer and rooted in heated propagating case, 18°C (65°F).
Seeds sown early to mid-spring and germinated in heated propagating case, 18°C (65°F).

OSTROWSKIA P
Seeds sown as soon as ripe and germinated in a garden frame. Plants take three to four years to reach flowering size.
Root cuttings taken during early or mid-winter and rooted in a garden frame. But bear in mind established plants do not like disturbance.

OSTRYA (Hop hornbeam) T
Seeds sown as soon as harvested (collect slightly green) during autumn in outdoor seed bed. Purchased seeds sown in spring after winter stratification.

OTHONNOPSIS S P
Semi-ripe cuttings taken during early or mid-summer and rooted in a heated propagating case, 15.5°C (60°F) or garden frame.
Division during early or mid-spring, replanting the portions immediately in their flowering positions.

OURISIA (Mountain foxglove) P
Division during early or mid-spring, replanting the portions immediately in flowering positions.
Seeds sown early or mid-spring and germinated in a cool greenhouse or garden frame.

OXALIS Mainly P, often B
Division between early and mid-spring, replanting the portions immediately in their flowering positions.
Seeds sown early or mid-spring and germinated in a heated propagating case, 15.5°C (60°F).

OXYDENDRUM (Sorrel tree) T
Seeds sown during early or mid-spring in trays of acid sandy peaty compost and germinated in a cool greenhouse or garden frame.
Simple or air layering between mid-spring and late summer, outdoors.

PACHYSANDRA (Mountain spurge) S S/P
Division during early or mid-spring, replanting the portions immediately in permanent positions.
Semi-ripe cuttings late summer/early autumn, rooted in a garden frame.

PACHYSTEGIA S
Semi-ripe cuttings taken during late summer and rooted in heated propagating case, 15.5°C (60°F).

PAEONIA (Peony) P/S
Division for perennials during early or mid-spring, replanting the portions immediately in flowering positions. Peonies dislike disturbance.
Seeds sown as soon as ripe and germinated in a greenhouse or garden frame.
Simple layering for shrubby species, early to mid-spring, outdoors. Layers take up to two years to form a good root system.

PALIURUS S/T
Seeds sown as soon as ripe and placed in a garden frame. Transfer to heated green-

house in spring. Alternatively stratify seeds over winter and sow in spring.
Root cuttings early or mid-winter, heated propagating case, 18°C (65°F), or garden frame.
Simple layering between mid-spring and late summer, outdoors.

PANICUM (Panic grass) An G/P G
Seeds for annuals sown early spring to early summer outdoors where they are to flower.
Division for perennial species during early or mid-spring, replanting the portions immediately in permanent positions.

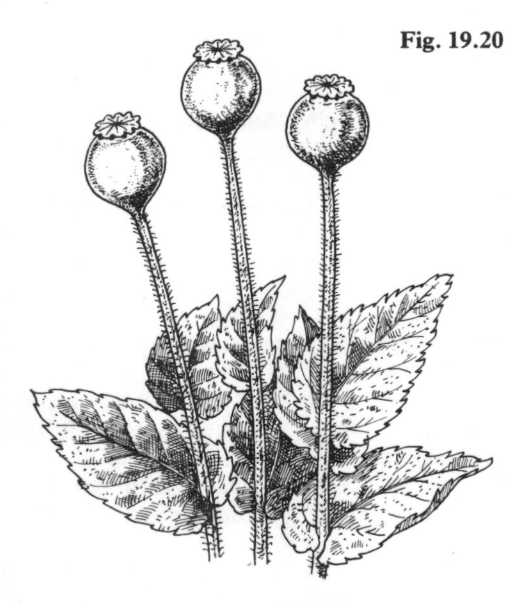

Fig. 19.20 The attractive seed pods of the opium poppy (papaver) contain myriad seeds. This and other annual poppies may be sown outdoors during spring in flowering positions.

PAPAVER (Poppy) An/P/A
Seeds for annuals, such as Shirley, Iceland and opium poppies, sown early to mid-spring outdoors where they are to flower (Fig. 19.20). Seeds of perennials and alpines can be sown in spring and germinated in a garden frame.
Division for perennials early to mid-spring, replanting the portions immediately in flowering positions.
Root cuttings for the perennial *P. orientale*, early to mid-winter, garden frame.

PARAHEBE S S/P
Softwood cuttings mid-spring to early summer, heated propagating case, 15.5°C (60°F).

Semi-ripe cuttings mid-summer to mid autumn, garden frame.

PARAQUILEGIA A
Seeds sown as soon as ripe or mid- to lat winter, in pots, and germinated outdoors.
Division early or mid-spring, replantin portions immediately.

PARNASSIA (Grass of Parnassus) P
Division early or mid-spring, replanting th portions immediately.
Seeds sown autumn or spring in moist soi outdoors.

PAROCHETUS (Shamrock pea, blu oxalis) P
Division early to mid-spring, replanting th portions immediately in permanent posi tion.
Seeds sown early to mid-spring, coo greenhouse or garden frame.
Cuttings mid-summer, garden frame.

PARROTIA (Iron tree) T
Simple layering mid-spring to late summer outdoors. Layers take two years to form good root system.
Seeds sown as soon as ripe and germinate in garden frame. Germination slow and erra tic.

PARROTIOPSIS T
Simple layering mid-spring to late summer outdoors. Layers take two years to form good root system.
Seeds sown as soon as ripe and germinate in a garden frame. Germination slow an erratic.

PARTHENOCISSUS (Includes Virgini creeper) Cl
Softwood cuttings between mid-spring an early summer, rooting them in a heate propagating case or under mist, 18–21°C (65–70°F).
Hardwood cuttings late autumn/earl winter, heated propagating case, 18–21°C (65–70°F).
Serpentine layering between mid-sprin and late summer, outdoors. Layers take on year to root.
Seeds sown during autumn in outdoor see

bed. Purchased seed should be stratified before sowing.

PASSIFLORA (Passion flower) Cl
Leaf-bud cuttings late spring or early summer, rooting them in a heated propagating case, 18°C (65°F).
Serpentine layering between mid-spring and late summer, outdoors. Layers take a year to root.
Seeds sown early or mid-spring and germinated in a heated propagating case, 18–21°C (65–70°F).
Suckers removed early spring and replanted elsewhere.

PAULOWNIA T
Root cuttings early to mid-winter, heated propagating case, 18°C (65°F).
Seeds sown early or mid-spring, bench in heated greenhouse. Surface sow and provide light.
Semi-ripe cuttings of side shoots, with heel, mid-summer, garden frame.

PELTANDRA (Arrow arum) Aq P
Division during mid-spring, replanting the portions immediately in permanent positions.

PELTIPHYLLUM (Umbrella plant) P
Division early or mid-spring, replanting the portions immediately in permanent positions.
Seeds sown early or mid-spring and germinated in garden frame.

PENNISETUM An G/P G
Seeds sown early or mid-spring and germinated in heated propagating case, 15.5°C (60°F).
Division for perennials, early or mid-spring, replanting the tufts immediately in permanent positions.

PENSTEMON (Beard tongue) P/A/S S
Softwood cuttings late spring or early summer, garden frame.
Semi-ripe cuttings late summer or early autumn, garden frame.
Division early or mid-spring, replanting the portions immediately in flowering positions.
Seeds sown early spring and germinated in heated propagating case, 15.5°C (60°F), or garden frame for alpines.

PERIPLOCA (Silk vine) Cl
Simple or serpentine layering late summer/early autumn, outdoors. Allow layers a year to root.
Semi-ripe cuttings mid- to late summer, heated propagating case, 15.5°C (60°F).
Division early or mid-spring, replanting the portions immediately in permanent positions.

PERNETTYA (Prickly heath) S
Semi-ripe cuttings mid- to late summer, heated propagating case, 15.5°C (60°F) or garden frame.
Layering (dropping) mid-spring outdoors. A plant is lifted and replanted to half its depth, spacing out the stems well. Each or most will then root.
Seeds sown as soon as ripe or early to mid-spring and germinated in a garden frame.
Division early or mid-spring, replanting the portions immediately in permanent positions.

PEROVSKIA P
Semi-ripe cuttings taken mid- to late summer and rooted in a heated propagating case, 15.5°C (60°F), or garden frame.

PETASITES (Butter bur, winter heliotrope) P
Division early or mid-spring, or as soon as flowering is over, replanting the portions immediately.

PETROPHYTUM (Rock spiraea) A
Soft cuttings early summer and rooted in a heated propagating case, 15.5°C (60°F).
Seeds sown as soon as ripe or during spring in a cool greenhouse.

PETTERIA S
Seeds sown late winter or early spring and germinated in a heated propgating case, 15.5°C (60°F). Soak seeds for 24 hours in warm water before sowing.

PHACELIA An/P
Seeds of annual species sown during early or mid-spring outdoors where they are to

flower. May also be sown in early autumn in mild areas if soil is well drained.

PHALARIS (Ribbon grass) P G/An G
Division for perennials, early or mid-spring, replanting the tufts immediately.
Seeds for annuals sown early or mid-spring outdoors where they are to flower.

PHELLODENDRON (Cork tree) T
Semi-ripe cuttings taken in mid-summer and rooted in a heated propagating case, 18°C (65°F).
Seeds sown early or mid-spring and germinated on bench of heated greenhouse, after winter stratification.
Root cuttings Remove roots in early winter and store in sand or sphagnum moss over winter in greenhouse. Then prepare and root in heated propagating case, 18°C (65°F).

PHILADELPHUS (Mock orange) S
Shrubs
Softwood cuttings late spring or early summer, heated propagating case, mist unit, 18–21°C (65–70°F) or garden frame.
Semi-ripe cuttings mid- to late summer, garden frame.
Hardwood cuttings early to mid-winter, garden frame.

PHILESIA S
Cuttings taken in mid-summer and rooted in a garden frame.
Suckers removed in mid-spring and immediately replanted in permanent positions.

PHILLYREA (Jasmine box) S/T
Simple layering between mid-spring and late summer, outdoors. Layers take a year to form a good root system.
Semi-ripe cuttings late summer, rooted in garden frame.

PHLOMIS P/S
Semi-ripe cuttings for shrubs, mid- to late summer, heated propagating case, 15.5°C (60°F) or garden frame.
Seeds for perennials and shrubs sown early or mid-spring and germinated in heated propagating case, 15.5°C (60°F).
Division for perennials, early to mid-spring, replanting the portions immediately.

PHLOX P/A
Division for perennial border phlox, early or mid-spring, replanting portions immediately in flowering positions.
Root cuttings for perennial border species, early or mid-winter, heated propagating case, 15.5°C (60°F), or garden frame.
Soft basal cuttings for perennial border species, early or mid-spring, rooting them in a garden frame.
Semi-ripe cuttings for alpines, early to mid-summer, rooting them in a garden frame.
Seeds for alpine species, sown mid- to late winter, in pots, and germinated outdoors.

PHORMIUM (New Zealand flax) P
Division early or mid-spring, replanting the portions, each with a minimum of three or four leaves, immediately in permanent positions.
Seeds sown early or mid-spring and germinated in heated propagating case, 15.5–18°C (60–65°F).

PHOTINIA S/T
Semi-ripe cuttings mid- to late summer, heated propagating case, 18°C (65°F); alternatively early to mid-autumn, rooting them in a garden frame.
Simple layering between mid-spring and late summer, outdoors. Layers take 18 months to form a good root system.
Seeds sown early or mid-spring in an outdoor seed bed, after winter stratification.

PHUOPSIS P
Division early or mid-spring, replanting the portions immediately in flowering positions.
Soft basal cuttings taken in summer and rooted in garden frame.

PHYGELIUS (Cape figwort) S
Semi-ripe cuttings between early and late summer, rooting them in a heated propagating case, 15.5°C (60°F).
Division between early and mid-spring, replanting the portions immediately.
Seeds sown early or mid-spring and germinated in a heated propagating case, 15.5°C (60°F).

PHYLLITIS (Hart's tongue fern) F
Division during early or mid-spring, replanting the portions immediately in permanent positions.
Spores sown as soon as ripe, heated propagating case, 18–21°C (65–70°F); keep moist and shaded, but light needed for germination.
Bulbils for the sterile *P. scolopendrium* 'Crispum'. Use frond bases as cuttings, inserting them in unheated propagating case. They will form bulbils, which develop into plantlets.

PHYLLODOCE (Mountain heath) S
Semi-ripe cuttings between mid- and late summer inserted in a heated propagating case, 15.5°C (60°F), or garden frame.
Simple layering between mid-spring and late summer, outdoors.
Seeds sown early or mid-spring and germinated in a heated propagating case, 15.5°C (60°F), or garden frame.

PHYLLOSTACHYS (Bamboo) P G
Division between early and mid-spring, replanting the portions immediately in their permanent positions.
Cuttings of young rhizomes during late winter or early spring, 30 cm (12 in) long, and planted in nursery bed outdoors or in boxes under glass.

PHYSALIS (Chinese lantern) P
Division between early and mid-spring, replanting the portions immediately in permanent positions.
Root cuttings early or mid-winter, inserting them in a garden frame.
Seeds sown mid-spring in a garden frame or outdoor seed bed.

PHYSOCARPUS S
Semi-ripe cuttings taken early to mid-summer and rooted in a garden frame.
Hardwood cuttings taken late autumn/early winter and rooted in a garden frame.
Seeds sown early or mid-spring in an outdoor seed bed.

PHYSOSTEGIA (False dragonhead, obedient plant) P
Division early to mid-spring, replanting the portions immediately in their permanent positions.
Soft basal cuttings early to mid-spring, rooting them in a garden frame.

PHYTEUMA (Horned rampion) A/P
Division early to mid-spring, replanting the portions immediately in flowering positions.
Seeds sown early spring and germinated in a garden frame.

PHYTOLACCA (Pokeweed) P
Division of the fleshy roots early to mid-spring, replanting the portions immediately in flowering positions.
Seeds sown early to mid-spring in an outdoor seed bed.

PICEA (Spruce) C
Veneer grafting for cultivars, preferably late summer, or early spring, greenhouse, with bottom heat, 21°C (70°F), and high humidity. Rootstocks are *Picea abies* for cultivars of this species, and *P. abies* or *P. pungens* for *P. pungens* cultivars.
Semi-ripe cuttings mid- to late summer rooted in a mist-propagation unit, 21°C (70°F).
Seeds sown early or mid-spring in outdoor seed bed.

PICRASMA T
Seeds if available, sown early spring and germinated in heated propagating case, 15.5°C (60°F).
Root cuttings early winter, heated propagating case, 15.5°C (60°F).

PIERIS S/Sm T
Semi-ripe cuttings mid- to late summer (when extension growth is no longer taking place), and rooted in a heated propagating case or under mist, 18–21°C (65–70°F).
Simple or air layering between mid-spring and late summer, outdoors. Layers can take two years to form a good root system.
Seeds sown early to mid-spring and germinated in a garden frame.

PILEOSTEGIA Cl
Semi-ripe cuttings mid- to late summer, rooted in a heated propagating case, 15.5°C (60°F).

Simple or serpentine layering between mid-spring and late summer, outdoors.

PIMELIA S/S S
Semi-ripe cuttings taken with a heel during mid-summer and rooted in a garden frame.
Seeds sown early spring and germinated in a garden frame.

PIMPINELLA (Aniseed, anise) An
Seeds sown early to mid-spring outdoors where the plants are to flower.

PINGUICULA (Butterwort) Ca P
Seeds sown early to mid-winter, on surface of compost consisting of 50:50 peat and sand. Stand the pot in shallow water in well-ventilated garden frame. Exposure to frost helps germination.
Division for plants with several crowns during early or mid-spring, replanting or repotting the portions immediately.

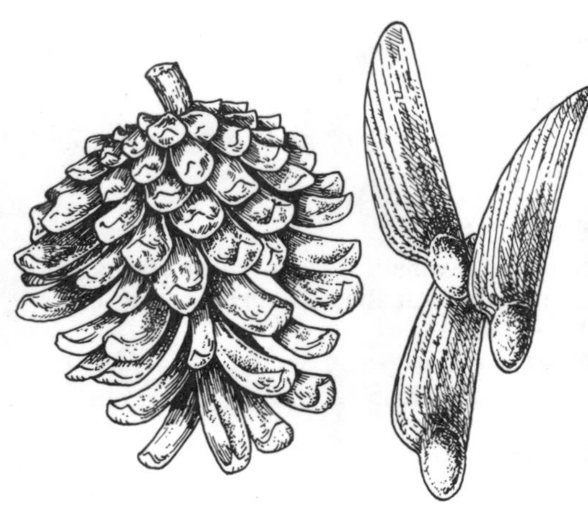

Fig. 19.21

Fig. 19.21 Pinus produce winged seeds in familiar cones. The seeds are sown during early to mid-spring in an outdoor seed bed where they generally germinate readily.

PINUS (Pine) C
Seeds sown early to mid-spring in an outdoor seed bed (Fig. 19.21).
Veneer grafting for cultivars of pinus, early spring, greenhouse, bottom heat, 15.5°C (60°F), and high humidity. The rootstocks are pinus species which should have the same number of needles as the scions. For example, use the two-needled *Pinus sylvestris* for two-needled pinus cultivars.

PIPTANTHUS S
Seeds sown early spring and germinated in a heated propagating case, 15.5°C (60°F). Soak seeds in warm water for one day before sowing, to soften the seed coats.

PISTACIA (Pistachio) T/S
Seeds sown early to mid-spring and germinated in heated propagating case, 15.5°C (60°F). Soak seeds in water for a day before sowing.
Simple layering between mid-spring and late summer, outdoors.

PITTOSPORUM T/S
Semi-ripe cuttings, wounded at the base by removing a thin slice of wood, mid- to late summer, heated propagating case, 18°C (65°F). Alternatively take cuttings in mid-autumn and root in a garden frame or heated propagating case.
Seeds sown early to mid-spring and germinated in heated propagating case.

PLANTAGO (Plantain) P
Division early to mid-spring, replanting immediately in permanent positions.
Seeds sown late spring in an outdoor seed bed.

PLATANUS (Plane) T
Seeds sown early to mid-spring in an outdoor seed bed.
Hardwood cuttings late autumn/early winter, rooting them in a garden frame.
Air layering between mid-spring and late summer, outdoors.

PLATYCARYA T
Air layering between mid-spring and late summer.
Seeds if available sown during spring in outdoor seed bed.

PLATYCODON (balloon flower) P
Seeds sown early spring and germinated in a cool greenhouse or garden frame.
Soft basal cuttings as soon as available in spring, rooting them in a heated propagating case, 15.5°C (60°F).
Division during early or mid-spring replanting the portions immediately. Slow to re-establish.

PLATYSTEMON (Cream cups) An
Seeds sown early or mid-spring outdoors where plants are to flower. Seeds may also be sown in early autumn in mild areas provided the soil is well drained.

POA (Meadow grass, bluegrass) An G/P G
Seeds sown during late spring outdoors where the plants are to grow.
Division for perennial species, early or mid-spring, replanting the tufts immediately in permanent positions.

PODOCARPUS C
Semi-ripe cuttings mid- to late summer, heated propagating case, 18°C (65°F).
Seeds sown early to mid-spring and germinated in heated propagating case, 15.5°C (60°F).

PODOPHYLLUM P
Division early to mid-spring, replanting the portions immediately in flowering positions.
Seeds sown during early spring, or when ripe in summer, and germinated in a garden frame.

POLEMONIUM (Jacob's ladder) P
Division early to mid-spring, replanting the portions immediately in flowering positions.
Seeds sown as soon as ripe, or in spring, outdoor seed bed.

POLYGALA (Milk wort) A
Soft or semi-ripe cuttings early to mid-summer, rooting them in a slightly heated propagating case or garden frame.
Division, if feasible, early to mid-spring, replanting the portions immediately in flowering positions.

POLYGONATUM (Solomon's seal) P
Division early or mid-spring, replanting the portions immediately in their flowering positions.
Rhizomes containing eyes or buds are removed in early or mid-spring and rooted in a heated propagating case, 15.5°C (60°F), or garden frame.
Seeds sown as soon as ripe and germinated in a garden frame.

POLYGONUM (Includes knotweeds and Russian vine) P/S Cl
Division for perennials, early to mid-spring, replanting the portions immediately in their flowering positions.
Semi-ripe cuttings for the trailing perennials, taken late summer and rooted in a garden frame.
Hardwood cuttings for the climbing species like *P. baldschuanicum* (Russian vine), early winter, heated propagating case, 15.5°C (60°F).

POLYPODIUM (Polypody) F
Division early to mid-spring, replanting the portions immediately in their permanent positions.
Spores sown as soon as ripe, heated propagating case, 18–21°C (65–70°F); keep moist and shaded, but light needed for germination.

POLYSTICHUM (Includes shield and holly ferns) F
Division early to mid-spring, replanting the portions immediately in their permanent positions.
Spores sown as soon as ripe, heated propagating case, 10°C (50°F); keep moist and shaded, but light needed for germination.
Bulbils on fronds. Remove fronds during early to mid-autumn, peg down on the surface of cutting compost and place on bench in cool greenhouse.

PONCIRUS (Japanese bitter orange) S
Seeds sown as soon as ripe and germinated in a garden frame.
Semi-ripe cuttings late summer or early autumn, heated propagating case, 18°C.

PONTEDERIA (Pickerel weed) Aq P
Division between mid-spring and early summer, replanting the portions immediately in their permanent positions.

POPULUS (Poplar) T
Hardwood cuttings taken late autumn or early winter and rooted in an outdoor nursery bed.
Suckers removed with roots attached during early or mid-spring and replanted in a nursery bed outdoors.

Fig. 19.22 *Primula denticulata* (drumstick primrose) produces comparatively thick fleshy roots which may be used as cuttings in winter by cutting them into 5 cm (2 in) long sections and rooting in a garden frame or cool greenhouse.

POTAMOGETON (Pondweed) Sub Aq
Cuttings taken between mid-spring and early summer. They can be bunched and held together near the base with a strip of lead. Plant direct in the pool.
Division of colonies between mid-spring and early summer, immediately replanting the portions in the pool.

POTENTILLA (Cinquefoil) S/P/A/An
Semi-ripe cuttings for shrubs, early to mid-autumn, rooting them in a garden frame or under a low polythene tunnel.
Division for perennials and alpines, early to mid-spring, replanting the portions immediately in permanent positions.
Seeds sown during early or mid-spring and germinated in a garden frame.
Soft basal cuttings for perennials, mid-spring, rooting them in a garden frame.

POTERIUM (Burnet) P
Division during early or mid-spring, replanting the portions immediately in their flowering positions.

PRATIA A
Alpines
Division early to mid-spring, replanting portions immediately in permanent positions.
Layering during spring, outdoors.
Semi-ripe cuttings early to mid-summer, rooted in heated propagating case 15.5°C (60°F).
Seeds sown early spring and germinated in garden frame.

PRIMULA P/A
Division for perennials and alpines (such as primroses, cowslips, auriculas and candelabra primulas), early or mid-spring, or immediately after flowering if they bloom in spring. Replant the portions immediately in their flowering positions.
Seeds for perennials and alpines, best sown as soon as ripe and germinated in a garden frame or cool greenhouse. Seeds of polyanthus, which are generally grown as biennials for bedding, can be sown between mid-winter and early spring and germinated in a heated propagating case, 15.5°C (60°F) or during mid-spring or late summer,

Fig. 19.22

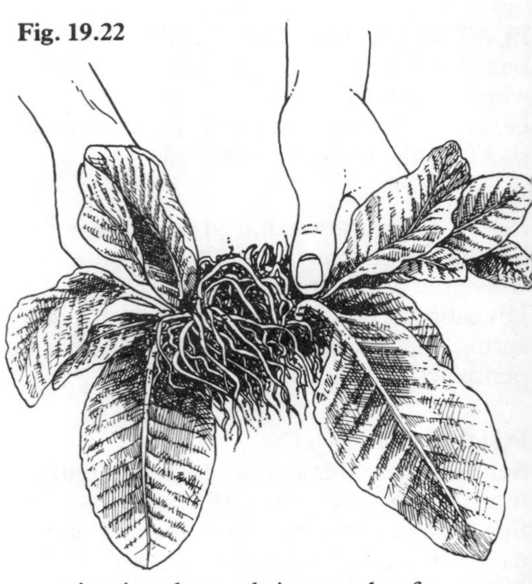

germinating the seeds in a garden frame.
Root cuttings for the perennial *Primula denticulata* (drumstick primrose), early to mid-winter, rooting them in a garden frame or cool greenhouse. (Fig. 19.22).

PRUNELLA (Self heal) P
Division of the mats during early or mid-spring, replanting the portions immediately in their permanent positions.

PRUNUS (Includes cherries, peaches and laurels) T/S
Whip and tongue grafting for cultivars of the trees, such as almonds, apricots, cherries, damsons, nectarines, peaches and plums, early spring, outdoors. Rootstocks: plum rootstock Brompton or St Julian A for almonds, apricots, nectarines and peaches; plum rootstocks Myrobalan B, St Julian A or 'Pixy' for plums and damsons; and *Prunus avium* (wild cherry) or 'Colt' for cherries.
Budding for the above, mid- to late summer, outdoors. Rootstocks the same.
Softwood cuttings for certain deciduous shrubs such as *P. glandulosa*, *P. pumila* (dwarf cherries), *P. triloba*, *P. tenella*, mid- to late spring, heated propagating case or mist unit, 18–21°C (65–70°F).
Semi-ripe cuttings for certain shrubs, particularly *P. laurocerasus* (cherry laurel) and *P. lusitanica* (Portugal laurel), early to mid-autumn, garden frame or low polythene

tunnel (also outdoors for cherry laurel).
Hardwood cuttings for *P. cerasifera* and cultivars, late autumn, outdoor nursery bed.
Seeds for all prunus, sown early to mid-spring in an outdoor seed bed after winter stratification.

PSEUDOLARIX (Golden larch) C
Seeds sown during autumn or early spring in an outdoor seed bed.

PSEUDOTSUGA (Includes Douglas fir) C
Seeds sown autumn or winter in an outdoor seed bed.

PTELEA (Hop tree) S/T
Seeds sown autumn, or in late spring after stratification, outdoor seed bed.
Simple layering between mid-spring and late summer, outdoors.

PTEROCARYA (Wingnut) T
Seeds sown as soon as ripe in outdoor seed bed. Bought seed should be sown as soon as possible after first soaking in water. Protect seedlings from frost.
Suckers removed with some roots attached during early spring and immediately replanted elsewhere.
Simple or air layering between mid-spring and late summer outdoors.

PULMONARIA (Lungwort) P
Division early to mid-spring or immediately after flowering, replanting the portions in permanent positions.
Seeds sown mid-to late spring in outdoor seed bed.

PULSATILLA (Includes pasque flower) P
Seeds sown as soon as ripe and germinated in a garden frame.
Root cuttings for *Pulsatilla vulgaris* (pasque flower), early to mid-winter, cool greenhouse.

PUSCHKINIA (Squill) B
Bulblets removed and replanted elsewhere in early autumn (planting time).
Seeds sown as soon as ripe, germinating them in a garden frame. Seedlings take two or three years to reach flowering size.

PYRACANTHA (Firethorn) S
Semi-ripe cuttings late summer, rooted in heated propagating case or under mist, 18–21°C (65–70°F).
Seeds sown early to mid-spring in outdoor seed bed, after winter stratification.

PYRETHRUM (CHRYSANTHEMUM) P
Division early spring, or after flowering, replanting the portions immediately. Discard any woody pieces.
Soft basal cuttings taken as soon as ready in spring and rooted in a heated propagating case, 15.5 °C (60 °F).
Seeds sown early spring and germinated in heated propogating case, 15.5 °C (60 °F).

PYRUS (Pear) T
Whip and tongue grafting early spring outdoors, for cultivars. Rootstock is usually Quince A or C, selections of *Cydonia oblonga* (common quince). But *Pyrus communis* (common pear) can also be used.
Budding mid-summer, outdoors. Rootstocks as above.
Seeds sown early to mid-spring in outdoor seed bed, after winter stratification.

QUERCUS (Oak) T
Seeds sown as soon as ripe in an outdoor seed bed. If sowing is not undertaken until spring, store seeds over winter in a cool place, in polythene bags containing moist peat.
Spliced side veneer grafting for cultivars, early spring, greenhouse, with bottom heat, 21°C (70°F), and humidity. Rootstocks are seedlings of the species.

Fig. 19.23

Fig. 19.23 Ramonda (rosette mullein) may be propagated from leaf cuttings in summer, rooting them in a garden frame or cool greenhouse. Use entire leaves complete with stalks.

RAMONDA (Rosette mullein) A
Leaf cuttings early to mid-summer, rooting them in a garden frame or cool greenhouse (Fig. 19.23).

Seeds sown during early spring and germinated in a garden frame.

RANUNCULUS (Includes buttercup and crowfoot) P/A/Aq
Division for perennials and alpines during early or mid-spring, for aquatic species between mid-spring and early summer, replanting the portions immediately in their permanent positions.
Seeds for perennials and alpines, sown as soon as ripe, or in late winter, and germinated in a garden frame.

RAOULIA P (for rock gardens)
Division of the mats in late summer, replanting the portions immediately in permanent positions. Small pieces can also be teased off from around the edge of a mat in spring and potted to grow on.

RESEDA (Mignonette) An
Seeds sown early or mid-spring outdoors where the plants are to flower. Seeds may also be sown during early autumn in mild areas with well-drained soil, for earlier flowering the following year.

RHAMNUS (Buckthorn) S/Sm T
Semi-ripe cuttings mid- to late summer, heated propagating case, 18 °C (65°F).
Seeds sown autumn, or early to mid-spring after winter stratification, in an outdoor seed bed.

RHAPHIOLEPIS S
Semi-ripe cuttings late summer, rooting them in a heated propagating case, 18–21°C (65-70°F).
Seeds sown as soon as ripe, autumn, and germinated in a heated propagating case, 18°C (65°F).

RHAZYA P
Seeds sown during early spring and germinated in a heated propagating case, 15.5°C (60°F).
Division early to mid-spring, replanting the portions immediately in flowering positions.

RHEUM (Rhubarb, including ornamental species) P
Division early to mid-spring, replanting the portions immediately in permanent positions. Each must contain at least one growth bud.
Seeds sown mid-spring in an outdoor seed bed.

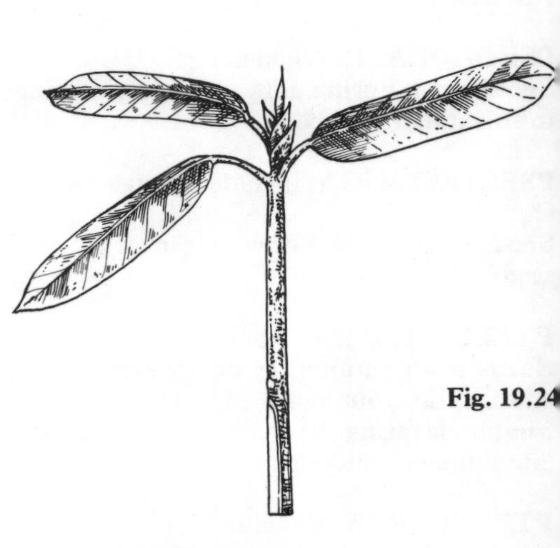

Fig. 19.24
Evergreen species and hybrids of rhododendron may be propagated from semi-ripe cuttings during autumn, rooting them in heat. First wound the base of cuttings by removing a sliver of bark to encourage rooting.

RHODODENDRON S/T
Semi-ripe cuttings for evergreen species and hybrids, early to mid-autumn, heated propagating case or mist unit, 21°C (70°F). Cuttings of evergreen azaleas, and dwarf rhododendrons, can be taken between early and late summer. Wound base of all cuttings by removing a sliver of bark (Fig 19.24).
Softwood cuttings for deciduous species and hybrids, including deciduous azaleas, mid-spring to early summer, heated propagating case or mist unit, 21°C (70°F). Lightly wound base of cuttings by removing a sliver of bark.
Simple layering for all rhododendrons and azaleas, between mid-spring and late summer, outdoors. Layers take two years to form a good root system.
Saddle or spliced side grafting for evergreen hybrids, early spring, greenhouse, with bottom heat, 18°C (65°F), and high humidity. Rootstocks are *Rhododendron ponticum* and *R.* 'Cunningham's White'. The latter is now preferred, as the former is too vigorous and produces suckers.
Seeds for species sown early to mid-spring on the surface of acid peaty compost, germinating them in a heated propagating case, 15.5°C (60°F).

RHODOHYPOXIS A
Division in spring, replanting or potting the portions immediately. Division must be carried out while plants are in growth.
Seeds sown during early spring in a cool greenhouse or garden frame.

RHODOTYPOS (White kerria) S
Semi-ripe cuttings early to mid-summer, heated propagating case, 18–21°C (65–70°F).
Hardwood cuttings early to late autumn, garden frame.

Fig. 19.25

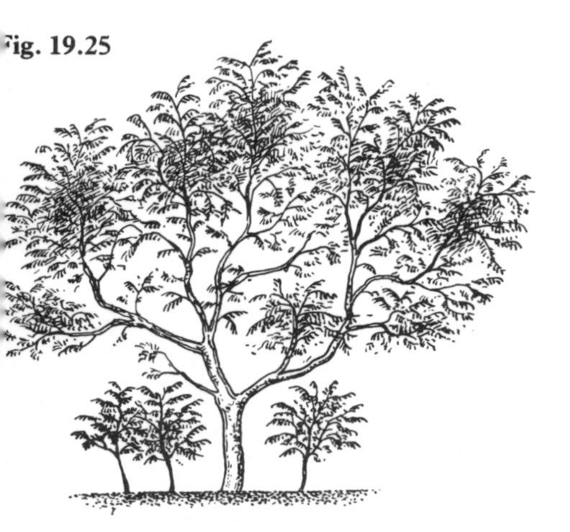

RHUS (Sumach) S/T
Root cuttings taken early to mid-winter and rooted in a garden frame.
Suckers removed with some roots attached, early to mid-spring, replanting immediately elsewhere (Fig. 19.25).

RIBES (Currants, including ornamental, and gooseberries) S
Shrub
Hardwood cuttings for black, red, white and flowering currants, and gooseberries, late autumn to early winter, outdoor nursery bed or garden frame. With red currant and gooseberry cuttings all buds except for three or four at the top should be cut out. This ensures the resultant bushes grow on a short leg, instead of producing growth from below soil level.

ROBINIA (False acacia, locust) T/S
Root cuttings taken in early or mid-winter and rooted in a garden frame.
Whip and tongue grafting for cultivars in early spring outdoors. The rootstocks are seedlings of the type; for instance, *R. pseudoacacia* for cultivars of that species.
Spliced side grafting late winter, greenhouse, with bottom heat, 21°C (70°F), and high humidity. Rootstocks as above.
Suckers with roots attached, removed and replanted elsewhere during early or mid-spring.
Seeds sown during early spring and germinated in an outdoor seed bed. First soak the seeds in warm water for two days to soften the hard seedcoat.

RODGERSIA P
Division early to mid-spring, replanting the portions immediately in their permanent positions.
Seeds sown early to mid-spring and germinated in a garden frame.

ROMNEYA (California tree poppy) P
Root cuttings taken early to mid-winter and rooted in a heated propagating case, 13–15.5°C (55–60°F).
Suckers removed with some roots attached during mid-spring and replanted elsewhere.
Seeds sown late winter or early spring and germinated in a heated propagating case, 13–15.5°C (55–60°F). Pot seedlings individually and establish in heated greenhouse. Overwinter in garden frame.

ROMULEA Cm
Offsets removed and replanted elsewhere during early autumn (planting time).
Seeds sown during early or mid-spring and germinated in a cool greenhouse.

ROSA (Rose) S/Cl
Budding for the hybrids (large-flowered, cluster-flowered, shrubs, climbers, ramblers and miniatures), between early summer and early autumn, outdoors. Rootstocks: *Rosa canina* (wild briar, dog rose), which suckers badly; improved commercial selections of this stock; *R. dumetorum* 'Laxa', a virtually thornless stock; and *R. rugosa* 'Hollandica' for standard roses. The latter two can be raised from hardwood cuttings.
Hardwood cuttings for strong-growing roses like ramblers and some rootstocks,

Fig. 19.25 Rhus (sumach) produce suckers from the roots. These can be used for propagation by removing them with some roots attached in early to mid-spring and planting them elsewhere.

Fig. 19.26 Roses (rosa) produce their seeds in familiar hips. Species may be sown during spring in an outdoor seed bed, after 12 to 18 months' stratification. Even so the germination rate may be poor.

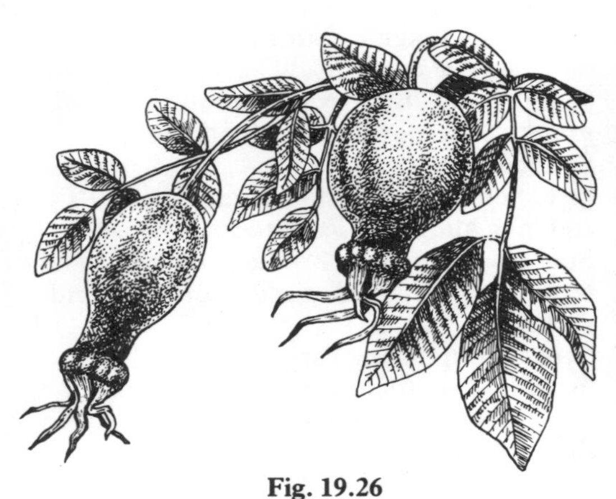

Fig. 19.26

taken late autumn or early winter and rooted in an outdoor nursery bed.
Seeds for species sown during early or mid-spring in an outdoor seed bed, after 12 to 18 months' stratification. Germination rate may be poor (Fig. 19.26).

ROSCOEA P
Division during spring as soon as growth is underway, replanting the portions immediately in their permanent positions. Root disturbance is disliked.
Seeds sown as soon as ripe and germinated in a garden frame.

ROSMARINUS (Rosemary) S
Semi-ripe cuttings taken early or mid-autumn and rooted in a garden frame or under a low polythene tunnel. Alternatively take cuttings mid-to late summer.

RUBUS (Bramble) S
Tip layering for blackberries and loganberries, mid-to late summer, outdoors. Layers will be well rooted by autumn.
Leaf-bud cuttings for blackberries and loganberries, late summer, rooting them in a garden frame.
Suckers with roots attached for raspberries, removed during late autumn or early winter and replanted in permanent positions.
Division for ornamental brambles which form colonies, such as *R. odoratus, R. spectabilis* and *R. parviflorus,* early to mid-spring, replanting the portions immediately in their permanent positions.
Root cuttings for the ornamental bramble *R. cockburnianus,* early to mid-winter garden frame or cool greenhouse.
Simple layering suitable for many species summer, outdoors.
Semi-ripe cuttings for most species, lat summer or early autumn, rooting them in garden frame.

RUDBECKIA (Coneflower) P/An
Division for perennials, early to mid-spring replanting the portions immediately in thei flowering positions.
Seeds for annuals and perennials, sow early spring and germinated in a heate propagating case, 15.5°C (60°F). Seeds o perennials may also be sown during lat spring in an outdoor seed bed.

RUMEX (Dock, sorrel) P
Seeds sown in early spring in an outdoo seed bed.

RUSCUS (Butcher's broom) S
Division early to mid-spring, replanting th portions immediately in their permanen positions.
Seeds sown as soon as ripe and germinate in a garden frame. Germination slow, as i growth of seedlings.

SAGINA (Pearlwort) P
Division for ornamental kinds such as *S subulata* 'Aurea', autumn, potting th portions and overwintering in a garde frame; or spring, replanting the portion immediately in permanent positions.

SAGITTARIA (Arrowhead) Aq P
Division between mid-spring and earl summer, replanting the portion immediately in their permanent positions.

SALIX (Willow, sallow) T/S
Hardwood cuttings late autumn or earl winter, outdoors.

SALVIA (Sage) P/S S/An
Division for perennials, between early an mid-spring, replanting the portion immediately in their flowering positions.
Semi-ripe cuttings for sub-shrubs, early t mid-summer, rooting them in a heate propagating case, 15.5°C (60°F).

The popular carnivorous perennial *Dionaea muscipula* (Venus fly trap) is most easily increased by division but alternative propagation methods are seeds and leaf cuttings.

Erythrina crista-galli (coral tree) grows remarkably quickly from seeds and will soon be producing its spectacular pea flowers. Sow early to mid-spring in heat.

The perennial *Mimosa pudica* (sensitive plant), usually grown as an annual, is easily raised from seeds sown during spring in heat.

The long-flowering evergreen shrub *Pachystachys lutea* (lollipop plant) can be raised from cuttings during spring or summer, rooting them in a heated propagating case.

Portulaca grandiflora (sun plant) is an annual bedding plant which is raised from seeds, sown either under glass or outdoors in flowering position.

Seedlings of *Strelitzia reginae* (bird of paradise flower) can take at least five years to produce their spectacular flowers. Can also be propagated by division.

The annual asparagus pea (*Lotus edulis*), whose young pods are cooked and eaten whole, can be sown during spring outdoors or under glass.

Celeriac, which has swollen celery-flavoured roots used as a winter vegetable, can be raised from seeds in spring, either under glass or outdoors.

Seeds of Florence fennel are sown during spring, preferably in a greenhouse. It produces bulb-like leaf bases which are used as a vegetable.

Okra, a tender annual, is grown for its edible pods. In cool temperate climates it has to be raised and grown in a greenhouse.

Propagate the giant moisture-loving perennial, *Gunnera manicata*, by division, removing small lateral crowns with roots attached during mid-spring. Start them off in pots.

Seeds of sub-shrubs and perennials may be sown during early or mid-spring and germinated in a heated propagating case, 15.5 °C (60 °F). Seeds of hardy annuals may be sown n the same period, but outdoors where the plants are to flower.

SAMBUCUS (Elder) S
Semi-ripe cuttings mid- to late summer, heated propagating case, 18°C (65°F).
Hardwood cuttings late autumn or early winter, garden frame or outdoor nursery bed.
Seeds sown early to mid-spring in an outdoor seed bed.

SANGUINARIA (Bloodroot) P
Division as soon as flowering is over, late pring or early summer, replanting the portions immediately in their flowering positions. The fleshy roots are easily damaged so handle plants carefully.

SANGUISORBA (Burnet) P
Division early to mid-spring, replanting the ortions immediately in their flowering ositions.
eeds sown mid-spring, garden frame.

ANTOLINA (Lavender cotton) S
emi-ripe cuttings taken between early and mid-autumn and rooted in a garden frame or nder a low polythene tunnel.
Hardwood cuttings taken late autumn/early winter and rooted in an outdoor nursery ed.

ANVITALIA (Creeping zinnia) An/P
eeds for the annual *S. procumbens,* early to mid-spring where the plants are to flower. lternatively sow during early autumn in mild areas with well-drained soil for early owers the following year.

APIUM S/T
hrubs and trees
emi-ripe cuttings taken during summer nd rooted in a heated propagating case.

APONARIA (Soapwort) An/Bi/P
ivision for perennials, early to mid-spring, planting the portions immediately in their owering positions.
Seeds for annuals, early to mid-spring, in flowering positions. Alternatively sow during early autumn in mild areas with well-drained soil, for earlier flowers the following year.
Softwood cuttings for perennials, especially rock-garden kinds, early summer, heated propagating case, 15.5°C (60°F), or garden frame.

SARCOCOCCA (Sweet box) S
Division early to mid-spring, replanting the portions immediately in their permanent positions.
Semi-ripe cuttings mid-autumn, rooting them in a garden frame. Alternatively take cuttings in mid-spring and root under mist, 21°C (70°F) – a quicker process.
Seeds sown early to mid-spring in an outdoor seed bed.

SASA (Bamboo) P G
Division during early or mid-spring, replanting the portions immediately in their permanent positions.

SASSAFRAS T
Suckers removed with some roots attached between autumn and spring and immediately replanted elsewhere.
Root cuttings taken during early or mid-winter, garden frame or cool greenhouse.
Seeds sown late winter or early spring and germinated in a heated propagating case, 15.5°C (60°F).

SAXIFRAGA (Saxifrage) A
Division early to mid-spring, or after flowering, replanting the portions immediately in their flowering positions.
Cuttings of single rosettes taken early to mid-summer and rooted in a cool greenhouse or garden frame.
Seeds sown mid- to late winter in pots and germinated outdoors.

SCABIOSA (Scabious, pincushion flower) P/An
Division for perennials, early to mid-spring, replanting the portions in their flowering positions.
Seeds for perennials, sown late spring or early summer in an outdoor seed bed. Seeds

for annuals, sown early to mid-spring outdoors where the plants are to flower. Alternatively sow annuals during early autumn in mild areas with well-drained soil.
Soft basal cuttings early to mid-spring, rooting them in a garden frame.

SCHISANDRA Tw S
Semi-ripe cuttings late summer, heated propagating case, 15.5–18°C (60–65°F). Alternatively root under mist, 18–21°C (65–70°F) in spring or summer.
Seeds sown as soon as ripe, generally mid-autumn, and germinated in a garden frame.
Serpentine layering between mid-spring and late summer, outdoors. Layers will be well rooted after a year.

SCHIZOPHRAGMA Cl
Semi-ripe cuttings taken late summer and rooted in a heated propagating case, 15.5°C (60°F).
Simple or serpentine layering between mid-spring and late summer, outdoors. Layers should be well rooted within a year.
Seeds sown during early or mid-spring and germinated in a heated propagating case, 18°C (65°F).

SCHIZOSTYLIS (Kaffir lily) R P
Division during mid-spring, replanting the portions immediately in their flowering positions. Ensure each clump contains about six young shoots.
Seeds sown during mid- or late spring and germinated in a heated propagating case.

SCIADOPITYS (Umbrella pine) C
Seeds sown early winter to ensure adequate chilling prior to germination, outdoor seed bed.

SCILLA (Squill) B
Bulblets removed and replanted elsewhere during late summer or early autumn (planting time).
Seeds sown as soon as ripe, or during mid- to late winter, in pots and germinated out of doors. Bulbs take from three to five years to reach flowering size.

SCIRPUS (Club rush) Wetland P
Division during mid-spring to early summer, replanting the tufts immediately in their permanent positions.

SCROPHULARIA (Figwort) P
Division early to mid-spring, replanting the portions immediately in their flowering positions.
Soft basal cuttings taken as soon as available in late spring or early summer and rooted in a cool greenhouse or garden frame.

SCUTELLARIA (Skull cap) A
Division during early or mid-spring, replanting the portions immediately in their flowering positions.
Seeds sown during mid-spring and germinated in a garden frame.

SEDUM (Stonecrop) P/A/An
Division during early or mid-spring, replanting the portions immediately in their flowering positions.
Soft cuttings taken early to mid-summer, garden frame or cool greenhouse.
Leaf cuttings taken early to mid-summer and rooted in a garden frame or cool greenhouse. Use whole leaves.
Seeds for the annual species, sown during early or mid-spring in flowering positions. May also self-sow freely. Seeds of perennials may be sown during mid-spring and germinated in a garden frame.

SEMIAQUILEGIA A
Seeds sown during autumn and germinated in a garden frame.

SEMPERVIVUM (Houseleek) A
Division during early or mid-spring, replanting the portions immediately in their permanent positions.
Offsets carefully removed with some roots attached and immediately replanted elsewhere.

SENECIO S/P
Semi-ripe cuttings for shrubs, taken during early to mid-autumn and rooted in a garden frame or under a low polythene tunnel.
Division for perennials, early to mid-spring, replanting the portions immediately in their flowering positions.
Seeds for shrubs and perennials, sown

luring early to mid-spring and germinated n a heated propagating case, 15.5°C (60°F), r garden frame.

EQUOIA (Redwood) C
eeds sown early to mid-spring in an utdoor seed bed.
emi-ripe cuttings taken early to mid-utumn and rooted in a garden frame. hoose upright-growing shoots for cuttings.

EQUOIADENDRON (Giant sequoia, iant redwood) C
eeds sown early to mid-spring in an utdoor seed bed.

ERRATULA P/A
ivision during early or mid-spring, eplanting the portions immediately.
eeds sown during spring and germinated a garden frame.

HIBATAEA (Bamboo) P G
ivision during early or mid-spring, eplanting the portions immediately in their ermanent positions.

HORTIA P
ivision early summer, after flowering, eplanting the portions immediately in their owering positions. Need cossetting to re-stablish sucessfully.
eeds (not often available) sown during id-spring and germinated in a garden ame.

IDALCEA (Greek mallow) P
ivision early to mid-spring, replanting the ortions immediately in their flowering ositions.
oft basal cuttings taken as soon as avail-le in spring and rooted in a heated prop-ating case, 15.5°C 60°F).
eds sown during mid-spring and germi-ted in a garden frame.

DERITIS P/S S
eds sown during and germinated in a rden frame.
mi-ripe cuttings summer, rooted in a rden frame or heated propagating case at .5°C (60°F).

SILENE (MELANDRIUM) (Catchfly, campion) P/A/An
Division for perennials and alpines, early to mid-spring, replanting the portions immediately in their flowering positions.
Soft basal cuttings for perennials and alpines, mid- to late spring, rooting them in a garden frame or cool greenhouse.
Seeds for annuals, sown early to mid-spring outdoors where they are to flower. Alternatively sow during early autumn in mild areas with well-drained soil. Seeds of perennials can be sown in an outdoor seed bed in late spring.

SILYBUM (Milk thistle) Bi
Seeds sown during late spring or early summer outdoors where the plants are to flower.

SISYRINCHIUM (Satin flower) P
Division early to mid-spring, replanting the portions immediately in their flowering positions.
Seeds sown during early autumn or early spring in pots and germinated in a garden frame. Self-sown seedlings may also appear around plants.

SKIMMIA S
Semi-ripe cuttings early to mid-autumn rooted under a low polythene tunnel or in a garden frame.
Semi-ripe cuttings mid-summer rooted in a mist-propagation unit, 21 °C (70 °F).
Simple layering between mid-spring and late summer, outdoors.
Seeds sown early to mid-spring after winter stratification and germinated in a garden frame. A slow method of producing plants.

SMILACINA (False Solomon's seal) P
Division mid-autumn, replanting the portions immediately in their permanent positions.

SOLANUM Scr S
Semi-ripe cuttings mid- to late summer, rooting them in a heated propagating case, 18°C (65°F). Cuttings may also be rooted under mist but remove them as soon as rooted.

SOLDANELLA A
Division during early summer (after flowering), replanting the portions immediately in their permanent positions. Or they can be established in pots in a garden frame and planted out the following spring.
Seeds sown as soon as ripe and germinated in a garden frame.

SOLIDAGO (Golden rod) P
Division early to mid-spring, replanting the portions immediately in their flowering positions.
Seeds sown during mid-spring and germinated in a garden frame.

× **SOLIDASTER (ASTER × SOLIDAGO** P
Division early to mid-spring, replanting the portions immediately in their flowering positions.

SOPHORA T/S
Trees and shrubs
Seeds sown during late winter in a heated propagating case, 18°C (65°F). Soak seeds in warm water for several hours prior to sowing, especially purchased seeds.

SORBARIA S
Shrubs
Hardwood cuttings taken late autumn or early winter and rooted in a garden frame.
Semi-ripe cuttings taken with a slight heel during early or mid-summer and rooted in a garden frame.
Root cuttings taken early to mid-winter and rooted in a garden frame.
Suckers removed with some roots attached any time between autumn and spring and immediately replanted elsewhere.

SORBUS (Includes mountain ash and whitebeam) T
Seeds sown as soon as ripe in an outdoor seed bed. Alternatively stratify over winter (especially purchased seeds) and sow during early spring.
Whip and tongue grafting for cultivars during early spring, outdoors. Rootstocks are *S. aucuparia* for cultivars of this and similar species, and *S. aria* or *S. intermedia* for whitebeam cultivars.
Budding mid-summer outdoors. Rootstocks as above.

SPARAXIS Cm
Cormlets removed during late autumn (planting time) and immediately replanted elsewhere. They may take two years to reach flowering size.
Seeds sown during late summer or early autumn, or in early spring, and germinated in a heated propagating case, 15.5°C (60°F). Grow on in a garden frame or greenhouse (frost-free) until flowering size is attained (up to three years) then plant in the garden.

SPARTINA P G
Division during early or mid-spring, replanting the tufts immediately in their permanent positions.

SPARTIUM (Spanish broom) S
Seeds sown during late winter and germinated in a heated propagating case, 15.5°C.

SPECULARIA (LEGOUSIA) (Include Venus's looking glass) An
Seeds sown during early to mid-spring outdoors where the plants are to flower. Alternatively sow during early autumn in mild areas with well-drained soil.

SPIRAEA S
Softwood cuttings early to mid-summer rooting them in a heated propagating case 15.5°C (60°F) or garden frame.
Semi-ripe cuttings mid-summer, rooted in a heated propagating case, 15.5°C (60°F).
Hardwood cuttings late autumn or early winter, rooting them in a garden frame.
Division is possible with some species, early to mid-spring, replanting the portions immediately in their permanent positions.
Suckers with roots attached can be removed from some species during early spring and immediately replanted elsewhere.

STACHYS (Includes woundwort and lamb's tongue) P
Division during early to mid-spring replanting the portions immediately in their flowering positions.
Seeds sown during mid-spring and germinated in a garden frame.

TACHYURUS S
imple layering between mid-spring and ate summer, outdoors. Layers take one or wo years to form a good root system.
emi-ripe cuttings with a heel mid- to late ummer, heated propagating case, 18°C 55°F). Rooting may be rather slow.
eeds sown during mid-autumn and germi-ated in a garden frame.

TAPHYLEA (Bladder nut) S/Sm T
imple layering during mid- or late ummer, outdoors. Layers take at least a ear to form a good root system.
uckers removed with some roots attached uring early spring and immediately eplanted elsewhere.
emi-ripe cuttings with a slight heel taken nid- to late summer and rooted in a heated ropagating case or under mist, 18–21°C 55–70°F).
eeds sown early or mid-spring and germi-ated in a heated propagating case.

TAUNTONIA Cl
emi-ripe cuttings taken during summer nd rooted in a garden frame.

TEPHANANDRA S
emi-ripe cuttings mid-summer, rooted in garden frame.
ardwood cuttings late autumn, rooted in n outdoor nursery bed.
ivision during early or mid-spring, eplanting the portions immediately in their ermanent positions.

TERNBERGIA B
ulblets removed during late summer or arly autumn (planting time) and replanted sewhere. They take one or two years to art flowering.
eeds sown as soon as ripe, or during mid- late winter, in pots, and germinated utdoors.

TEWARTIA (STUARTIA) S/Sm T
imple or air layering between mid-spring nd late summer, outdoors. Rooting takes at ast one year.
oftwood cuttings early summer, rooted in heated propagating case, 18°C (65°F).
eeds sown during mid-autumn and germi-nated in a garden frame. Germination may be very slow.

STIPA (Feather grass) P G
Division early to mid-spring, replanting the tufts immediately in their permanent posi-tions.
Seeds sown during mid-spring in an outdoor seed bed.

STOKESIA (Stokes' aster) P
Division during early or mid-spring, replanting the portions immediately in their flowering positions.
Root cuttings taken late autumn or early winter and rooted in a garden frame.
Seeds sown during early or mid-spring and germinated in a garden frame.

STRANVAESIA S/Sm T
Simple layering between mid-spring and late summer, outdoors.
Seeds sown early to mid-spring in an outdoor seed bed, after stratification during the winter.
Semi-ripe cuttings wounded at the base, taken during late summer or autumn and rooted in a heated propagating case or under mist, 18–21°C (65–70°F).

STRATIOTES (Water soldier) Aq
Offsets removed when sufficiently developed, during spring or summer, and replaced in the water.

STYRAX (Snowbell, storax) T/S
Simple or air layering betwen mid-spring and late summer, outdoors.
Semi-ripe cuttings taken early to mid-summer and rooted in a heated propagating case, 18°C (65°F). Expect low percentage rooting.
Seeds sown early to mid-spring and germi-nated in a heated propagating case, 18°C (65°F).

SUCCISA (Devil's bit) P
Seeds sown late spring or early summer in an outdoor seed bed.
Division early to mid-spring, replanting the portions immediately in their flowering positions.

SYCOPSIS S/T
Simple or air layering between mid-spring and late summer, outdoors.
Seeds sown during early or mid-spring and germinated in a heated propagating case, 18°C (65°F).
Semi-ripe cuttings mid-summer, heated propagating case, 21°C (70°F).

SYMPHORICARPOS (Snowberry) S
Division during early or mid-spring, replanting the portions immediately in their permanent positions.
Suckers removed with some roots attached during early or mid-spring and immediately replanted in nursery bed or permanent positions.
Hardwood cuttings taken late autumn or early winter and rooted in a garden frame or outdoors.
Seeds sown during early or mid-spring in an outdoor seed bed.

SYMPHYANDRA (Ring bellflower) P
Seeds sown during autumn and germinated in a garden frame. Plants may also self-sow. Ideally grow as biennials.

SYMPHYTUM (Comfrey) P
Root cuttings taken during early or mid-winter and rooted in a garden frame.
Division during early or mid-spring, replanting the portions immediately in their flowering positions.
Seeds sown late spring in an outdoor seed bed.

SYMPLOCOS S/T
Simple or air layering between mid-spring and late summer, outdoors.
Semi-ripe cuttings with a heel taken during mid-summer and rooted in a heated propagating case, 15.5°C (60°F).
Seeds sown as soon as ripe or during early spring and germinated in a garden frame or cool greenhouse. Germination of purchased seeds may be very slow.

SYRINGA (Lilac) S
Softwood cuttings taken during late spring or early summer and rooted under mist or in a heated propagating case, 18–21°C (65–70°F).
Semi-ripe cuttings with a heel taken mid- t late summer and rooted in a heated prop agating case or under mist, 18–21°C (65 70°F).
Root cuttings when dormant, garde frame.
Simple layering between mid-spring an late summer, outdoors. Layers should b well rooted within a year.
Suckers removed early spring and replante elsewhere.
Spliced side veneer grafting for cultivar during early spring, in a greenhouse bottom heat, 21°C (70°F) and high humid ity. The rootstock is *Ligustrum vulgare* (com mon privet). Do not use *Syringa vulgar* (common lilac) as it suckers badly.
Seeds sown during autumn, or early to mic spring after winter stratification, in a outdoor seed bed.

TAMARIX (Tamarisk) S/Sm T
Hardwood cuttings taken during lat autumn or early winter and rooted in garden frame or outdoors.

TANACETUM P
Division where possible, early spring, after flowering, replanting the portion immediately in flowering positions. Discar woody pieces.
Soft basal cuttings as soon as available i spring, rooting them in a heated propaga ing case, 15.5°C (60°F).
Seeds sown early spring and germinated i heated propagating case, 15.5°C (60°F).

TAXODIUM (Swamp cypress) C
Hardwood cuttings taken during lat autumn or early winter and rooted in heated propagating case, 18–21°C (65 70°F).
Seeds sown during early or mid-spring in garden frame or outdoor seed bed.

TAXUS (Yew) C
Semi-ripe cuttings taken between early an mid-autumn and rooted in a heated prop agating case, 18°C (65°F) or garden frame.
Veneer grafting for cultivars (particular golden-leaved kinds), early spring in greenhouse, with bottom heat, 24°C (75°F and high humidity. The rootstock is *Tax*

baccata (common yew).
Seeds sown between early and mid-spring in an outdoor seed bed, after stratifying for two winters. Some seeds may not germinate until the following year.

TECOPHILAEA (Chilean crocus) Cm
Cormlets removed and replanted elsewhere in early autumn. They take up to four years to attain flowering size.
Seeds sown during autumn and germinated in a cool greenhouse. Seedlings also take about four years to reach flowering size. Tecophilaeas are generally grown in pots or pans in cool greenhouses, except in very mild areas.

TELLIMA P
Division during early or mid-spring, replanting the portions immediately in their flowering positions.
Seeds sown during early spring and germinated in a garden frame. Purple-leaved cultivar will not come true from seeds.

TEUCRIUM (Germander) S/P
Division if feasible, early to mid-spring, replanting the portions immediately in their flowering positions.
Semi-ripe cuttings with a heel for shrubby species, taken between early and late summer and rooted in a heated propagating case, 15.5°C (60°F).
Seeds sown during early or mid-spring and germinated in a heated propagating case, 15.5°C (60°F) or garden frame.

THALICTRUM (Meadow rue) P
Division early to mid-spring, replanting the portions immediately in their flowering positions. Divisions are slow to re-establish.
Soft basal cuttings mid- to late spring, rooting them in a garden frame.
Seeds sown during early spring and germinated in a garden frame. Double forms do not produce seeds and are propagated vegetatively.

THELYPTERIS F
Division early to mid-spring, replanting the portions immediately.
Spores sown as soon as ripe, heated propagating case, 18–21°C (65–70°F); keep moist and shaded, but light needed for germination.

THERMOPSIS (False lupin) P
Division early to mid-spring, replanting the portions immediately in their flowering positions. Some species like *T. lupinoides* do not like to be disturbed and divisions are slow to re-establish.
Seeds sown during mid-spring in an outdoor seed bed.

THLASPI (Penny cress) A
Seeds sown in pots during early to mid-spring and germinated outdoors or in a garden frame.
Soft cuttings mid- to late spring, rooted in a garden frame.

THUJA (Arbor-vitae) C
Semi-ripe cuttings early to mid-autumn rooted in a garden frame or heated propagating case, 18°C (65°F). Alternatively root during mid-spring under mist or in a heated propagating case, 18–21°C (65–70°F).
Seeds sown during early to mid-spring in an outdoor seed bed.

THUJOPSIS (Hiba arbor-vitae) C
Semi-ripe cuttings early to mid-autumn rooted in a garden frame, or heated propagating case at 18°C (65°F).
Seeds sown early to mid-spring in an outdoor seed bed.

THYMUS (Thyme) S/A
Division early to mid-spring, replanting the portions immediately in their permanent positions.
Softwood cuttings of shrubby species early summer, rooted in a heated propagating case, 15.5°C (60°F) or garden frame.
Seeds sown during early spring and germinated in a garden frame.

TIARELLA (Foam flower) P
Division early to mid-spring, replanting the portions immediately in their flowering positions.
Seeds, especially for species not easily divided, such as *T. wherryi*, sown early spring and germinated in a garden frame.

TILIA (Lime, linden) T
Seeds sown as soon as ripe (or even better in the green stage) in an outdoor seed bed. Germination can be very erratic. Purchased seed should be subjected to a high temperature (warm greenhouse) followed by chilling in refrigerator (both for four months) prior to sowing.
Simple or air layering between mid-spring and late summer, outdoors. Layers usually take two years to form a good root system, although some species root within a year.
Cuttings for some species such as *T. europaea* and *T. cordata,* early summer, under mist, 21°C (70°F).

TOLMIEA (Pick-a-back plant) P
Division early to mid-spring, replanting the portions immediately.
Leaves with well-developed plantlets removed and rooted in a garden frame or cool greenhouse. Alternatively peg down leaves in the soil. Sometimes plantlets root naturally into the soil.

TORREYA C
Semi-ripe cuttings late summer or early autumn, rooted in a heated propagating case, 18°C (65°F), or garden frame.
Seeds sown during early or mid-spring and germinated in a heated propagating case, 18°C (65°F). Germination slow and erratic.

TOVARA P
Division during early or mid-spring, replanting the portions immediately in their flowering positions.
Cuttings taken during mid-summer (non-flowering shoots) and rooted in a garden frame.

TRACHELIUM (Throatwort) P
Soft cuttings taken as soon as available during spring and rooted in a garden frame or heated propagating case, 15.5°C (60°F).
Seeds sown early to mid-spring and germinated in a heated propagating case, 15.5°C (60°F).

TRACHELOSPERMUM Cl
Simple or serpentine layering between mid-spring and late summer, outdoors. Layers take a year to form a good root system.
Semi-ripe cuttings late summer, heated propagating case or mist unit, 21°C (70°F).

TRACHYCARPUS (Chusan palm) T
Seeds sown during early spring and germinated in a heated propagating case, 24°C (75°F). Grow on young plants in pots in a heated greenhouse, minimum 10°C (50°F) for two years.
Suckers, with about three leaves, removed with some roots attached between mid- and late spring. Pot them and grow on in a heated greenhouse as for seedlings.

TRADESCANTIA (Spiderwort) P
Division early to mid-spring, replanting the portions immediately in their flowering positions.
Seeds sown early to mid-spring and germinated in a garden frame, but seedlings of cultivars will be variable.

TRICYRTIS (Toad lily) P
Division early to mid-spring, replanting the portions immediately in their flowering positions.
Seeds sown during early or mid-spring and germinated in a garden frame.

TRIFOLIUM (Clover, trefoil) P
Division early to mid-spring, replanting the portions immediately in their permanent positions.
Seeds sown during spring in an outdoor seed bed.

TRILLIUM (Wood lily) P
Division late summer or early spring, replanting the portions immediately in their flowering positions. Each portion must contain at least one bud. Trilliums are slow to re-establish after division.
Seeds sown as soon as ripe and germinated in a garden frame. Seeds can take a year and a half or more to germinate. Grow on seedlings in pots in a garden frame for two years. They take five or six years to start flowering.

TROCHODENDRON (Wheel tree) T
Air or simple layering between mid-spring and late summer, outdoors.

Semi-ripe cuttings mid- to late summer rooted in a heated propagating case, 18°C (65°F).

TROLLIUS (Globe flower) P
Division early to mid-spring, replanting the portions immediately in their flowering positions.
Seeds sown as soon as ripe and germinated in a garden frame. Stored or purchased seeds can take 12 months or more to germinate.

TROPAEOLUM (Nasturtium) An/P
Seeds for annuals, sown during mid-spring where the plants are to flower. Seeds of perennial species may be sown during early or mid-spring and germinated in a heated propagating case or garden frame.
Soft basal cuttings of perennial species taken during mid- or late spring and rooted in a heated propagating case, 15.5°C (60°F).
Division for some perennials like *T. speciosum* and *T. polyphyllum* during early spring, replanting the portions immediately in their permanent positions.
Tubers for species which produce them, separating them when lifting in autumn, or at planting time (early to mid-spring).

TSUGA (Hemlock) C
Semi-ripe cuttings mid-summer to early autumn, rooted in a garden frame.
Seeds sown during autumn or winter in an outdoor seed bed.

TULIPA (Tulip) B
Bulblets removed mid- to late autumn (planting time) and replanted elsewhere. Large bulblets may flower the following year, smaller ones within two years.
Seeds sown as soon as ripe, in pots, and germinated in a garden frame. Seedlings can take from three to seven years to reach flowering size, depending on species.

TYPHA (Reedmace) Aq P
Division between mid-spring and early summer, cutting through the rhizomes with a sharp knife, replanting the portions immediately in their permanent positions.

ULEX (Gorse) S
Seeds sown during early or mid-spring and germinated in a garden frame or outdoor seed bed.
Semi-ripe cuttings taken during early autumn and rooted in a garden frame.
Softwood cuttings taken early to mid-summer and rooted in a garden frame.

ULMUS (Elm) T
Softwood cuttings taken early summer and rooted in individual pots in a heated propagating case or under mist, 21°C (70°F).
Hardwood cuttings taken late autumn or early winter and rooted in a heated propagating case, 21°C (70°F), or garden frame.
Suckers removed with some roots attached during early spring and immediately replanted elsewhere.
Simple or air layering between mid-spring and late summer, outdoors.
Root cuttings taken early to mid-winter and rooted in individual pots in a garden frame.
Whip and tongue grafting during early spring outdoors. Rootstocks are seedlings of *U. glabra* (wych elm).
Budding for cultivars, using the chip method, mid-summer, outdoors. Rootstocks as above.
Seeds sown in the green stage or as soon as ripe in an outdoor seed bed.

UMBELLULARIA (California laurel) T
Seeds sown early to mid-spring in heated propagating case, 15.5°C (60°F).
Semi-ripe cuttings with a heel taken mid- to late summer and rooted in a garden frame.
Simple or air layering between mid-spring and late summer, outdoors.

UTRICULARIA (Bladderwort) Ca Aq
Division between mid-spring and early summer, replacing the portions immediately in water.

UVULARIA (Bellwort) P
Division during early or mid-spring, replanting the portions immediately in their flowering positions.

VACCINIUM (Includes blueberry, cranberry, bilberry) S
Semi-ripe cuttings late summer, rooting them in a garden frame.

Layering (dropping) in mid-spring, outdoors. A plant is lifted and replanted to half its depth, spacing out the stems well. Each or most will then root.
Simple layering between mid-spring and late summer, outdoors. Layers take one or two years to form good root systems.
Division during early spring, replanting the portions immediately in their permanent positions.
Seeds sown during autumn and germinated in a garden frame.

VALERIANA (Valerian) P
Division during early or mid-spring, replanting the portions immediately in permanent positions.
Soft basal cuttings taken as soon as available in mid- to late spring and rooted in a garden frame, or heated propagating case at 15.5°C (60°F).
Seeds sown during mid-spring in an outdoor seed bed.

VERATRUM (False hellebore) P
Division during early spring, replanting the portions immediately in their permanent positions.
Seeds sown as soon as ripe and germinated in a garden frame. Seedlings attain flowering size in about four years.

VERBASCUM (Mullein) P/Bi/S S
Division for perennials during early or mid-spring, replanting the portions immediately in their flowering positions.
Root cuttings taken early to mid-winter and rooted in a greenhouse or garden frame.
Cuttings of basal shoots for sub-shrubs, taken late summer and rooted in a garden frame.
Seeds sown during mid-spring in an outdoor seed bed.

VERBENA (Vervain) P
Seeds sown mid-winter to early spring and germinated in a heated propagating case, 18–21°C (65–70°F). Germination rate and time variable.
Semi-ripe cuttings taken during late summer and rooted in a garden frame or heated propagating case, 15.5°C (60°F).

VERNONIA (Ironweed) P
Division early to mid-spring, replanting the portions immediately in their flowering positions.
Seeds sown during early or mid-spring and germinated in a garden frame.

VERONICA (Speedwell) P/A
Division during early or mid-spring replanting the portions immediately in their flowering positions.
Soft basal cuttings taken as soon as available in spring and rooted in a garden frame or heated propagating case, 15.5°C (60°F).
Cuttings of side shoots of alpine species mid- to late summer, rooting them in a garden frame.
Seeds sown during spring in an outdoor seed bed.

VIBURNUM S
Softwood cuttings for deciduous species rooted in a heated propagating case or mist unit, 21°C (70°F) during spring.
Semi-ripe cuttings mid-summer to early autumn, rooted in a heated propagating case, 21°C (70°F), or garden frame.
Hardwood cuttings late autumn to early winter, rooting them in a garden frame.
Simple layering between mid-spring and late summer, outdoors.
Seeds sown as soon as ripe and germinated in a garden frame. Germination can be very slow, as can growth of seedlings.

VINCA (Periwinkle) S S
Division during early or mid-spring, replanting the portions immediately in their flowering positions.
Semi-ripe cuttings taken between late summer and mid-autumn and rooted in a garden frame or under a low polythene tunnel.
Serpentine layering of long stems between mid-spring and late summer, outdoors. Generally they take root naturally.

VIOLA (Pansy, violet) P/A
Seeds for species sown during early or mid-spring and germinated in a garden frame or cool greenhouse. Summer-flowering pansies used for bedding can be sown between late winter and early spring and germinated in a

eated propagating case, 15.5°C (60°F), for owering the same year; or early to mid-ummer in a garden frame to flower the ollowing year. Winter- and spring-flower-ng bedding pansies are sown during early ummer and germinated in a garden frame.
Division for perennials early to mid-spring, eplanting the portions immediately in their owering positions.
Cuttings of basal shoots for perennials, late ummer, rooting them in a garden frame.
Plantlets pegged down until rooted for *V. dorata*.

ITIS (Vines, including grape vines) Cl
Eye cuttings early to mid-winter, rooted in heated propagating case, 18°C (65°F).
Semi-ripe cuttings with a heel, mid- to late ummer, rooted in a heated propagating ase, 18°C (65°F).
Hardwood cuttings late autumn, garden rame.
Serpentine layering between mid-spring nd late summer, outdoors.
Seeds sown during autumn and germinated n a garden frame.

WAHLENBERGIA P/A/An
Division early to mid-spring, replanting the portions immediately in their flowering positions.
Soft basal cuttings as soon as available in spring, rooting them in a garden frame.
Seeds sown mid- to late winter in pots and germinated outdoors.

WALDSTEINIA P
Division early to mid-spring, replanting the portions immediately in permanent positions.

WATTAKAKA Cl
Seeds sown during spring, garden frame.
Semi-ripe cuttings during summer rooted in a garden frame.

WEIGELA S
Softwood cuttings taken during early summer and rooted in a garden frame.
Semi-ripe cuttings between early and late summer, rooting them in a heated propagating case, 15.5°C (60°F) or garden frame.
Hardwood cuttings late autumn or early winter, rooting them in a garden frame.

WISTERIA Cl
Serpentine or simple layering between mid-spring and late summer, outdoors. The layers take a year to form a good root system.
Softwood cuttings early to mid-summer, using side shoots with short internodes, inserted under mist, 21°C (70°F).

WOODWARDIA (Chain fern) F
Division early to mid-spring, replanting the portions immediately.

WULFENIA R-G P
Seeds sown during early spring and germinated in a garden frame.
Division early to mid-spring, replanting the portions immediately in their flowering positions.

XANTHOCERAS S/T
Seeds sown during spring in an outdoor seed bed, after winter stratification.
Root cuttings taken early to mid-winter and rooted in a heated propagating case, 18°C (65°F).

XANTHORHIZA S
Division in autumn or early spring, replanting the portions immediately in their permanent positions.

XERANTHEMUM (Immortelle) An
Seeds sown during early or mid-spring outdoors where the plants are to flower.

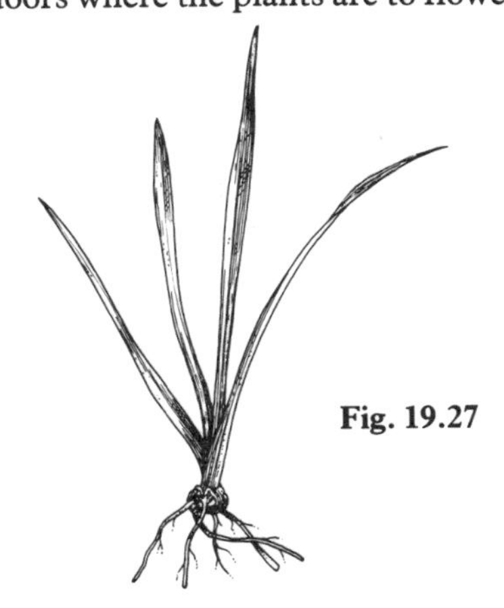

Fig. 19.27

Fig. 19.27 A young yucca plant produced from a bud or 'toe' planted in an outdoor nursery bed or garden frame during spring.

YUCCA S/T
Suckers with roots attached removed during early or mid-spring and replanted elsewhere (Fig. 19.27).
Buds or 'toes' situated at base of stems and on rhizomes. In mid- to late spring lift plant, remove some buds and plant them 12 mm (½ in) deep in outdoor nursery bed or garden frame.
Seeds sown during early or mid-spring and germinated in a heated propagating case, 18–21°C (65–70°F).

ZANTEDESCHIA (Arum lily) P
Offsets for the hardy *Z. aethiopica*, carefully removed during spring and immediately replanted in flowering positions. Alternatively divide established clumps.
Seeds sown during early or mid-spring and germinated in a heated propagating case, 15.5–18°C (60–65°F).

ZANTHOXYLUM (Prickly ash) S/T
Root cuttings taken during early or mid-winter and rooted in a heated propagating case, 18°C (65°F).
Seeds sown as soon as ripe in autumn and germinated in a heated propagating case, 18°C (65°F).
Suckers with roots attached removed during early or mid-spring and immediately replanted elsewhere.

ZAUSCHNERIA (California fuchsia) P
Division during early or mid-spring, replanting the portions immediately in their flowering positions.
Soft basal cuttings taken during late spring and rooted in a heated propagating case, 15.5–18°C (60–65°F).
Semi-ripe cuttings taken late summer or early autumn and rooted in a heated propagating case, 15.5–18°C (60–65°F) or garden frame.
Seeds sown during early spring and germinated in a heated propagating case.

ZELKOVA T
Seeds sown as soon as ripe in the autumn in an outdoor seed bed. Unripe or green seeds can also be sown during early autumn.
Air or simple layering between mid-spring and late summer, outdoors.

ZENOBIA S
Semi-ripe cuttings taken during summer and rooted in a heated propagating case, 15.5°C (60°F).
Simple layering between mid-spring and late summer, outdoors. Layers take about 18 months to form a good root system.
Seeds sown during early to mid-spring and germinated in a garden frame.

ZEPHYRANTHES (Zephyr lily) B
Bulblets removed and replanted in their flowering positions during mid-spring (planting time).
Division of established clumps during mid-spring, replanting immediately in their permanent positions.
Seeds sown during autumn or winter, in pots, and germinated in a cool greenhouse or garden frame.

CHAPTER 20

Tender Ornamental and Fruiting Plants

KEY TO PLANT TYPES

An = Annuals
An G = Annual grasses
Aq = Aquatics
B = Bulbs
Bi = Biennials
C = Conifers
Ca P = Carnivorous perennials
Cc = Cacti
Cl = Climbers
Cl P = Climbing perennials
Cl S = Climbing shrubs
Cm = Corms
F = Ferns
Fl Aq = Floating aquatic
O = Orchids
P = Perennials
P G = Perennial grasses
Pm = Palms
R P = Rhizomatous perennials
S = Shrubs
S S = Sub-shrubs
Su = Succulents
Su An = Succulent annuals
Sub Aq = Submerged aquatics
T = Trees
T F = Tree ferns
Tu P = Tuberous perennials

ABUTILON S
Seeds early to mid-spring, heated propagating case, 18°C (65°F).
Semi-ripe cuttings mid- to late summer, heated propagating case, 18°C (65°F).

ACACIA S/T
Seeds sown as soon as ripe, or in spring, germinating them in a heated propagating case, 15.5°C (60°F). To soften the hard seedcoats, place in hot water and leave in the cooling water until they have swelled, for several days if necessary.
Semi-ripe cuttings with a heel mid- to late summer, rooting them in a heated propagating case, 15.5–18°C (60–65°F).

ACALYPHA S
Softwood cuttings taken during mid-spring and rooted in a heated propagating case, 21°C (70°F).
Semi-ripe cuttings with heel taken during summer and rooted in a heated propagating case, 21°C (70°F).

ACHIMENES (Hot-water plant) P
Seeds sown late winter or early spring and germinated in a heated propagating case, 21°C (70°F).
Softwood cuttings, in spring, rooted in heated propagating case, 21°C (70°F).
Leaf cuttings as for softwood cuttings.
Rhizomes can be separated and potted separately during early spring (planting time), heated greenhouse.

ACIDANTHERA Cm
Cormlets removed during mid-spring (planting time) and potted separately, cool greenhouse. They take about three years to flower.
Seeds sown late winter or early spring and germinated in a cool greenhouse. Seedlings will flower in two or three years.

ADA O
Pseudobulbs, mature and without leaves, removed while repotting in spring and potted separately.

ADIANTUM (Maidenhair fern) F
Division during early or mid-spring, repotting the portions immediately in the smallest possible pots.
Spores sown as soon as ripe, heated propagating case, 21°C (70°F), with good light but shade from sun, and constant moisture.

ADROMISCHUS Su
Leaf cuttings early to mid-summer, heated propagating case.
Stem cuttings as above.

AECHMEA (Urn plant) P
Offsets removed with some roots attached when about 15 cm (6 in) high, between mid-spring and mid-summer, potted and established in a heated propagating case.
Seeds sown when ripe or during early or mid-spring and germinated in a heated propagating case, 21°C (70°F). Seedlings are very slow growing

AEONIUM Su
Leaf cuttings taken between mid-spring and mid-summer and rooted in a heated propagating case, 18–21°C (65–70°F).
Stem cuttings taken between mid-spring and mid-summer and rooted in a heated propagating case, 15.5–21°C (60–70°F).
Seeds sown during early to mid-spring and germinated in a heated propagating case, 21°C (70°F).

AERIDES O
Cuttings prepared from lower shoots when 10 cm (4 in) in length, mid-to late spring, inserted in the potting compost, heated propagating case, 18–21°C (65–70°F).

AESCHYNANTHUS Cl or Tr P
Softwood cuttings taken from shoot tips, mid- to late spring, heated propagating case, 18–21°C (65–70°F).
Simple layering between mid-spring and mid-summer, where the plant is growing.

AGAPETES S/Cl
Semi-ripe cuttings late summer or early autumn, rooted in a heated propagating case, 21°C (70°F).
Seeds sown during early spring and germinated in a heated propagating case, 21°C (70°F).

AGAVE (Century plant) Su
Offsets removed during spring and established in pots in a heated propagating case.
Seeds sown early to mid-spring and germinated in a heated propagating case, 21°C (70°F).

AGERATUM (Floss flower) An/Bi
Seeds sown between late winter and mid-spring and germinated in a heated propagating case, 15.5–18°C (60–65°F).

AGLAONEMA (Chinese evergreen) P
Division during mid-spring, repotting th portions and providing normal greenhous or room conditions.
Semi-ripe cuttings taken from base c plants, mid- to late summer, rooted in heated propagating case, 21°C (70°F).
Stem section cuttings taken mid- to lat summer and rooted in a heated propagatin case, 21°C (70°F).

AICHRYSON Su
Cuttings taken between late spring and mid summer, rooted in a heated propagatin case, 15.5–21°C (60–70°F).
Seeds sown early to mid-spring, germinat ing them in a heated propagating case, 21°C

ALLAMANDA Cl
Softwood cuttings taken from tips of sten or side shoots during spring, rooted in heated propagating case, 21–24°C (70-75°F).

ALOCASIA R P
Cuttings of the rhizomes taken during earl spring and rooted in a heated propagatin case, 21°C (70°F).
Cuttings prepared from the stems durin spring and rooted in a heated propagatin case, 21°C (70°F).
Suckers removed with some roots attache in spring and immediately potted. Establis in heated propagating case.

ALOE Su
Offsets removed during mid-spring an immediately potted.
Leaf cuttings taken between late spring an mid-summer and rooted in a heated prop-agating case, 15.5–21°C (60–70°F).
Seeds sown early to mid-spring, heated propagating case, 21°C (70°F).

ALONSOA (Mask flower) P
Seeds sown during early spring and germinated in a heated propagating case, 15.5°C (60°F). Grown as annuals outdoors or under glass

ALPINIA (Ginger lily) P
Division of clumps in late spring, potting the portions immediately.

ALTERNANTHERA P
Division during mid-spring, potting the portions immediately
Semi-ripe cuttings taken from shoot tips during late summer, rooting them in a heated propagating case.

AMARANTHUS An
Seeds sown during early spring, either in a heated propagating case, 15.5°C (60°F), or direct in flowering positions outdoors.

AMMOBIUM (Sand immortelle) An
Seeds sown during early spring and germinated in a heated propagating case, 15.5°C.

ANANAS (Pineapple) P
Offsets or suckers removed during mid-spring when well developed, potted and established in a heated propagating case.

ANGRAECUM O
Cuttings of side shoots if available, late spring to early summer, rooted in heated propagating case, 21°C (70°F).

ANIGOZANTHUS (Kangaroo paw) P
Division mid- to late spring, potting the portions immediately.
Seeds sown as soon as ripe during late summer and germinated in heated propagating case, 18°C (65°F). Germination slow.

ANTHURIUM P
Division early spring, potting the portions immediately.
Seeds sown as soon as ripe and germinated in a heated propagating case, 21–24°C (70–75°F).

ANTIRRHINUM (Snapdragon)
P, grown as An
Seeds sown during late winter or early spring and germinated in a heated propagating case, 15.5–18°C (60–65°F), for summer bedding. Sow from mid-summer to early autumn for winter/spring pot plants under glass.

APHELANDRA S
Softwood cuttings of side shoots mid-spring to mid-summer, heated propagating case, 21°C (70°F).
Leaf-bud cuttings as above.
Seeds sown during spring, heated propagating case, 21°C (70°F).

APOROCACTUS (Includes rat-tail cactus)
Cc
Cuttings taken late spring to mid-summer, heated propagating case, 15.5–21°C (60–70°F).
Seeds sown early to mid-spring, heated propagating case, 21°C (70°F).

ARACHIS (Peanut) An
Seeds sown during spring in heated propagating case, 24°C (75°F).

ARAUCARIA (Norfolk Island pine) C
Seeds sown during early spring and germinated in a heated propagating case, 10–13°C (50–55°F). A slow process.
Cuttings of young shoots from cut-back plants, spring, heated propagating case, 15.5°C (60°F).

ARCTOTIS (African daisy) P/S
Seeds sown during early to mid-spring and germinated in heated propagating case, 18°C (65°F).
Semi-ripe cuttings for perennials, late summer or early autumn, heated propagating case, 15.5°C (60°F).

ARDISIA T/S
Trees and shrubs
Soft or semi-ripe cuttings with heel obtained from side shoots produced by cut-back plants, between early spring and early autumn, heated propagating case, 18°C.
Seeds sown during mid-spring, heated propagating case, 18°C (65°F).

ARGYRANTHEMUM (Marguerite)
P, often grown as An
Seeds sown during late winter, germinating them in a heated propagating case, 13°C (55°F).
Cuttings taken early autumn and rooted in heated propagating case, 15.5°C (60°F).

ARGYRODERMA Su
Seeds sown early to mid-spring and germinated in heated propagating case, 21°C (70°F).

Division for tufted species, early summer, potting the portions immediately.

ARIOCARPUS (Living rock) Cc
Seeds sown early to mid-spring and germinated in heated propagating case, 21°C (70°F).

ASCLEPIAS (Milkweed) P
Seeds sown during early spring and germinated in heated propagating case, 15.5–18°C (60–65°F).
Softwood cuttings taken from shoot tips, late spring, heated propagating case.

ASPARAGUS P, some Cl
Division during mid-spring, potting the portions immediately.
Seeds sown early to mid-spring and germinated in heated propagating case, 15.5°C (60°F).

ASPIDISTRA (Cast-iron plant) P
Division mid-spring, potting the portions immediately.

Fig. 20.1

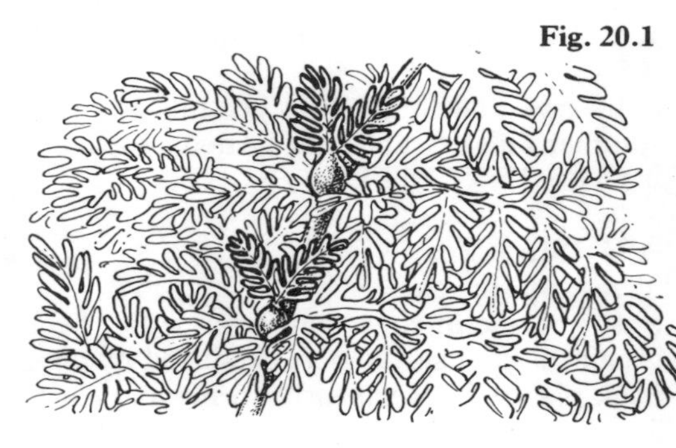

Fig. 20.1 *Asplenium bulbiferum* (spleenwort) produces 'bulbils' on its leaves which develop into plantlets. If it is desired to increase this species, pick off plantlets when well developed and plant in a seed tray.

Fig. 20.2 Tubers of begonias may be divided in early spring when potting. Cut tubers into sections, each with a minimum of one shoot.

ASPLENIUM (Spleenwort) F
Division mid-spring, potting the portions immediately.
Plantlets produced on leaves. Pick off when well developed and plant in seed tray (Fig. 20.1).
Spores sown as soon as ripe, heated propagating case, 21°C (70°F), with good light but shade from sun, and constant moisture.

ASTROPHYTUM (Star cactus) Cc
Seeds sown early to mid-spring and germinated in heated propagating case, 21°C (70°F).

ASYSTASIA (MACKAYA) S
Softwood cuttings between mid-spring and early summer, heated propagating case.
Semi-ripe cuttings summer, heated propagating case.

BABIANA Cm
Seeds sown mid-spring and germinated in heated propagating case, 15.5°C (60°F).
Cormlets removed and potted during mid autumn.

BANKSIA S/T
Semi-ripe cuttings mid-summer to mid autumn, unheated but close propagating case.
Seeds sown as soon as ripe, heated propagating case, 15.5–18°C (60–65°F).

BAROSMA (AGATHOSMA) S
Semi-ripe cuttings with a heel, late summer/early autumn, heated propagating case, 15.5°C (60°F).
Seeds early to mid-spring, germinated in heated propagating case, 15.5°C (60°F).

BEAUCARNEA (NOLINA)
(Includes pony tail) P
Offsets if produced, removed during spring when re-potting and immediately potted in individual pot.
Seeds sown early spring and germinated in heated propagating case, 18°C (65°F).

Fig. 20.2

BEGONIA P
Seeds sown mid- to late winter, including *B. semperflorens* and tuberous begonias for summer bedding, germinating them in a heated propagating case, 15.5°C (60°F).
Soft cuttings taken mid-spring to early summer and rooted in a heated propagating case, 18–21°C (65–70°F), especially for fibr-

ous-rooted species and winter-flowering begonias.
Division during mid-spring, especially for rhizomatous begonias, potting the portions immediately.
Leaf cuttings for some species, including *B. masoniana* and *B.rex*, between mid-spring and early summer, heated propagating case, 18–24°C (65–75°F).
Division of tubers when potting in early spring. Cut tubers into sections, each with a minimum of one shoot (Fig. 20.2).

BELOPERONE (JUSTICIA, DREJERELLA) (Shrimp plant) S
Soft or semi-ripe cuttings taken between early and late summer and rooted in a heated propagating case, 18°C (65°F).

BERTOLONIA P
Cuttings taken during spring or summer and rooted in a heated propagating case, 21–24°C (70–75°F).
Leaf cuttings as above.

BIGNONIA (DOXANTHA) (Cross vine) Cl
Cuttings taken during summer and rooted in a heated propagating case.
Simple or serpentine layering between mid-spring and late summer.

BILLARDIERA Cl
Semi-ripe cuttings late summer or early autumn, rooted in a heated propagating case, 15.5°C (60°F).
Seeds sown early to mid-spring and germinated in heated propagating case, 15.5°C (60°F).

BILLBERGIA P
Division mid-spring, potting the portions immediately.
Seeds sown early to mid-spring and germinated in heated propagating case, 21°C (70°F), with high humidity.

BLECHNUM F
Division mid-spring, potting the portions immediately.
Spores sown as soon as ripe, heated propagating case, 21°C (70°F), with good light but shade from sun, and constant moisture.

BLETILLA O
Division early to mid-spring, potting the portions immediately.

BOMAREA Tw Cl
Division during mid-spring, potting or replanting the portions immediately.
Seeds sown during mid-spring and germinated in a heated propagating case, 18°C (65°F).

BORONIA S
Shrubs
Semi-ripe cuttings with heel, late summer or early autumn, rooted in a heated propagating case.
Seeds sown early to mid-spring and germinated in a heated propagating case, 15.5°C (60°F).

BORZICACTUS Cc
Seeds sown early to mid-spring in heated propagating case, 21°C (70°F).

BOUGAINVILLEA (Paper flower) S/Cl
Semi-ripe cuttings mid- to late summer, heated propagating case, 21–24°C (70–75°F).
Hardwood cuttings mid-winter, heated propagating case, 18°C (65°F).

BOUVARDIA S/P
Softwood cuttings between mid-spring and early summer, heated propagating case, 21°C (70°F).

BRACHYGLOTTIS S/T
Semi-ripe cuttings taken late summer and rooted in heated propagating case.

BRASSAVOLA O
Division of well-established plants during spring, immediately replanting or potting the portions.
Pseudobulbs, mature and without leaves, removed while repotting in spring and potted separately. Leave at least six pseudobulbs on parent plant.

BRASSIA O
Division of well-established plants during spring, immediately replanting or potting the portions.

Pseudobulbs, mature and without leaves, removed while repotting or replanting in spring, and potted or planted separately.

BROMELIA P
Offsets or suckers removed with roots during mid-spring when well developed, potted and established in a heated propagating case.

BROWALLIA (Bush violet) An/P
Seeds sown during early spring and germinated in a heated propagating case, 18°C (65°F). Also sow during late summer for winter-flowering pot plants. Grown as annuals.

BRUNFELSIA S
Semi-ripe cuttings, but not too woody, taken between early and late summer and rooted in a heated propagating case, 21°C (70°F).

BULBOPHYLLUM O
Division during spring or immediately after flowering, potting the portions individually.
Pseudobulbs, mature and without leaves, removed while repotting in spring or after flowering and potted separately.

CAESALPINIA T/S/Cl
Seeds sown during spring and germinated in a heated propagating case, 21°C (70°F). First soak in warm water for several hours to assist germination.
Simple layering between mid-spring and early summer.

CALADIUM (Angel wings) Tu P
Division of dormant plants during late winter or early spring, potting them individually. Small tubers can also be separated and potted.

CALANDRINA (Rock purslane)
P, grown as An
Seeds sown during spring and germinated in heated propagating case, 15.5–18°C (60–65°F).

CALANTHE O
Division of clumps during repotting, potting the portions immediately.

Pseudobulbs, mature and without leaves removed while repotting and potted separately.

CALATHEA P
Division during mid-spring, potting th portions immediately and establishing in heated propagating case, 18°C (65°F).

CALCEOLARIA (Slipper wort) A/Bi/P/S
Seeds sown late winter to early spring fo bedding plants, and early summer for po plants, heated propagating case, 18°C (65°F).
Softwood cuttings for shrubby species mid-spring, rooted in a heated propagatin case, 15.5°C (60°F).
Semi-ripe cuttings with heel for shrubb species, late summer or early autumn rooted in heated propagating case, 15.5°C (60°F).

CALLISIA P
Cuttings, softwood or semi-ripe, using tip of shoots, taken between mid-spring an early autumn and rooted in a heated prop agating case.

CALLISTEMON (Bottle brush) S/T
Softwood or semi-ripe cuttings, ideall with heel of older wood, early to lat summer, heated propagating case, 18–21°C (65–70°F).
Seeds sown early spring, heated propaga ing case, 15.5–18°C (60–65°F).

CALLISTEPHUS (China aster) An
Seeds for bedding plants sown during earl spring in heated propagating case, 15.5°C (60°F). Alternatively sow during early t mid-spring and germinate in a garde frame. Seeds may also be sown during mid spring direct in flowering positions. For po plants, sow late spring or early summer an germinate in heated propagating case.

CALOCEPHALUS An/S/P
Semi-ripe cuttings during summer, roote in heated propagating case, 15.5°C (60°F) for the shrubby kinds grown as bedding c pot plants.

CAMPANULA (Bell flower) P/Bi

ɔft cuttings for *C. isophylla*, between mid-ring and early summer, rooting them in a ·ated propagating case, 15.5°C (60°F).
ɛeds for *C. isophylla* and *C. pyramidalis*, wn during early spring and germinated in ated propagating case, 15.5°C (60°F).

ANNA (Indian shot) R P
ivision of the rhizomes during mid-spring ıen shoots are visible, potting or replant-g the portions immediately.
ɛeds sown during late winter and germi-ted in a heated propagating case, 21°C 0°F), after soaking in warm water for one y. Alternatively chip or file the hard seed-at.

APSICUM (Ornamental pepper) grown as An
ɛeds sown during mid-spring and germi-ted in a heated propagating case, 18°C 5°F).

ARMICHAELIA (Butterfly broom) S
ɛmi-ripe cuttings with heel taken during :e summer and rooted in heated propagat-g case, 15.5°C (60°F).
ɛeds sown during spring and germinated heated propagating case, 15.5°C (60°F).

ARNEGIEA (Saguaro) Cc
ɛeds sown early to mid-spring and germi-ted in heated propagating case, 21°C)°F).

ARPOBROTUS Su
ıttings late spring to mid-summer, heated opagating case, 15.5–21°C (60–70°F).

ARYOTA (Fishtail palm) Pm
ɛeds sown during spring and germinated a heated propagating case, 27°C (80°F).
ıckers removed with roots attached ıring spring and immediately potted.

ASSIA T/S
ɛmi-ripe cuttings with a heel, late ımmer, rooted in heated propagating case, .5–18°C (60–65°F).
ɛeds sown early to mid-spring and germi-ted in heated propagating case, 15.5°C)°F).

CATHARANTHUS (Includes Madagascar periwinkle) S
Seeds sown during spring and germinated in a heated propagating case, 15.5–18°C (60–65°F).
Semi-ripe cuttings taken during summer and rooted in heated propagating case, 15.5–18°C (60–65°F)

CATTLEYA O
Division when repotting during spring, potting the portions immediately.
Pseudobulbs, mature and without leaves, removed while repotting in spring and potted separately. Leave at least four pseudobulbs on parent plant.

CELOSIA (Cockscomb) An
Seeds sown during late winter or early spring and germinated in a heated propagating case, 18°C (65°F).

CELSIA An/Bi/P
Seeds sown during mid-spring and germinated in a heated propagating case, 15.5°C (60°F). Sow biennials and perennials during early to mid-summer and germinate in a garden frame.
Cuttings for perennials, spring or autumn, rooting them in a garden frame.

CENTAUREA P
Seeds sown during winter or early spring and germinated in heated propagating case, 15.5°C (60°F).
Soft cuttings taken between late summer and early autumn and rooted in a heated propagating case, 15.5°C (60°F).

CEPHALOCEREUS (Old man cactus) Cc
Seeds sown during early or mid-spring and germinated in a heated propagating case, 21°C (70°F).

CEPHALOTUS (West Australian pitcher plant) P
Division of established plants during spring, potting the portions immediately.
Root cuttings taken mid- to late spring and rooted in peat and sand on greenhouse bench.
Leaf cuttings. Remove entire leaves and

root in sphagnum moss in an unheated propagating case with high humidity,

CEREUS Cc
Seeds sown early to mid-spring and germinated in a heated propagating case, 21°C (70°F).
Cuttings, if branching occurs, taken late spring to mid-summer and rooted in a heated propagating case, 15.5–21°C (60–70°F).

Fig. 20.3

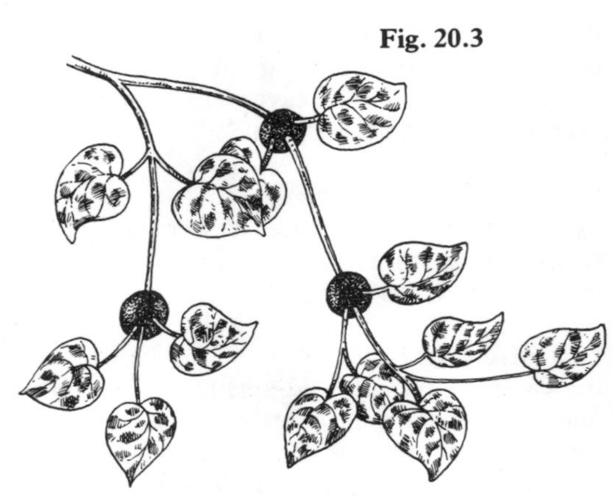

Fig. 20.3 *Ceropegia woodii* (hearts entangled) produces tubers on its stems which may be removed and rooted in heat, when they will develop into new plants.

CEROPEGIA (Hearts entangled)
Cl or Tr Su
Seeds sown early to mid-spring and germinated in a heated propagating case, 21°C (70°F).
Cuttings taken late spring to mid-summer and rooted in a heated propagating case, 15.5–21°C (60–70°F).
Tubercles on stems, if formed, removed and rooted in a heated propagating case, 15.5–21°C (60–70°F) (Fig. 20.3).

CESTRUM S/T
Shrubs and trees
Semi-ripe cuttings taken late summer or early autumn and rooted in heated propagating case, 21–24°C (70–75°F).
Seeds sown early to mid-spring and germinated in heated propagating case, 21–24°C (70–75°F).

CHAMAECEREUS (Peanut cactus) Cc
Cuttings late spring to mid-summer, rooted in a heated propagating case, 15.5–21°C (60–70°F).

Seeds sown early to mid-spring and germi nated in heated propagating case, 21° (70°F).

CHAMAEDOREA Pm
Suckers, if produced, removed with roo and potted during mid-spring.
Seeds sown during early or mid-spring an germinated in a heated propagating cas 25–27°C (77–80°F).

CHAMAEROPS Pm
Suckers removed with roots and potte during mid-spring.
Seeds sown during early or mid-spring an germinated in a heated propagating cas 24°C (75°F).

CHLIDANTHUS B
Bulblets removed and potted individuall during mid-spring.

CHLOROPHYTUM (Spider plant) P
Division mid-spring, potting the portion immediately.
Plantlets layered into pots during sprin and summer.

CHORIZEMA S/Cl
Soft or semi-ripe cuttings taken durin early summer or early autumn and rooted i a heated propagating case, 18°C (65°F).
Seeds sown during early to mid-spring an germinated in a heated propagating cas 18°C (65°F).

CHRYSALIDOCARPUS Pm
Seeds sown during spring in heated prop agating case, 27°C (80°F).

CHRYSANTHEMUM P
Soft basal cuttings for greenhouse chrysa themums, between mid-winter and ear spring, heated propagating case, 15.5° (60°F).
Seeds for charm and cascade chrysa themums, sown between mid-winter an early spring and germinated in heated prop agating case, 15.5°C (60°F). Seeds of C *parthenium* are sown during late winter early spring for summer bedding, an germinated in a heated propagating cas 15.5° (60°F).

IBOTIUM T F
pores sown during spring and germinated . a heated propagating case, 21°C (70°F).

ISSUS (Includes grape ivy and kangaroo ne) Cl/Su S
oft or semi-ripe cuttings between mid-ring and late summer, heated propagating se, 15.5–18°C (60–65°F).
uttings of succulent species late spring to id-summer, heated propagating case, 21°C 0°F).
eeds for succulent species, early to mid-ring, heated propagating case, 21°C 0°F).

ITRUS (Includes calamondin, citron, apefruit, lemon, lime, orange and shad-ock) T/S
emi-ripe cuttings taken late summer or rly autumn and rooted in heated propagat-g case, 18°C (65°F).
eeds sown early to mid-spring and germi-ted in a heated propagating case, 15.5°C 0°F).

LEISTOCACTUS Cc
uttings between late spring and mid-mmer, heated propagating case, 15.5–°C (60–70°F).
eeds sown early to mid-spring and germi-ted in a heated propagating case, 21°C 0°F).

LEOME (Spider flower) An/S
eeds for the annual species which is rmally grown, early to mid-spring, heated opagating case, 18°C (65°F).

LERODENDRUM S/Cl
oftwood or semi-ripe cuttings taken etween mid-spring and late summer and oted in a heated propagating case, 18–°C (65–70°F)
eeds sown early to mid-spring and germi-ted in a heated propagating case, 21°C 0°F).

LETHRA S/T
emi-ripe cuttings with heel, late summer, eated propagating case, 15.5°C (60°F).
imple layering between mid-spring and te summer.

Seeds sown during early or mid-spring and germinated in a heated propagating case, 15.5°C (60°F).

CLIANTHUS (Includes glory pea and parrot's bill) S
Semi-ripe cuttings late summer or early autumn, heated propagating case.
Seeds sown early to mid-spring, heated propagating case, 18–21°C (65–70°F).

CLITORIA (Butterfly pea) P/S, often Tw
Seeds sown during spring and germinated in heated propagating case, 21°C (70°F), after chipping and soaking in warm water for a day.
Softwood cuttings taken during summer and rooted in heated propagating case.

CLIVIA R P
Division when flowering is over, is a matter of taking care to avoid damaging the fleshy rhizomes, replanting or potting the portions immediately.
Offsets removed with some roots attached, spring or summer, potting them individually.
Seeds sown as soon as ripe and germinated in a heated propagating case, 15.5°C (60°F).

COBAEA (Cup and saucer vine) S Cl
Seeds sown individually in small pots during early or mid-spring and germinated in a heated propagating case, 18°C (65°F). Generally grown as annuals.

COCOS (Coconut palm) Pm
Seeds sown early to mid-spring and germinated in a heated propagating case, 27–29°C (80–85°F).

CODIAEUM (Croton) S
Soft or semi-ripe cuttings taken from shoot tips between mid-spring and late summer. Dip bases in powdered charcoal, then root in a heated propagating case, 24°C (75°F).
Air layering between mid-spring and late summer, normal greenhouse conditions.

COELOGYNE O
Division during spring when repotting, potting the portions immediately.

Pseudobulbs, mature and without leaves, removed while repotting in spring and potted separately.

COFFEA (Coffee) S
Semi-ripe cuttings mid- to late summer, rooted in a heated propagating case, 24°C (75°F).
Seeds sown between early and mid-spring and germinated in a heated propagating case, 24°C (75°F).

COLEUS (Includes flame nettle) An/P/SS
Soft cuttings taken from shoot tips between early and mid-spring and rooted in a heated propagating case, 15.5–18°C (60–65°F).
Seeds sown late winter or early spring and germinated in a heated propagating case, 18°C (65°F).

COLOCASIA (Includes taro) Tu or R P
Division in spring, replanting or potting the portion immediately.

COLQUHOUNIA S
Semi-ripe cuttings mid- to late summer, heated propagating case.
Seeds sown early to mid-spring and germinated in heated propagating case.

COLUMNEA P/S S
Soft or semi-ripe cuttings taken between mid-spring and late summer and rooted in a heated propagating case, 18–21°C (65–70°F).
Seeds sown early to mid-spring and germinated in a heated propagating case, 18°C (65°F).

COMMELINA (Day flower) P
Division early to mid-spring, potting the portions immediately.
Soft or semi-ripe cuttings taken early to late summer and rooted in a heated propagating case.
Seeds sown early to mid-spring and germinated in a heated propagating case, 21°C (70°F).

CONOPHYTUM (Cone plant) Su
Division mid-summer, potting the portions immediately.
Seeds sown early to mid-spring and germinated in a heated propagating case, 21° (70°F).

COPIAPOA Cc
Offsets removed with some roots betwe early and late summer and individual potted.
Seeds sown early to mid-spring and germ nated in a heated propagating case, 21° (70°F).

COPROSMA S/T
Semi-ripe cuttings taken mid- to la summer and rooted in a heated propagati case.
Seeds sown early to mid-spring and germ nated in heated propagating case, 15.5° (60°F).

CORDYLINE S/T
Cuttings consisting of stem sections, la spring or early summer, rooted in a heat propagating case, 21°C (70°F).
Suckers removed early or mid-spring ar potted individually.
Seeds sown mid-spring and germinated in heated propagating case, 18°C (65°F).

CORONILLA (Crown vetch) S
Semi-ripe cuttings late summer/ear autumn, heated propagating case.
Seeds early to mid-spring, heated propaga ing case, 15.5°C (60°F).

CORREA S/T
Semi-ripe cuttings late summer, rooted heated propagating case.
Seeds sown early to mid-spring and germ nated in heated propagating case.

CORYPHANTHA Cc
Offsets removed with some roots attache between early and late summer, and pott individually.
Seeds sown early to mid-spring and germ nated in heated progagating case, 21° (70°F).

COSMOS (Cosmea) An/P
Seeds for annuals sown late winter or ear spring and germinated in a heated propaga ing case, 15.5°C (60°F).
Soft basal cuttings for the tuberous pere

nial, *C. atrosanguineus*, early to mid-spring, heated propagating case.

COSTUS (Spiral flag) R P
Cuttings (stem sections), late spring or early summer, heated propagating case, 21°C (70°F).
Division during mid-spring, replanting the portions immediately.

COTULA An
Seeds for the annual species, *C. barbata*, sown early spring and germinated in heated propagating case, 15.5°C (60°F).

COTYLEDON Su
Seeds sown early to mid-spring and germinated in heated propagating case, 21°C (70°F).
Stem cuttings late spring to mid-summer, heated propagating case, 15.5–21°C (60–70°).
Leaf cuttings, using whole leaves, late spring to mid-summer, heated propagating case, 15.5–21°C (60–70°F).

CRASPEDIA P, usually grown as An
Seeds sown during early spring and germinated in a heated propagating case, 15.5°C.

CRASSULA Su
Stem cuttings late spring to mid-summer, heated propagating case, 15.5–21°C (60–70°F).
Leaf cuttings, using whole leaves, late spring to mid-summer, heated propagating case, 15.5–21°C (60–70°F).
Seeds sown early to mid-spring and germinated in heated propagating case, 21°C (70°F).

× CRINODONNA (× **AMARCRINUM**) (**AMARYLLIS** × **CRINUM**) B
Division of the clumps in mid-spring, potting the portions immediately.
Bulblets removed mid-spring and potted individually.

CRINUM B
Division of clumps during mid-spring, potting the portions individually.
Bulblets removed mid-spring and potted individually. Should flower in three years.

Seeds sown as soon as ripe, or during spring, germinated in a heated propagating case, 21°C (70°F). Seedlings can take five years to flower.

CROSSANDRA P/S S/S
Soft or semi-ripe cuttings taken between mid-spring and late summer and rooted in a heated propagating case, 21°C (70°F).
Seeds sown early to mid-spring and germinated in a heated propagating case, 18°C (65°F).

CROTALARIA S/P
Seeds sown during spring after soaking in warm water, heated propagating case.
Semi-ripe cuttings mid-to late summer, heated propagating case.

CRYPTANTHUS (Earth star) P
Offsets or suckers with roots attached, removed when well developed, early summer, immediately potted and established in a heated propagating case.

× CRYPTBERGIA (CRYPTANTHUS × BILLBERGIA) P
As for *Cryptanthus*

CRYPTOCORYNE Aq
Division during spring or summer, replanting the portions immediately.

CTENANTHE P
Division during mid-spring, potting the portions immediately.

CUCUMIS (Melon) An
Seeds sown during early spring in a heated greenhouse. Sow them singly in small pots and germinate in a heated propagating case, 15.5°C (60°F). Plant in cropping positions when plants have formed five leaves. For raising and cropping in an unheated greenhouse sow at the end of mid-spring and germinate in above temperature. Melons for cropping in garden frames or cloches should be raised in a heated propagating case as above during mid-spring and planted in late spring.

CUCURBITA (Ornamental gourds) An
Seeds sown in individual pots during late

spring and germinated in a garden frame. Alternatively sow outdoors at the end of late spring.

CUPHEA P/An/S
Softwood or semi-ripe cuttings for shrubby species, between mid-spring and mid-summer, heated propagating case, 15.5–18°C (60–65°F).
Softwood cuttings for perennial species, early to mid-spring, heated propagating case, 15.5–18°C (60–65°F).
Seeds for perennial species sown mid- to late winter and germinated in heated propagating case, 15.5°C (60°F).

CUPRESSUS (Cypress) C
Semi-ripe cuttings for *C. cashmeriana*, late summer, heated propagating case.

CUSSONIA T/S
Air layering during spring under normal greenhouse conditions.

CYATHEA T F
Spores sown as soon as ripe or during spring, heated propagating case, 21°C (70°F), with good light but shade from sun, and constant moisture.

CYCAS P
Seeds sown early to mid-spring and germinated in heated propagating case, 21–24°C (70–75°F).
Suckers removed when well developed with roots attached and immediately potted. Establish in heated propagating case.

CYCLAMEN Tu P
Seeds for pot plants sown late summer or early autumn, or mid- to late winter, germinating them in a heated propagating case, 15.5°C (60°F).

Fig. 20.4 Cymbidiums may be propagated by removing mature leafless pseudobulbs while repotting in spring and potting them separately.

CYMBIDIUM O
Division early to mid-spring, immediately potting the portions.
Pseudobulbs, mature and without leaves, removed while repotting early to mid-spring and potted separately. Leave at least four pseudobulbs on parent plant (Fig. 20.4).

Fig. 20.5 The bract rosettes of cyperus (umbrella plant) may be used as cuttings in spring or summer, rooting them in heat. For convenience, the bracts may be reduced in length.

CYPERUS (Umbrella plant) P
Division mid-spring, potting the portions immediately.
Cuttings formed of bract rosettes, between mid-spring and late summer, heated propagating case, 15.5°C (60°F) (Fig. 20.5).
Seeds sown early to mid-spring and germinated in heated propagating case, 18–21°C (65–70°F).

CYPHOMANDRA (Tree tomato) T/S
Seeds sown early to mid-spring and germinated in heated propagating case, 18°C (65°F).

CYPRIPEDIUM (Slipper orchid) O
Division mid-spring, potting the portions immediately.

CYRTANTHUS B
Bulblets removed in spring and potted individually.
Seeds sown during spring and germinated in a heated propagating case, 15.5°C (60°F).

CYRTOMIUM F
Division during mid-spring, potting the portions immediately.
Spores sown as soon as ripe and germinated in a heated propagating case, 21°C (70°F) with good light but shade from sun, and constant moisture.

:YTISUS (Broom) S
;emi-ripe cuttings with heel late summer/ :arly autumn, heated propagating case, 5.5–18°C (60–65°F).
;eeds sown early to mid-spring and germi-ıated in heated propagating case, 15.5°C 60°F).

)AHLIA Tu P
)ivision of the dormant tubers when plant-ng outdoors during early or mid-spring.
;oft basal cuttings taken from plants tarted into growth under glass, late winter o early spring, heated propagating case, 5.5–18°C (60–65°F).
;eeds sown late winter or early spring and erminated in a heated propagating case, 5.5°C (60°F).

)ARLINGTONIA (Cobra lily) Ca P
)ffshoots removed when well developed, vhen repotting in spring and potted indi-idually.
;eeds sown as soon as ripe in the normal otting compost, placing the pan in shallow vater to keep compost constantly moist; ool greenhouse.

)ASYLIRION (Bear grass) P
;eeds sown during spring in heated prop-gating case, 15.5°C (60°F).
'uckers removed during spring and potted ndividually.

)ATURA (Angel's trumpets) S/T/An
'oft or semi-ripe cuttings for shrubby pecies, late spring to early autumn, heated ropagating case, 15.5–18°C (60–65°F).
eeds for annual species, sown early to mid-pring and germinated in a heated propagat-ıg case, 15.5°C (60°F).

)AVALLIA F
)ivision mid-spring, potting the portions ıdividually.
pores sown as soon as ripe, heated prop-gating case, 21°C (70°F), with good light ut shade from sun, and constant moisture.
.hizome sections 8 cm (3 in) long with əme leaves, pinned down into compost, əring.

DENDROBIUM O
Division after flowering, potting the portions immediately.
Pseudobulbs, mature and without leaves, removed while repotting after flowering and potted separately. Leave at least four pseudobulbs on parent plant. Long pseudobulbs can be cut into sections with two nodes and used as cuttings. Root in heated propagating case.

DIANTHUS (Carnation) P
Softwood cuttings for perpetual-flowering carnations, taken early spring and rooted in heated propagating case, 15.5–18°C.
Seeds for annual carnations used for summer bedding, and for perpetual-flowering carnations, sown mid- to late winter and germinated in a heated propagating case, 15.5°C (60°F).

DICHORISANDRA P
Division during early spring, replanting or potting the portions immediately.
Cuttings taken during summer and rooted in heated propagating case, 21°C (70°F).

DICKSONIA Tf
Spores sown during spring and germinated in heated propagating case, 15.5°C (60°F).

DIEFFENBACHIA (Dumb cane) P
Cuttings consisting of stem sections, early summer, heated propagating case, 21–24°C (70–75°F). The poisonous sap must be kept away from mouth and eyes.
Air layering spring or summer, normal greenhouse conditions.

DIMORPHOTHECA (African daisy, Cape marigold) An/P/S
Seeds for annuals, and other types, sown early spring and germinated in heated propagating case, 18°C (65°F).
Semi-ripe cuttings for perennials and shrubby species, late summer, heated propagating case, 15.5°C (60°F), or garden frame.

DIONAEA (Venus fly trap) Ca P
Division mid-spring, potting the portions immediately in individual pots.
Seeds sown early to mid-spring in the

normal potting compost. Stand the container in water to keep moist. Unheated propagating case. Seedlings can take up to seven years to flower.
Leaf cuttings spring or summer, entire leaves, heated propagating case, 21–26°C (70–80°F), and high humidity. Root them in sphagnum moss.

DIOSCOREA Tu P
Division early or mid-spring, potting the portions immediately.
Tubers cut into pieces in spring, potted and established in heated propagating case.
Soft basal cuttings as soon as available in spring and rooted in heated propagating case.

DISA O
Division when plants are dormant, immediately potting the portions. Only divide plants with six new growths – split into two equal portions.

DIZYGOTHECA S/T
Air layering mid-spring to late summer, normal greenhouse conditions.
Seeds sown early to mid-spring and germinated in heated propagating case, 21°C (70°F).
Cuttings in the form of stem sections, summer, heated propagating case, 21°C (70°F).

DOROTHEANTHUS (Includes Livingstone daisy) Su An
Seeds sown late winter or early spring and germinated in heated propagating case, 15.5°C (60°F).

DRACAENA S/T
Cuttings in the form of stem sections, early to mid-summer, heated propagating case, 21–24°C (70–75°F).
Leaf-bud cuttings as above.
Seeds sown early to mid-spring and germinated in heated propagating case, 26°C (80°F).
Air layering between mid-spring and late summer, normal greenhouse conditions.

DROSERA (Sundew) Ca P
Division early to mid-spring, potting the portions immediately.
Seeds sown during early to mid-spring o[n] the normal potting compost. Keep th[e] compost constantly moist. Germinate i[n] unheated propagating case.
Leaf cuttings late spring or early summe[r], unheated propagating case, high humidit[y]. Lay flat on sphagnum moss and light[ly] cover with moss.
Root cuttings, spring, inserted in spha[g]num moss and rooted in heated propagatin[g] case, 21°C (70°F).

DURANTA S/T
Softwood cuttings mid- to late sprin[g], heated propagating case, 21°C (70°F).

DYCKIA P
Offsets removed in spring when we[ll] developed, potted and established in heate[d] propagating case.

ECCREMOCARPUS (Chilean glory vine) Cl
Seeds sown late winter or early spring an[d] germinated in heated propagating cas[e], 15.5°C (60°F).

Fig. 20.6 To propagate the succulent echeverias from leaf cuttings, remove entire leaves between late spring and mid-summer and root them in heat.

ECHEVERIA Su
Division if feasible, mid-spring, potting th[e] portions in individual pots.
Offsets removed with some roots attache[d], late spring to mid-summer, and potted ind[i]vidually.
Leaf cuttings late spring to mid-summe[r], heated propagating case, 15.5–21°C (6[0]–70°F). Use entire leaves (Fig. 20.6).

Seeds sown early to mid-spring and germinated in heated propagating case, 21°C (70°F).

ECHINOCACTUS Cc
Seeds sown early to mid-spring and germinated in heated propagating case, 21°C (70°F).

ECHINOCEREUS Cc
Seeds sown early to mid-spring and germinated in heated propagating case, 21°C (70°F).
Cuttings for branching kinds, late spring to mid-summer, heated propagating case, 15.5–21°C (60–70°F).

ECHINOFOSSULOCACTUS Cc
Seeds sown early to mid-spring and germinated in heated propagating case, 21°C (70°F).

ECHINOPSIS Cc
Seeds sown early to mid-spring and germinated in heated propagating case, 21°C (70°F).
Offsets removed with some roots attached during mid-spring and potted into individual pots.

EDGEWORTHIA S
Semi-ripe cuttings early to late summer, heated propagating case, 15.5°C (60°F), or garden frame.

EICHHORNIA (Water hyacinth) Fl Aq
Young plants which form on runners are removed during summer and replaced in the water.

EMILIA An/P
Seeds sown during spring and germinated in a heated propagating case.

ENCEPHALARTOS S/T
Seeds sown during spring and germinated in heated propagating case, 29°C (85°F). Seeds not always viable.

EPACRIS S
Semi-ripe cuttings early to late summer, heated propagating case, 15.5°C (60°F).

EPIDENDRUM O
Division of established plants when repotting in spring.
Cuttings of stem-like pseudobulbs taken after flowering and rooted in heated propagating case.

EPIPHYLLUM (Orchid cactus) Cc
Cuttings consisting of stem sections, late spring to mid-summer, heated propagating case, 15.5–21°C (60–70°F).
Seeds sown early to mid-spring and germinated in heated propagating case, 21°C (70°F).

EPIPREMNUM (SCINDAPSUS)
(Includes devil's ivy) Cl
Cuttings consisting of stem sections or tips, early to late summer, heated propagating case, 21–24°C (70–75°F).
Leaf-bud cuttings as above.

EPISCIA (Includes flame violet) P
Division mid-spring, potting the portions in individual pots.
Cuttings taken early to late summer and rooted in a heated propagating case, 24°C (75°F).
Plantlets at ends of stolons removed, potted and rooted in heated propagating case, 24°C (75°F).
Leaf cuttings, using entire leaves complete with stalk, spring or summer, heated propagating case, 24°C (75°F).

ERANTHEMUM S/P
Softwood cuttings taken during late spring and rooted in heated propagating case, 21°C (70°F).

ERICA (Cape heath) S
Softwood cuttings taken between mid-spring and early summer and rooted in a heated propagating case, 15.5°C (60°F).

ERYTHRINA (Coral tree) T/S/P
Seeds sown early to mid-spring and germinated in a heated propagating case, 21°C (70°F).
Softwood cuttings taken with a heel during spring or early summer and rooted in a heated propagating case, 21°C (70°F).

ESPOSTOA Cc
Seeds sown early to mid-spring and germinated in a heated propagating case, 21°C (70°F).

EUCALYPTUS (Gum tree) T
Seeds sown early to mid-spring and germinated in a heated propagating case, 15.5°C (60°F).

EUCHARIS B
Bulblets removed and potted individually in spring.
Seeds sown during spring and germinated in heated propagating case, 21°C (70°F).

EUPHORBIA S/P/Su
Softwood cuttings for the species *E. fulgens, E. pulcherrima* (poinsettia) and *E. splendens* (crown of thorns), late spring or early summer, heated propagating case, 21°C (70°F).
Seeds for succulent species, early to mid-spring, heated propagating case, 21°C (70°F).
Cuttings for succulent species, late spring to mid-summer, heated propagating case, 15.5–21°C (60–70°F).

EUSTOMA An/P
Seeds sown during late winter and germinated in a heated propagating case, 21–24°C (70–75°F).

EXACUM An/Bi/P
Seeds sown early spring and germinated in a heated propagating case, 18°C (65°F). A sowing may also be made during late summer if winter warmth is available.

FASCICULARIA P
Offsets removed during spring, when well developed and with roots, potted individually and established in heated propagating case.

FAUCARIA (Tiger's jaw) Su
Seeds sown early to mid-spring and germinated in heated propagating case, 21°C (70°F).
Cuttings late spring to mid-summer, heated propagating case, 15.5–21°C (60–70°F).

FEIJOA S
Semi-ripe cuttings with a slight heel, mid to late summer, heated propagating case 15.5–18°C (60–65°F).
Simple layering between mid-spring and late summer, *in situ*.
Seeds sown early to mid-spring, heated propagating case, 15.5°C (60°F).

FELICIA (Includes kingfisher daisy)
An/S S
Seeds for annuals, sown late winter or early spring and germinated in heated propagating case, 15.5°C (60°F).
Softwood cuttings for sub-shrubs, late spring or early summer, heated propagating case, 15.5°C (60°F).

FENESTRARIA Su
Seeds sown early to mid-spring and germinated in heated propagating case, 21°C (70°F).

FEROCACTUS Cc
Seeds sown early to mid-spring and germinated in heated propagating case, 21°C (70°F).

FICUS (Includes figs and rubber plants)
T/S/Cl
Soft or semi-ripe cuttings between mid spring and mid-summer, heated propagating case, 21–24°C (70–75°F).
Leaf-bud cuttings, especially for *F. elastica* (rubber plant), as above.
Air layering between mid-spring and late summer, normal greenhouse conditions.
Seeds sown early to mid-spring and germinated in heated propagating case, 21–24°C (70–75°F). Light needed for germination which may be slow.

FITTONIA (Net leaf, nerve plant) Cr P
Division mid-spring, potting the portions and establishing them in a heated propagating case, 21°C (70°F).
Cuttings taken early to late summer and rooted in heated propagating case, 21°C (70°F).

FRANCOA P
Seeds sown mid-spring and germinated in heated propagating case, 15.5°C (60°F).

Division mid-spring, potting the portions immediately.

FREESIA Cm
Seeds sown thinly in flowering containers (pots or boxes) from early spring to early summer, in succession if desired, germinating them in a cool greenhouse or garden frame. First soak them in water for 24 hours to soften hard seed coats. Thin out surplus seedlings.
Cormlets removed during late summer or early autumn (planting time) and replanted in pots or boxes. Grow on in cool greenhouse. They may bloom the following year.

FUCHSIA S
Softwood cuttings as soon as available in spring, heated propagating case, 15.5°C (60°F).
Semi-ripe cuttings late summer, as above.
Seeds sown early or mid-spring and germinated in heated propagating case, 15.5°C (60°F).

GARDENIA (Includes Cape jasmine) S/T
Softwood cuttings with a heel in spring as soon as available, heated propagating case, 18–21°C (65–70°F).
Semi-ripe cuttings late summer, as above.

GASTERIA Su
Leaf cuttings late spring to mid-summer, heated propagating case, 15.5–21°C (60–70°F).
Division or offsets mid-spring, potting the portions immediately.
Seeds sown early to mid-spring and germinated in heated propagating case, 21°C (70°F).

GAZANIA P, usually grown as An
Seeds sown late winter or early spring and germinated in heated propagating case, 15.5°C (60°F).
Semi-ripe cuttings late summer, heated propagating case, 15.5°C (60°F).

GEOGENANTHUS (Seersucker plant) P
Cuttings taken during spring, heated propagating case, 21°C (70°F).

GERBERA (Barberton daisy) P
Division mid-spring, potting the portions immediately.
Cuttings of side shoots spring or summer, heated propagating case, 15.5–18°C (60–65°F).
Seeds sown early or mid-spring and germinated in heated propagating case, 15.5–18°C (60–65°F).

GESNERIA P/S
Seeds sown early to mid-spring and germinated in heated propagating case, 21°C (70°F).
Cuttings, using non-flowering shoots, spring or summer, heated propagating case, 21°C (70°F).

GIBBAEUM Su
Seeds sown early to mid-spring and germinated in heated propagating case, 21°C (70°F).
Division of established clumps in spring, potting the portions immediately.

GLADIOLUS Cm
Cormlets, planting them in outdoor nursery bed during mid-spring. They can take two or three years to attain flowering size. Lift in autumn, dry off and store frost-free over winter.
Seeds sown early to mid-spring and germinated in heated greenhouse, 7–13°C (45–55°F). Dry off the small corms in autumn, then treat as above. They take two or three years to reach flowering size.

GLECHOMA (ground ivy) P
Division during spring, potting the portions immediately.
Cuttings taken between mid-spring and late summer and rooted in heated propagating case, 15.5°C (60°F).

GLORIOSA (Climbing lily) Tu Cl
Offsets removed and potted during late winter or early spring, when starting parents into growth.
Seeds sown late winter or early spring, individually in small pots, and germinated in heated propagating case, 21–24°C (70–75°F). Plants can take up to four years to come into flower.

GLOTTIPHYLLUM Su
Cuttings late spring to mid-summer, heated propagating case, 15.5–21°C (60–70°F).
Seeds sown early to mid-spring and germinated in heated propagating case, 21°C (70°F).

GOMPHRENA (Globe amaranth) An
Seeds sown early spring and germinated in heated propagating case, 15.5–18°C (60–65°F).

GRAPTOPETALUM Su
Leaf cuttings late spring to mid-summer, heated propagating case, 15.5–21°C (60–70°F).
Seeds sown early to mid-spring and germinated in heated propagating case, 21°C (70°F).

GREVILLEA S/T
Seeds sown early to mid-spring and germinated in a heated propagating case, 15.5°C (60°F), after soaking in water for 48 hours.
Semi-ripe cuttings early to mid-summer, heated propagating case, 15.5–18°C (60–65°F).

GUZMANIA P
Offsets late spring or summer, well developed and with roots, potted individually and established in a heated propagating case.

GYMNOCALYCIUM Cc
Offsets removed in summer and potted individually.
Seeds sown early to mid-spring and germinated in heated propagating case, 21°C (70°F).

GYNURA P/S/Cl
Cuttings taken between late spring and late summer and rooted in heated propagating case, 21°C (70°F).

HABRANTHUS B
Bulblets removed while repotting in spring, and potted individually,
Seeds sown during spring and germinated in heated propagating case, 15.5°C (60°F).

HAEMANTHUS (Blood lily) B
Bulblets removed during early spring while repotting and potted individually.
Seeds sown as soon as ripe and germinated in a heated propagating case, 15.5–18°C (60–65°F).

HAKEA S/Sm T
Seeds sown in spring and germinated in a heated propagating case, 15.5°C (60°F).
Semi-ripe cuttings with a heel taken in early or mid-summer and rooted in a heated propagating case.

HAWORTHIA Su
Offsets removed late spring to mid-summer and potted individually.
Division during spring, potting the portions immediately.
Seeds sown early to mid-spring and germinated in heated propagating case, 21°C (70°F).

HECHTIA P
Offsets, well developed, removed with roots attached during spring, potted and established in heated propagating case.

HEDERA (Ivy) Cl
Soft or semi-ripe cuttings early to mid-summer, heated propagating case, 15.5°C (60°F).
Leaf-bud cuttings mid-spring to early summer, heated propagating case, 15.5°C.
Simple or serpentine layering late spring or early summer, *in situ*.

HEDYCHIUM (Ginger lily) P
Division in mid-spring, potting or planting the portions immediately.

HELIAMPHORA (Sun pitcher) Ca P
Division of the rhizomes during spring, taking care to avoid damaging the brittle roots. Each portion should contain several pitchers.

HELICHRYSUM P/An/S S/S
Seeds for annuals, sown late winter or early spring and germinated in heated propagating case, 18°C (65°F).
Semi-ripe cuttings for shrubby kinds during summer, heated propagating case, 15.5°C (60°F).

IELICONIA (Lobster claws) P
ffsets removed during late spring and nmediately potted or replanted.

IELIOCEREUS Cc
uttings late spring to mid-summer, heated ropagating case, 15.5–21°C (60–70°F).
eeds sown early to mid-spring and germiated in a heated propagating case, 21°C.

IELIOTROPIUM (Heliotrope) S
oftwood cuttings as soon as available in oring, heated propagating case, 15.5–18°C 50–65°F).
emi-ripe cuttings late summer, heated ropagating case, 15.5–18°C (60–65°F).
eeds sown late winter or early spring and erminated in heated propagating case, 5.5–18°C (60–65°F).

IEMIGRAPHES P
uttings taken in spring or summer and ooted in a heated propagating case, 21°C 70°F).

IIBISCUS (Shrubby mallow) S
oftwood or semi-ripe cuttings mid-spring late summer, with a heel, heated propgating case, 18°C (65°F).

IIPPEASTRUM B
ulblets removed and potted individually in utumn, when repotting.
eeds sown as soon as ripe or in early spring nd germinated in heated propagating case, 5.5–18°C (60–65°F).

IOWEA (KENTIA) Pm
eeds sown during spring and germinated n heated propagating case, 27°C (80°F).

IOYA (Wax flower) Cl/S
emi-ripe cuttings in summer, heated propgating case, 15.5–18°C (60–65°F).

IUERNIA Su
eeds sown early to mid-spring and germiated in heated propagating case, 21°C 70°F).
uttings late spring to mid-summer, heated ropagating case, 15.5–21°C (60–70°F).

HUMEA Bi
Seeds for *H. elegans* sown during mid-summer and germinated in a garden frame or cool greenhouse.

HUNNEMANNIA P, grown as An
Seeds sown during early spring and germinated in heated propagating case, 15.5°C (60°F). Alternatively sow outdoors during mid-spring where the plants are to flower.

HYDRANGEA S
Softwood cuttings mid-spring and rooted in a heated propagating case, 15.5°C (60°F).
Semi-ripe cuttings late summer, as above.

HYLOCEREUS Cc
Cuttings late spring to mid-summer, heated propagating case, 15.5–21°C (60–70°F).

HYMENOCALLIS (Spider lily) B
Bulblets removed in mid-spring when repotting and potted individually. They will flower within two to three years.

HYPOCYRTA (ALLOPLECTUS) (Clog plant) P
Semi-ripe cuttings early to late summer, heated propagating case, 21°C (70°F).

HYPOESTES (Polka dot plant) P/S S
Seeds sown early to mid-spring and germinated in heated propagating case, 18–21°C (65–70°F).
Cuttings taken spring or summer and rooted in heated propagating case, 18–21°C (65–70°F).

IMPATIENS (Busy lizzy, balsam) An/P
Seeds for perennials and annuals, sown early spring and germinated in heated propagating case, 15.5–18°C (60–65°F).
Cuttings for perennials, mid-spring to late summer, heated propagating case, 15.5°C (60°F).

IPOMOEA (Morning glory) An/P Cl
Seeds sown early to mid-spring and germinated in heated propagating case, 18°C (65°F). Seeds should first be soaked in tepid water for half a day or so, to speed up germination.

Fig. 20.7 Some kalanchoe (bryophyllum) species produce plantlets on their leaves. These may be removed when developed and rooted in pots or trays in heat.

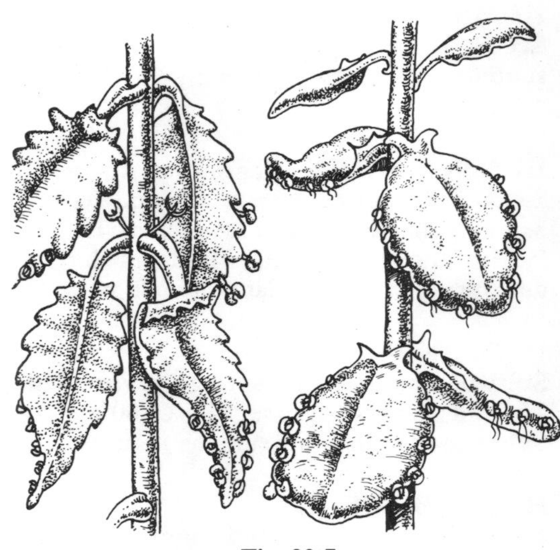

Fig. 20.7

IPOMOPSIS P/Bi
Seeds sown early spring and germinated in heated propagating case.

IRESINE P
Soft cuttings taken mid-spring to early summer and rooted in heated propagating case, 18–21°C (65–70°F).

IXORA S
Semi-ripe cuttings during summer, heated propagating case, 21–24°C (70–75°F).

JACARANDA T
Semi-ripe cuttings with heel, early to mid-summer, heated propagating case, 21°C (70°F).
Seeds sown early to mid-spring and germinated in heated propagating case, 21–24°C (70–75°F).

JASMINUM (Jasmine) S/Cl
Semi-ripe cuttings with heel, late summer or early autumn, heated propagating case, 15.5°C (60°F).

JOVELLANA P/S S
Semi-ripe cuttings during summer, heated propagating case, 15.5°C (60°F).

JUBAEA (Chilean wine palm) Pm
Seeds sown during spring and germinated in heated propagating case, 27°C (80°F).

JUSTICIA (JACOBINIA) S/P
Softwood cuttings mid-spring to earl summer, heated propagating case, 18–21°((65–70°F).

KALANCHOE (BRYOPHYLLUM) Su P/S
Seeds sown early to mid-spring and germi nated in heated propagating case, 21°C.
Cuttings late spring to mid-summer, heate propagating case, 15.5–21°C (60–70°F).
Leaf cuttings as above.
Plantlets produced on leaves, remove when developed and rooted in pots or trays heated propagating case, 15.5°C (60°F (Fig.20.7).

KOCHIA (Burning bush, summer cypress) A
Seeds sown early spring and germinated i heated propagating case, 15.5°C (60°F). May also be sown outdoors during mi spring, in permanent positions.

KOHLERIA P
Division of rhizomes during winter or earl spring while repotting. Each piece mus have a growth bud. Establish in heate propagating case, 21°C (70°F).
Cuttings in summer, heated propagatin case, 18°C (65°F).

LACHENALIA (Cape cowslip) B
Bulblets removed during late summer whil repotting and potted individually.
Seeds sown as soon as ripe or during earl spring and germinated in heated propaga ing case, 15.5°C (60°F).

LAELIA O
Division during spring or after flowering potting the portions immediately.
Pseudobulbs, mature and without leaves removed while repotting in spring an potted separately. Leave at least fou pseudobulbs on parent plant.

LAGERSTROEMIA (Crape myrtle) S/T
Semi-ripe cuttings in summer, heated prop agating case, 18°C (65°F).
Seeds sown during autumn or spring an germinated in heated propagating case, 21 24°C (70–75°F).

Root cuttings early to mid-winter, heated ›ropagating case, 18°C (65°F).

.AMPRANTHUS Su
Cuttings late summer, heated propagating ase, 15.5–21°C (60–65°F).
eeds sown early to mid-spring and germiated in heated propagating case, 21°C.

.ANTANA S
oftwood cuttings mid-spring to early ımmer, heated propagating case, 15.5–8°C (60–65°F).
emi-ripe cuttings late summer, as above.
eeds sown late winter and germinated in eated propagating case, 15.5°C (60°F).

.APAGERIA (Chilean bell flower) Cl
imple layering between mid-spring and ırly summer, *in situ.*
eeds sown as soon as ripe, or in spring, eated propagating case, 15.5–18°C (60–5°F). Seeds are slow to germinate. Soak in 'ater for 48 hours before sowing.

.APEIROUSIA Cm
ormlets removed in early spring while ›potting and potting separately.
eeds sown during spring and germinated heated propagating case, 15.5°C (60°F).

ATANIA (Latan palm) Pm
eeds sown during spring and germinated heated propagating case, 27°C (80°F).

EONOTIS (Lion's ear) S
emi-ripe cuttings early to mid-summer, eated propagating case, 15.5–18°C (60–5°F).

EPTOSPERMUM S/T
emi-ripe cuttings mid-to late summer, eated propagating case, 15.5°C (60°F). ʼound at base by removing a thin slice of ırk to expose wood.
eeds sown early to mid-spring in a heated ropagating case, 18°C (65°F). May be slow germinate.

EUCADENDRON (Silver tree) T
eeds sown during spring and germinated heated propagating case, 15.5°C (60°F).

LEUCOCORYNE B
Bulblets removed during autumn and potted separately.

LIMONIUM (Sea lavender, statice) An/P
Seeds of annual species sown late winter or early spring and germinated in a heated propagating case, 15.5°C (60°F).

LITHOPS (Living stones) Su
Seeds sown early to mid-spring and germinated in heated propagating case, 21°C (70°F).
Division in summer, potting plant bodies individually.

LIVISTONA (Fountain palm) Pm
Seeds sown during spring and germinated in heated propagating case, minimum 24°C (75°F).

LOBELIA P/S/An
Seeds for annuals (used for bedding schemes) and perennials, sown late winter or early spring, heated propagating case, 15.5–18°C (60–65°F).
Division for perennials early to mid-spring, replanting the portions immediately.
Soft or semi-ripe cuttings for shrubby species, mid-spring to late summer, heated propagating case.

LOBIVIA Cc
Seeds sown early to mid-spring, heated propagating case, 21°C (70°F).
Offsets with or without roots removed in spring or early summer and potted individually.

LOPHOPHORA (Dumpling cactus, peyote) Cc
Seeds sown early to mid-spring, heated propagating case, 21°C (70°F).

LOTUS P
Softwood cuttings early to mid-summer, heated propagating case, 21°C (70°F).
Seeds sown during spring and germinated in heated propagating case, 18°C (65°F).

LUCULIA S
Semi-ripe cuttings with heel, mid-to late summer, heated propagating case, 18°C.

Seeds sown during spring and germinated in heated propagating case, 18°C (65°F).

LUFFA (Loofah) An Cl
Seeds sown singly in small pots, early to mid-spring, and germinated in heated propagating case, 24–29°C (75–85°F).

LYCASTE O
Division when repotting or replanting in spring.
Pseudobulbs, mature and without leaves, removed while repotting or replanting in spring and potted separately. Leave at least four pseudobulbs on parent plant.

LYCOPODIUM (Club moss) P
Division during mid-spring, potting the portions immediately.
Spores sown as soon as ripe and germinated in heated propagating case, 21°C (70°F).

LYCORIS B
Bulblets removed in late summer and potted individually.
Seeds sown as soon as ripe and germinated in heated greenhouse.

MAMMILLARIA Cc
Division, or removal of offsets, summer, potting the portions individually.
Seeds sown early to mid-spring, heated propagating case, 21°C (70°F).

MANDEVILLA (DIPLADENIA) Cl
Semi-ripe cuttings early to mid-summer, heated propagating case, 18–21°C (65–70°F).
Seeds sown early to mid-spring and germinated in heated propagating case, 18°C (65°F).

MANETTIA Cl
Softwood or semi-ripe cuttings during summer, heated propagating case, 18–21°C (65–70°F).

MARANTA (Includes prayer plant) P
Division during mid-spring, potting the portions immediately.
Cuttings with about three leaves attached, late spring or summer, heated propagating case, 21°C (70°F).

MARTYNIA (Unicorn plant) An
Seeds sown early spring and germinated in heated propagating case, 18°C (65°F).

MASDEVALLIA O
Division of well-established plants during spring while repotting, ensuring each portion has four to six stems.

MAXILLARIA O
Division during spring when repotting.
Pseudobulbs, mature and without leaves removed while repotting during spring and potted separately. Leave at least four pseudobulbs on parent plant.

MEDINILLA (Includes rose grape) S/Cl
Softwood or semi-ripe cuttings between mid-spring and late summer, heated propagating case, 24°C (75°F).

MELALEUCA (Bottle brush) S/T
Semi-ripe cuttings mid-to late summer heated propagating case, 18°C (65°F).

MELIA (Bead tree) T
Seeds sown as soon as ripe and germinated in heated propagating case, 18°C (65°F).
Semi-ripe cuttings during summer, heated propagating case, 18°C (65°F).

MELIANTHUS (Honey bush) S/P
Semi-ripe cuttings mid-to late summer heated propagating case, 18–21°C (65–70°F).
Seeds sown mid-spring and germinated in heated propagating case, 18–21°C (65–70°F).
Division or suckers mid-spring, potting or replanting immediately.

METROSIDEROS T/S
Semi-ripe cuttings in summer, heated propagating case, 18–21°C (65–70°F).
Simple layering during spring, *in situ.*
Seeds sown early to mid-spring and germinated in heated propagating case, 18°C.

MICHELIA T/S
Semi-ripe cuttings late summer, heated propagating case, 18–21°C (65–70°F).
Simple layering mid-spring to summer, *in situ.*

IKANIA Cl/S/P
uttings taken in summer and rooted in ated propagating case, 18°C (65°F).

ILLA B
ulblets removed in spring when repotting d potted individually.

ILTONIA O
ivision during autumn or early spring, tting the portions immediately.
eudobulbs, mature and without leaves, moved while repotting in autumn or early ring, and potted separately. Leave at least ur pseudobulbs on parent plant.

IMOSA (Sensitive plant) P, grown as An
eds sown early to mid-spring and germi- ted in heated propagating case, 18–21°C 5–70°F).

INA Cl
eds sown early to mid-spring and germi- ted in heated propagating case, 18°C 5°F). First soak seeds in tepid water for lf a day.

IRABILIS (Marvel of Peru) An/P
eds sown early to mid-spring and germi- ted in heated propagating case, 18°C 5°F).
vision in early spring for tuberous peren- als.

ITRARIA Cl S
mi-ripe cuttings taken during summer d rooted in heated propagating case, .5°C (60°F).
vision mid-spring, potting or replanting portions immediately.

ONSTERA (Includes Swiss cheese plant)

af-bud cuttings early to late summer, ated propagating case, 24–27°C (75– F).
ttings of shoot tips as above.
nple layering spring or summer, *in situ.*

ORAEA Cm
rmlets removed and potted separately ring early autumn.
eds sown early to mid-spring and germi- nated in heated propagating case, 15.5–18°C (60–65°F).

MUCUNA Cl
Simple layering during summer, *in situ.*
Seeds sown during spring and germinated in heated propagating case, 18–24°C (65–75°F). Seeds should be soaked in warm water for two hours before sowing.

MUSA (Banana) P
Division late spring to mid-summer, potting the portions immediately.
Suckers or offsets removed with roots attached late spring to mid-summer and immediately potted.
Seeds sown mid-spring and germinated in heated propagating case, 21°C (70°F).

NELUMBO (Lotus) Aq
Division mid-spring, replanting the portions immediately.
Seeds sown singly in small pots during spring, shallowly submerged in water and germinated at 18–24°C (65–75°F). Hard seed coat should be filed.

NEMATANTHUS P/S
Softwood or semi-ripe cuttings during summer, rooted in heated propagating case, 21°C (70°F).

NEMESIA An
Seeds sown late winter or early spring and germinated in heated propagating case, 15.5°C (60°F).

NEOREGELIA P
Offsets removed when well developed and rooted, potted individually and established in a heated, humid propagating case.
Seeds sown early to mid-spring and germi- nated in heated propagating case, 21°C (70°F), with high humidity.

NEPENTHES (Pitcher plant) Ca P
Cuttings prepared from lateral shoots or stem tips, spring, heated propagating case with high humidity, 27–30°C (80–85°F).
Seeds sown during spring and germinated in heated propagating case with high humid- ity, minimum 24°C (75°F).

NEPHROLEPIS (Ladder fern) F
Division mid-spring, replanting the portions immediately in small pots.

NERINE (Includes Guernsey lily) B
Bulblets removed during late summer when repotting and potted individually.
Seeds sown as soon as ripe (these cannot be stored) and germinated in cool greenhouse.

NERIUM (Oleander) S
Semi-ripe cuttings mid-to late summer, heated propagating case, 15.5–18°C (60–65°F).
Seeds sown early to mid-spring and germinated in heated propagating case, 18–21°C (65–70°F).

NICOTIANA (Ornamental tobacco plant) An/P
Seeds sown late winter or early spring and germinated in heated propagating case, 18°C (65°F).

NIDULARIUM P
Offsets removed between mid-spring and mid-summer, when well developed and rooted, potted individually and established in heated propagating case.
Seeds sown early to mid-spring and germinated in heated propagating case, 21°C (70°F), with high humidity.

NOLANA P/S S, usually grown as An
Seeds sown during spring and germinated in heated propagating case, 15.5°C (60°F).

NOTOCACTUS Cc
Seeds sown early to mid-spring and germinated in a heated propagating case, 21°C (70°F).

NYMPHAEA (Tropical water lilies) Aq
Seeds sown as soon as ripe in pans, slightly submerged in water, temperature 24–27°C (75–80°F). Grow on seedlings throughout winter in same temperature.
Young plants formed on leaves. Remove and pot individually when roots have formed. Grow on throughout winter in water maintained at 15.5–18°C (60–65°F).
Small tubers formed at base of crown should be stored over winter in damp sand to perpetuate the plants.

OCHNA S/T
Semi-ripe cuttings with a heel, mid-to la summer, heated propagating case, 18°C.
Seeds sown early to mid-spring and gern nated in heated propagating case, 15.5 (60°F).

ODONTOGLOSSUM O
Division during late summer/early autum or early to mid-spring.
Pseudobulbs, mature and without leave removed while repotting and potted sep rately. Leave at least four pseudobulbs parent plant.

OLEA (Olive) T
Semi-ripe cuttings with a heel mid-to la summer, heated propagating case, 18–21 (65–70°F).
Seeds sown early to mid-spring and gern nated in a heated propagating case, 18–21 (65–70°F).

ONCIDIUM O
Division early to mid-spring, potting t portions immediately.
Pseudobulbs, mature and without leave removed while repotting in spring a potted separately. Leave at least fo pseudobulbs on parent plant.

OPLISMENUS (Basket grass) P G
Division spring or summer, potting t portions immediately.
Cuttings taken in spring or summer a rooted in heated propagating case, 15.5 (60°F).

OPUNTIA (Prickly pear) Cc
Cuttings of single mature pads late spring mid-summer, heated propagating cas 15.5–21°C (60–70°F) (Fig.20.8).
Seeds sown early to mid-spring and gern nated in a heated propagating case, 21 (70°F). First soak seeds in tepid water for hours to soften hard seed coats.

ORNITHOGALUM (Chincherinchee)
Bulblets removed when repotting in early mid-autumn and potted separately.
Seeds sown as soon as ripe or in early spri and germinated in a cool greenhouse.

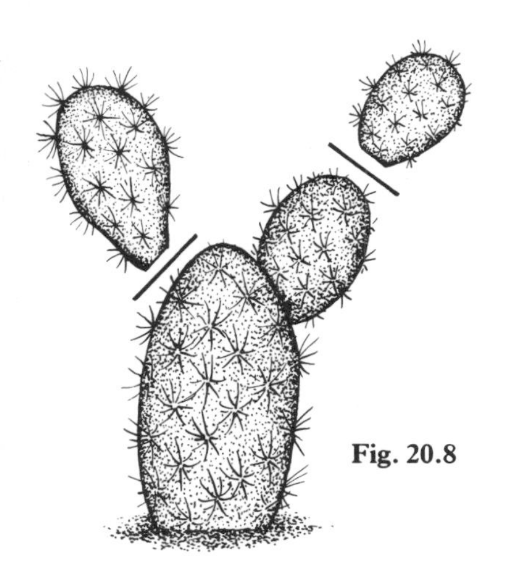
Fig. 20.8

Fig. 20.8 Opuntia species (prickly pears) are easily propagated from cuttings or single pads during spring or summer, rooting them in a heated propagating case.

STEOSPERMUM P
uttings taken during mid-summer and oted in a heated propagating case, 15.5°C 0°F).

XALIS P, inc Su/Tu
ivision between early and mid-spring, tting the portions immediately.
eds sown early or mid-spring and germi- ted in a heated propagating case, 15.5°C 0°F).
ittings for succulent species, late spring to id-summer, rooting them in a heated opagating case, 15.5–21°C (60–70°F).

XYPETALUM (TWEEDIA) H Cl
ittings late spring or summer, heated opagating case, 18–21°C (65–70°F).
eds sown during spring and germinated a heated propagating case, 15.5°C (60°F).

ACHYCEREUS Cc
eds sown early to mid-spring and germi- ted in a heated propagating case, 21°C 0°F).

ACHYPHYTUM Su
ittings late spring to mid-summer, heated opagating case, 15.5–21°C (60–70°F).
af cuttings, using entire leaves, as above.
eds sown early to mid-spring and germi- ted in a heated propagating case, 21°C 0°F).

PACHYSTACHYS (Includes lollipop plant) S/P
Cuttings taken during spring or summer and rooted in a heated propagating case, 18–21°C (65–70°F).

PAMIANTHE B
Bulblets removed during late winter and potted individually.
Seeds sown as soon as ripe and germinated in a heated propagating case.

PANCRATIUM B
Bulblets removed during late summer or early autumn and immediately potted into individual pots.
Seeds sown as soon as ripe and germinated in a heated propagating case, 15.5–18°C (60–65°F).

PANDANUS (Screw pine) T/S
Offsets or suckers removed with roots in mid-spring and potted individually. They should be established in a heated propagating case, 24°C (75°F).

PANDOREA (Includes wonga-wonga vine) Cl
Semi-ripe cuttings mid- to late summer rooted in a heated propagating case, 21°C (70°F).
Seeds sown early to mid-spring and germinated in a heated propagating case, 18°C (65°F).
Simple layering during summer, *in situ*.

PAPHIPOPEDILUM (Slipper orchid) O
Division of established plants, with at least six new growths, between late winter and late spring, potting the portions immediately. Split the plant in half so that each portion has three new growths.

PARODIA Cc
Seeds sown early to mid-spring and germinated in a heated propagating case, 21°C (70°F).

PASSIFLORA (Passion flower) Cl
Leaf-bud cuttings late spring or early summer, rooting them in a heated propagating case, 18°C (65°F).
Simple or serpentine layering between

mid-spring and late summer, *in situ*.
Seeds sown early or mid-spring and germinated in a heated propagating case, 18–21°C (65–70°).

PAVONIA S/P
Soft of semi-ripe cuttings taken between mid-spring and late summer and rooted in a heated propagating case, 21°C (70°F).
Seeds sown early to mid-spring and germinated in a heated propagating case.

PEDILANTHUS Su P
Cuttings prepared from the tips of stems during spring, allowing them to dry out for a day before inserting in a heated propagating case, 21°C (70°F). Keep only barely moist during rooting.

PELARGONIUM P/S/S S
Semi-ripe cuttings for perennials and shrubby species, including *P. domesticum* (regal pelargonium), *P. × hortorum* (zonal pelargonium) and *P. peltatum* (ivy-leaf pelargonium), late summer or early autumn (mid-summer if to be trained as standards), rooting them on the greenhouse staging or in a garden frame.
Seeds for all species and for the seed-raised strains of zonal pelargonium for summer bedding, mid- to late winter, germinating them in a heated propagating case, 21°C (70°F), avoiding temperature fluctuations which cause erratic germination.

PELLAEA (Cliffbrake) F
Division during mid-spring, potting the portions immediately. Each piece must have a portion of rhizome and at least two fronds.
Spores sown as soon as ripe and germinated in a heated propagating case, 21°C (70°F), with good light but shade from sun, and constant moisture.

PELLIONIA P
Division during mid-spring, potting the portions immediately.
Cuttings taken from the shoot tips during spring or early summer and rooted in a heated propagating case, 21–24°C (70–75°F).

PENTAS (Star cluster) S/P
Softwood cuttings taken between m
spring and early summer and rooted ir
heated propagating case, 18°C (65°F).
Seeds sown early to mid-spring and ger
nated in a heated propagating case, 18–21
(65–70°F).

PEPEROMIA (Pepper elder) P
Leaf cuttings, using mature entire lea
complete with petioles, between mid-spri
and late summer, rooting them in a hea
propagating case, 18–21°C (65–70°F).
Stem cuttings are possible with so
species, rooted as above.
Division in mid-spring, potting the portic
immediately.
Seeds sown early to mid-spring and gern
nated in a heated propagating case, 18–21
(65–70°F).

PERESKIA Cc
Stem cuttings taken late spring to m
summer and rooted in a heated propagati
case, 15.5–21°C (60–70°F).
Seeds sown early to mid-spring and gern
nated in a heated propagating case, 21
(70°F).

PERILLA An
Seeds sown during late winter or ea
spring and germinated in a heated propag
ing case, 18°C (65°F).

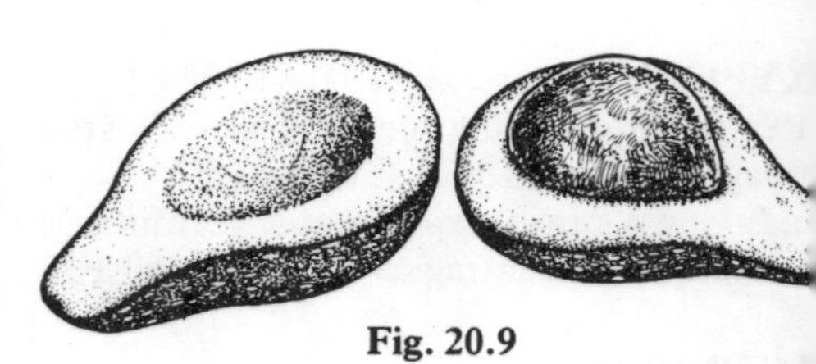
Fig. 20.9

Fig. 20.9 The avocado pear (persea) produces a single large seed ('stone'). This should be sown in a pot as soon as available, lightly covered with compost and germinated in a heated propagating case.

PERSEA (Avocado pear) T
Seeds sown as soon as available and gern
nated in a heated propagating case, 18–21
(65–70°F). Sow seeds in individual pots a
lightly cover with compost (Fig.20.9).

PETREA S/Cl
Semi-ripe cuttings taken in summer a
rooted in a heated propagating case, 1
21°C (65–70°F).

PETUNIA P/An
Seeds sown during early spring and germinated in a heated propagating case, 15.5°C (60°F).

PHAIUS O
Division when repotting after flowering, potting the portions immediately.
Pseudobulbs, mature and without leaves, removed while repotting after flowering and potted separately. Leave at least four pseudobulbs on parent plant.

PHALAENOPSIS (Moth orchid) O
Division in late spring, potting the portions immediately and keeping them only slightly moist until established. Not a reliable method as plants do not often provide the opportunity for division.
Stem cuttings taken after flowering and rooted in a heated propagating case. This is not an easy method of propagation.
Plantlets may develop on flowering stems. These can be removed when rooted, potted and established in a heated propagating case.

PHASEOLUS (Snail flower) P Cl
Seeds sown during early spring and germinated in a heated propagating case.

PHILODENDRON S/Cl
Cuttings of stem sections or stem tips, taken between early and late summer and rooted in a heated propagating case, 21–24°C (70–75°F).
Leaf-bud cuttings as above.
Air layering between mid-spring and late summer, in normal greenhouse conditions.
Simple layering between mid-spring and late summer, *in situ*.
Seeds sown during mid-spring and germinated in a heated propagating case, 24–27°C (75–80°F).

PHLOX (Annual phlox) An
Seeds of *P. drummondii* sown during early spring for summer bedding and germinated in a heated propagating case, 15.5°C (60°F). Seeds may also be sown during early autumn to provide greenhouse pot plants for spring flowering.

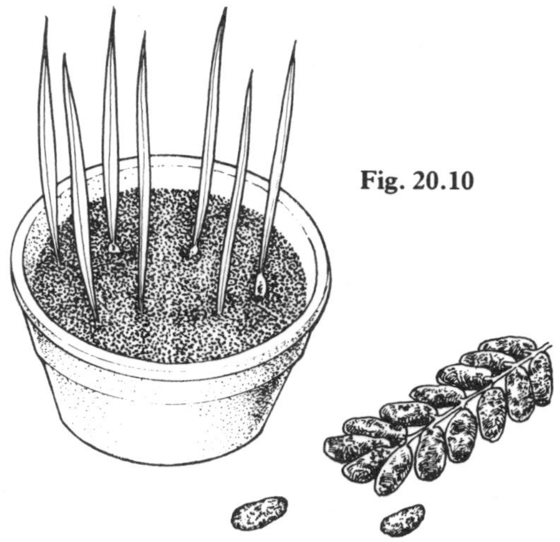

Fig. 20.10

Fig. 20.10 The date palm (phoenix) can be raised from seeds sown in spring and germinated in a minimum temperature of 24 °C (75 °F). Seeds initially produce a single leaf.

PHOENIX (Includes date palm) Pm
Seeds sown during early or mid-spring and germinated in a heated propagating case, minimum 24°C (75°F) (Fig.20.10).
Suckers removed when well developed and rooted during early summer and potted individually.

PHRAGMIPEDIUM O
Division of established plants during spring, potting the portions immediately. Only divide a plant with at least six new growths, splitting it into two equal portions.

PHYSALIS (Cape gooseberry)
P, grown as An
Seeds of *P. peruviana* sown during mid-spring and germinated in a heated propagating case, 18°C (65°F).

PILEA P
Soft or semi-ripe cuttings taken between mid-spring and late summer and rooted in a heated propagating case, 18–21°C (65–70°F). Best to regularly renew plants as pileas quickly deteriorate after a few years.
Division during mid-spring, potting the portions immediately.
Seeds sown during early or mid-spring and germinated in a heated propagating case, 18–21°C (65–70°F).

PINGUICULA (Butterwort) Ca P
Seeds sown early to mid-winter, on surface

of compost consisting of 50:50 peat and sand. Stand the pot in shallow water in cool greenhouse.
Division for plants with several crowns during early or mid-spring, potting the portions immediately.

PIPER (Pepper) Cl/S/Sm T
Semi-ripe cuttings taken in summer and rooted in a heated propagating case, 18–24°C (65–75°F).

PISONIA S/T
Cuttings obtained from tips of shoots during spring or summer and rooted in a heated propagating case, 18–21°C (65–70°F).

PISTIA (Water lettuce) Fl Aq
Division by separating plantlets during spring or summer and immediately replacing them in the water.

PITCAIRNIA P
Offsets removed when well developed and rooted during spring or summer, potted individually and established in a heated propagating case with high humidity.

PITTOSPORUM T/S
Semi-ripe cuttings, wounded at base by removing a thin slice of wood, mid- to late summer, and rooted in a heated propagating case, 18°C (65°F). Alternatively take cuttings in mid-autumn and root in heated propagating case.
Seeds sown early to mid-spring and germinated in a heated propagating case, 18°C (65°F).

PLATYCERIUM (Stag's horn fern) F
Division during mid-spring, planting or potting the portions immediately.
Spores sown as soon as ripe and germinated in a heated propagating case.
Plantlets removed and planted or potted during mid-spring.

PLECTRANTHUS P/S S
Cuttings taken from shoot tips in spring or early summer and rooted in a heated propagating case, 18°C (65°F). Best to raise new plants on a regular basis as young specimens are the most attractive.
Division mid-spring, potting the portions immediately.
Seeds sown early to mid-spring and germinated in a heated propagating case, 18°C (65°F).

PLEIONE O
Division when flowering is over, potting the portions immediately.
Pseudobulbs, mature and without leaves removed while repotting after flowering and potted separately. Leave at least four pseudobulbs on parent plant.
Pseudobulblets may form on the tops of older pseudobulbs. These can be collected when they drop off and in mid- or late spring planted to half their depth in pans of the normal potting compost.

PLEOMELE (DRACAENA) P
Cuttings in the form of stem sections for *P. reflexa*, early to mid-summer, heated propagating case, 21–24°C (70–75°F).
Leaf-bud cuttings as above.
Air layering between mid-spring and late summer in normal greenhouse conditions.

PLUMBAGO (Leadwort) S
Semi-ripe cuttings with a heel for *P. auriculata (P.capensis)*, early to mid-summer and rooted in a heated propagating case, 18°C (65°F).
Soft basal cuttings for *P. indica (P. rosea)*, mid- to late spring, rooting them in a heated propagating case, 18–21°C (65–70°F).
Seeds sown early to mid-spring and germinated in a heated propagating case, 21°C (70°F). Seeds should only be very lightly covered.

PLUMERIA (Frangipani) S/T
Cuttings taken from the tips of stems, before leaves appear, in spring, rooting them in a heated propagating case, 18–21°C (65–70°F).

POLIANTHES (Tuberose) Tu P
Offsets removed during autumn and repotted. However, these may not flower well in cool temperate climates as they do not ripen sufficiently, so it is the usual practice to buy imported tubers on an annual basis.

'OLYPODIUM F
)ivision early to mid-spring, potting the ortions immediately.
.hizomes used as cuttings and rooted in a eated, humid propagating case. Peg the cm (3 in) lengths of rhizome onto the .urface of potting compost.
pores sown in spring and germinated in a eated propagating case.

OLYSCIAS T/S
uttings, either stem sections or tips of ems, taken during mid- to late summer ıd rooted in a heated propagating case, 21–4°C (70–75°F).

ORTULACA (Includes sun plant) An/P
eeds of *P. grandiflora* (sun plant) sown uring early spring and germinated in a eated propagating case, 18°C (65°F). Seeds ay also be sown outdoors during mid- to te spring where the plants are to flower. hey should be barely covered with fine il.

RIMULA P, grown as An or Bi
eeds sown according to type and germiated in a heated propagating case, 15.5°C 0°F). Keep shaded and moist. Sow *P.* × *wensis* in late winter or early spring; *P. alacoides* (fairy primrose) late spring or rly summer; *P. obconica* early to mid-ring; and *P. sinensis* (Chinese primrose) te spring to early summer. To grow olyanthus and coloured primroses as pot ants sow the seeds during late spring or rly summer and germinate in a shaded rden frame or cool greenhouse. Keep oist at all times.

ROSTANTHERA (Mint bush) S
emi-ripe cuttings taken mid- to late mmer and rooted in a heated propagating se, 18–21° (65–70°F).
eeds sown during spring and germinated a heated propagating case, 18°C (65°F).

ROTEA S/T
emi-ripe cuttings taken mid- to late mmer and rooted in a heated propagating se, 18–21°C (65–70°F).
eeds sown early to mid-spring and germiated in a heated propagating case, 18–21°C (65–70°F). Sow one seed per 8 cm (3 in) pot to avoid pricking out with its inevitable root disturbance.

PSEUDOPANAX T/S
Seeds sown as soon as ripe and germinated in a garden frame or cool greenhouse.
Semi-ripe cuttings in summer, heated propagating case.

PTERIS (Table fern) F
Division during mid-spring, potting the portions individually.
Spores sown as soon as ripe or in early spring and germinated in a heated propagating case, 15.5°C (60°F).

PUNICA (Pomegranate) S/T
Semi-ripe cuttings taken with a heel in mid-summer and rooted in a heated propagating case, 15.5–18°C (60–65°F).
Seeds sown during early or mid-spring and germinated in a heated propagating case, 15.5°C (60°F).

PUYA P
Offsets removed when they are well developed, with some roots if possible, spring or summer, potted and then established in a heated propagating case.
Seeds sown early to mid-spring and germinated in a heated propagating case, 18–21°C.

PYCNOSTACHYS P
Cuttings taken during early summer and rooted in a heated propagating case 18–21°C (65–70°F).
Seeds sown during spring and germinated in a heated propagating case, 18°C (65°F).

QUESNELIA P
Offsets removed when well developed during spring or summer, potted, then established in a heated propagating case with high humidity.

QUISQUALIS S/Cl
Semi-ripe cuttings taken during summer and rooted in a heated propagating case, 24–27°C (75–80°F).
Seeds sown early to mid-spring and germinated in a heated propagating case.

RADERMACHERA T/S
Seeds sown early to mid-spring and germinated in a heated propagating case.
Air layering between mid-spring and late summer in normal greenhouse conditions.

RAVENALA (Travellers' tree) T
Seeds sown during spring and germinated in a heated propagating case, 21°C (70°F).
Suckers, if produced, may be removed in spring or summer, potted and then allowed to become established in a heated propagating case, minimum 24°C (75°F).

REBUTIA Cc
Seeds sown early to mid-spring and germinated in a heated propagating case, 21°C (70°F).
Offsets removed during spring or summer and potted individually in suitable-sized pots.

REHMANNIA P
Seeds sown during mid-spring and germinated in a heated propagating case, 15.5°C (60°F). It is best to raise new plants each year, treating them as biennials and discarding the old flowered plants.

REINWARDTIA S S
Soft basal cuttings taken during mid- to late spring and rooted in a heated propagating case, 15.5°C (60°F). Reinwardtias can be grown as annuals, discarding the old plants after flowering. Plants are best discarded anyway after two or three years by which time they will be deteriorating.

RHAPHIDOPHORA (Shingle plant) Cl
Cuttings prepared from tips of stems or stem sections, each containing one leaf, early to late summer, rooting them in a heated propagating case, 24–27°C (75–80 °F).
Simple layering spring or summer, *in situ.*

RHAPHIOLEPIS S
Semi-ripe cuttings taken during mid- or late summer and rooted in a heated propagating case, 18–21°C (65–70 °F).

RHAPIS (Includes ground rattan) Pm
Suckers removed during spring and potted individually.
Division of estalished plants in spring feasible, potting the portions immediately.
Seeds sown early to mid-spring and germ nated in a heated propagating case, 27° (80°F).

RHIPSALIDOPSIS (Includes East cactus) Cc
Cuttings consisting of one to three joint late spring to mid-summer, rooted in heated propagating case, 15.5–21°C (60 70°F). It is the usual practice to pot thr rooted cuttings in one pot to make a bett display.
Seeds sown early to mid-spring and germ nated in a heated propagating case, 21° (70°F).

RHIPSALIS Cc
Cuttings taken late spring to mid-summ and rooted in a heated propagating cas 15.5–21°C (60–70°F).
Seeds sown early to mid-spring and germ nated in a heated propagating case, 21° (70°F).

RHODOCHITON (Purple bell vine) Cl
Seeds sown during early spring and germ nated in a heated propagating case, 15.5° (60°F).

RHODODENDRON S
Seeds for the tender species, sown early mid-spring and germinated in a heat propagating case, 15.5°C (60°F).
Semi-ripe cuttings for species and cultivar including cultivars of *R. simsii* (India azalea), early to mid-summer, rooting the in a heated propagating case, 18°C (65°F).
Simple layering for all tender rhododendrons between mid-spring and late summe *in situ.*

RHOEO (Boat lily) P
Cuttings of basal shoots taken during ear or mid-summer and rooted in a heated pro agating case, 18°C (65°F). Plants should discarded when two or three years old ar replaced with young specimens, which a more attractive.
Seeds sown early to mid-spring and germ nated in a heated propagating case, 18–21° (65–70°F).

RHOICISSUS (Grape ivy) Cl
Semi-ripe cuttings taken early to late summer and rooted in a heated propagating case, 18°C (65°F).
Leaf-bud cuttings taken early to late summer and rooted in a heated propagating case, 18°C (65°F).
Seeds sown during early or mid-spring and germinated in a heated propagating case, 18°C (65°F).

RICINUS (Castor-oil plant)
S, generally grown as An
Seeds sown during late winter or early spring and germinated in a heated propagating case, 21°C (70°F). First soak the seeds in tepid water for 24 hours to encourage faster germination. Sow one seed per small pot.

RIVINA (Bloodberry) P
Cuttings taken during spring and rooted in a heated propagating case, 18–21°C (65–70°F).
Leaf cuttings, using complete leaves with stalks, taken during spring and rooted in a heated propagating case, 18–21°C (65–70°F).
Seeds sown during early or mid-spring and germinated in a heated propagating case, 15.5–18°C (60–65°F). Plants are best raised annually from seeds or cuttings, discarding the old plants after flowering.

ROCHEA Su
Cuttings taken during mid-spring from plants which have been cut back and rooted in a heated propagating case, 15.5–21°C (60–70°F).
Seeds sown early to mid-spring and germinated in a heated propagating case, 21°C (70°F).

ROYSTONIA (Royal palm) Pm
Seeds sown during spring and germinated in a heated propagating case with a minimum temperature of 27°C (80°F).

RUELLIA P/S S
Cuttings mid- to late spring from plants which have been cut back, rooted in a heated propagating case, 18–21°C (65–70°F). Best to raise new plants every two years as young specimens are the most attractive.
Division, if feasible, during mid-spring, immediately potting the portions into suitable-size pots.
Seeds sown early to mid-spring and germinated in a heated propagating case, 18–21°C (65–70°F).

SABAL Pm
Seeds sown during spring and germinated in a heated propagating case, 27°C (80°F).
Suckers removed during spring, with roots if possible, potted immediately then established in a heated propagating case.

SACCHARUM (Sugar cane) P G
Suckers removed in spring and immediately replanted or potted.
Cuttings of stems taken during spring, rooting them in a heated propagating case, 21–27°C (70–80°F).

SAINTPAULIA (African violet) P
Leaf cuttings consisting of entire leaves complete with stalks, between early and late summer and rooted in a heated propagating case, 18–21°C (65–70°F). Insert one cutting in a small pot.
Division during mid-spring, potting the portions immediately.
Seeds sown early to mid-spring and germinated in a heated propagating case, 18–21°C (65–70°F).

SALPIGLOSSIS (Painted tongue) An/Bi
Seeds sown during early spring for flowers in summer and autumn, and in early autumn for flowers during winter and spring, heated propagating case, 18°C (65°F). Seeds may also be sown outdoors in their flowering sites during late spring.

SALVIA (Sages, including scarlet sage)
An/Bi/P/S/S S
Seeds for *S. splendens* (scarlet sage) grown for summer bedding, and other half-hardy annuals, sown during early spring and germinated in a heated propagating case, 18°C (65°F). Seeds also for shrubby and perennial species, as above.
Softwood cuttings for shrubby species, mid- to late spring, rooting them in a heated

propagating case, 15.5°C (60°F).
Semi-ripe cuttings for shrubby species taken mid- to late summer and rooted in a heated propagating case, 15.5°C (60°F).
Division for perennial species, early to mid-spring, planting or potting the portions immediately.

SALVINIA Fl Aq
Division by separating the colonies during summer, immediately replacing the portions in water.

SANCHEZIA S
Cuttings taken between early spring and mid-summer and rooted in a heated propagating case, 18–21°C (65–70°F).

SANDERSONIA Tu P
Division by separating tubers during repotting in spring, potting them individually to grow on to flowering size.
Seeds sown during early spring and germinated in a heated propagating case, 24°C (75°F). Seedlings take two or three years to attain flowering size.

SANSEVIERIA (Includes mother-in law's tongue) P
Leaf cuttings, using sections of leaf about 5 cm (2 in) in length, taken between early and late summer and rooted in a heated propagating case, 21°C (70°F). Make sure the sections are inserted the right way up and to half their length. This method is not recommended for *S. trifasciata* 'Laurentii' as the resultant plants will not have the yellow leaf edge (leaves will be plain green). Propagate it from suckers.
Suckers with two or three leaves and well rooted removed during mid-spring and immediately planted in individual pots of suitable size.

SARRACENIA (Trumpet pitcher) P
Seeds (collected previous autumn) sown during late winter and germinated in a cool greenhouse. Use the growing compost (e.g. live sphagnum moss, peat and sand) and sow in a pan. Cover seeds lightly with very fine sphagnum peat. Place pan in shallow water.
Division during spring just as plants are starting into growth. Cut through rhizomes cleanly and ensure portions have roots.
Rhizome sections during spring, each with some roots, potting into growing compost with tops of sections exposed. Stand pots in shallow water in a cool greenhouse. Time of shoot production is extremely variable.

SAUROMATUM (Voodoo lily) Tu P
Offsets removed in early spring and immediately potted.

SAXIFRAGA (Mother of thousands) P
Plantlets can be removed when well developed and rooted in individual pots. Alternatively layer plantlets into pots and sever from parent when rooted. Normal greenhouse conditions for these methods.

SCHEFFLERA (BRASSAIA, HEPTAPLEURUM) S/T
Air layering between mid-spring and late summer, normal greenhouse conditions.
Semi-ripe cuttings during summer, rooting them in a heated propagating case, 18–21°C (65–70°F).
Seeds sown early to mid-spring and germinated in a heated propagating case, 21–24°C (70–75°F).

SCHINUS (Peruvian mastic tree, pepper tree) S/T
Semi-ripe cuttings taken during summer and rooted in a heated propagating case.
Seeds sown during spring and germinated in a heated propagating case.

SCHIZANTHUS (Poor man's orchid, butterfly flower) An
Seeds sown early spring to produce pot plants for flowering in summer and autumn, or late summer/early autumn for late winter/spring flowering, heated propagating case, 15.5°C (60°F). Seeds may also be sown outdoors during mid-spring where the plants are to flower.

SCHIZOPETALON An
Seeds sown during spring and germinated in a heated propagating case, 13°C (55°F). Pot seedlings as soon as large enough to handle as they dislike root disturbance.

Fig. 20.11

Fig. 20.11 The cuttings of schlumbergera (which includes the Christmas cactus) should consist of two or three joints. They are best rooted during spring or summer in a heated propagating case.

SCHLUMBERGERA (Includes Christmas cactus) Cc
Cuttings consisting of two or three joints taken between late spring and mid-summer and rooted in a heated propagating case, 15.5–21°C (60–70°F) (Fig. 20.11).

SCIRPUS (ISOLEPIS)
(Miniature bulrush) P
Division for *S. cernuus* (miniature bulrush) during mid-spring, potting the portions immediately. Keep the compost permanently moist by standing pots in shallow water.

SEDUM (Stonecrop) Su
Cuttings of stems taken late spring to mid-summer and rooted in a heated propagating case, 15.5–21°C (60–70°F).
Leaf cuttings, using entire leaves, taken late spring to mid-summer and rooted in a heated propagating case, 15.5–21°C (60–70°F).
Division if feasible during mid-spring, potting the portions immediately.
Seeds sown early to mid-spring and germinated in a heated propagating case, 21°C (70°F).

SELAGINELLA P
Cuttings taken during the spring or summer and rooted in a heated propagating case, 21°C (70°F).
Division in mid-spring, potting portions immediately.

SELENICEREUS (Queen of the night) Cc
Cuttings consisting of sections of stem, 10–15 cm (4–6 in) in length, taken between late spring and mid-summer and rooted in a heated propagating case, 15.5–21°C (60–70°F).
Seeds sown early to mid-spring and germinated in a heated propagating case, 21°C (70°F).

SENECIO (Includes cineraria)
An/P/Su/Cl/S
Seeds for *S.* × *hybridus* (*S. cruentus*) which is grown as a biennial pot plant, and annual species, sown between mid-spring and late summer and germinated in a heated propagating case, 15.5°C (60°F). Seeds of *S. bicolor cineraria* (*S. cineraria, Cineraria maritima*), often grown as an annual for summer bedding, are sown in late winter or early spring and germinated as above. Seeds of succulents are sown early to mid-spring and germinated in a temperature of 21°C (70°F).
Cuttings for perennials, shrubs, climbers and succulents, spring or summer, rooting them in a heated propagating case, 18–21°C (65–70°F).

SETCREASEA P
Cuttings taken between mid-spring and late summer and rooted in a heated propagating case, 15.5–18°C (60–65°F). Best to replace plants after three years as they will start to deteriorate.
Division during mid-spring, potting the portions immediately.

SIDERITIS S/S S/P
Semi-ripe cuttings taken during summer and rooted in a heated propagating case, 18°C (65°F).

SINNINGIA (Includes gloxinia) Tu P/S S
Leaf cuttings taken between early and late summer and rooted in a heated propagating case, 21°C (70°F). Lay an entire leaf on the compost after cutting through the veins on the underside.
Seeds sown late winter or early spring and germinated in a heated propagating case, 21°C (70°F).
Division of tubers during early spring, cutting them into sections and ensuring each portion has a shoot, pot them, using potting compost, and establish in a heated propagating case.
Soft basal cuttings early to mid-spring, with a small slice of tuber attached, rooting them in a heated propagating case, 21°C (70°F).

SMILAX Cl
Semi-ripe cuttings taken in summer and rooted in a heated propagating case, 18–21°C (65–70°F).
Division during mid-spring, potting the portions immediately.
Seeds sown early to mid-spring and germinated in a heated propagating case, 18–21°C.

SMITHIANTHA P
Leaf cuttings taken in late spring or early summer and rooted in a heated propagating case, 21°C (70°F). Use entire leaves complete with leaf stalk which can be reduced in length to 12 mm (½ in).
Seeds sown late winter or early spring and germinated in a heated propagating case, 21°C (70°F).
Division by separating rhizomes during repotting in late winter or early spring.

SOLANDRA Cl
Semi-ripe cuttings taken during summer and germinated in a heated propagating case, 18–21°C (65–70°F).

SOLANUM (Winter and Jerusalem cherries) S
Seeds for *S. capsicastrum* (winter cherry) and *S. pseudocapsicum* (Jerusalem cherry), which are grown as annual pot plants, sown late winter or early spring and germinated in a heated propagating case, 18°C (65°F).

SOLEIROLIA (HELXINE) (Mind-your-own-business) Pr P
Division can be carried out during spring, summer or autumn, immediately potting or planting the portions.

SOLLYA (Bluebell creeper) Cl
Cuttings taken during spring or summer and rooted in a heated propagating case, 18–21°C (65–70°F).

SONERILA P/S
Cuttings of the shoot tips taken during spring or summer and rooted in a heated propagating case, 18–21°C (65–70°F).
Seeds sown during early or mid-spring and germinated in a heated propagating case, 21–24°C (70–75°F).

SOPHRONITIS O
Division during early or mid-spring, each portion containing one or more pseudobulbs, potting immediately in suitable-sized pots using the normal compost.

SPARMANNIA (Includes African hemp) S/T
Softwood cuttings produced by plants which have been cut back, taken during spring and rooted in a heated propagating case, 15.5°C (60°F).

SPATHIPHYLLUM (Includes white sails) P
Division of established plants during mid-spring, immediately potting the portions individually.

SPREKELIA (Jacobean or Aztec lily) B
Bulblets removed and potted separately in early autumn while repotting.
Seeds sown during early or mid-spring and germinated in a heated propagating case, 18–21°C (65–70°F). Seedlings can take six or seven years to reach flowering size.

STANHOPEA O
Division during spring when repotting, planting the portions immediately into pots of suitable size.
Pseudobulbs, mature and without leaves, removed while repotting in spring and potted separately. Leave at least four pseudobulbs on parent plant.

STAPELIA (Carrion flowers) Su
Division during mid-spring, potting the portions immediately.
Cuttings of stems taken late spring to mid-summer and rooted in a heated propagating case, 15.5–21°C (60–70°F). Dry off the cuttings for four days before inserting them.
Seeds sown early to mid-spring and germinated in a heated propagating case, 21°C (70°F). Germination is normally very fast – a matter of a few days.

STENOTAPHRUM (St Augustine grass) P G
Division in mid-spring, potting the portions immediately.
Cuttings consisting of a tuft of leaves and a

ort length of stem, taken in spring or mmer and rooted in a heated propagating se, 15.5–18°C (60–65°F).

TEPHANOTIS (Wax flower) Cl
emi-ripe cuttings taken mid- to late mmer and rooted in a heated propagating se, 18–21°C (65–70°F).
mple or serpentine layering between id-spring and late summer *in situ*.
eds sown early to mid-spring and germi- ted in a heated propagating case, 21°C 0°F).

TRELITZIA (Bird of paradise flower) P
ivision during mid-spring, potting the rtions immediately.
eds sown early to mid-spring and germi- ted in a heated propagating case, 18–21°C 5–70°F). Seedlings can take at least five ars to reach flowering size.

TREPTOCARPUS (Cape primrose) P
eaf cuttings for tufted species, late spring mid-summer, rooting them in a heated opagating case, 18°C (65°F). Use small tire leaves; or cut long leaves into 8 cm in) long sections.
ttings of stems for appropriate species, ken mid- to late spring and rooted as for f cuttings.
ivision for tufted species during early ring, potting the portions immediately.
eds sown mid- to late winter for autumn wers, or early to mid-spring for summer ooms the following year, and germinated a heated propagating case, 18°C (65°F).

TREPTOSOLEN (Marmalade bush) S
ftwood cuttings taken during late spring early summer and rooted in a heated opagating case, 15.5–18°C (60–65°F).

TROBILANTHES P/S
ttings taken during spring and rooted in eated propagating case, 24°C (75°F).

TROMANTHE P
ivision during mid-spring, potting the rtions immediately.

YAGRUS (COCOS) (Dwarf coconut lm) Pm
Seeds for *S.weddelliana (Cocos weddelliana)* sown early to mid-spring and germinated in a heated propagating case, 24–27°C (75–80°F).

SYNGONIUM (Goose foot) Cl
Cuttings consisting or stem sections or tips taken between early and late summer and rooted in a heated propagating case, 21–24°C (70–75°F).

TAGETES (Includes African and French marigolds) An
Seeds sown early to mid-spring and germinated in a heated propagating case, 18°C (65°F). Seeds germinate within a few days.

TECOMA S/T
Semi-ripe cuttings taken in summer and rooted in a heated propagating case, 18–21°C (65–70°F).

TECOMARIA
(Includes Cape honeysuckle) Cl
Semi-ripe cuttings taken during summer and rooted in a heated propagating case, 18–21°C (65–70°F).

TELOPEA S
Simple layering early spring, *in situ*.
Seeds sown early spring and germinated in a heated propagating case, 15.5–18°C (60–65°F).

TETRANEMA (Mexican foxglove) P
Division during mid-spring, potting the portions immediately.
Seeds sown early to mid-spring and germinated in a heated propagating case, 15.5–18°C (60–65°F).

THUNBERGIA An/P Cl/S/P
Seeds sown early spring and germinated in a heated propagating case, 15.5 – 18°C (60 – 65°F).
Softwood cuttings for shrubby species consisting of sections of stem taken during late spring and rooted in a heated propagating case, 18 – 21°C (65 – 70°F).

TIBOUCHINA (Glory bush) S/S S/Cl/P
Softwood or semi-ripe cuttings taken between mid-spring and late summer and

rooted in a heated propagating case, 18 – 21°C (65 – 70°F).

TIGRIDIA (Tiger flower) B
Bulblets removed in mid-spring when repotting or planting and potted separately.
Seeds sown early to mid-spring and germinated in a heated propagating case, 15.5 – 18°C (60 – 65°F).

Fig. 20.12
Tillandsias produce offsets which may be removed in spring or summer when well developed, potted individually and established in a heated propagating case.

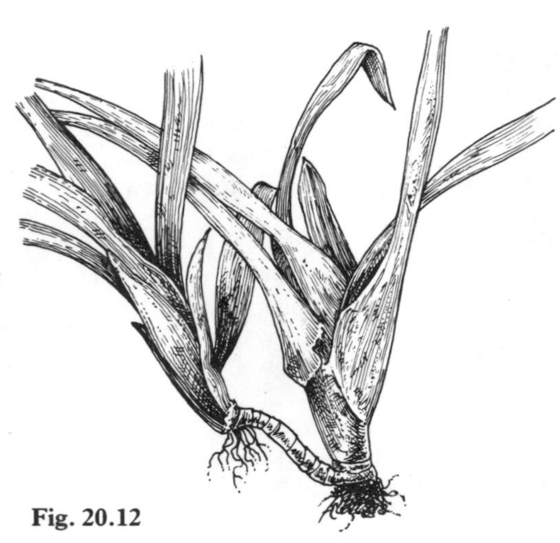

Fig. 20.12

TILLANDSIA P
Offsets removed late spring or summer, well developed and with roots, potted individually and established in a heated propagating case (Fig. 20.12). Atmospheric tillandsias must not be potted but mounted on wood, tree branch, etc.
Seeds sown early to mid-spring and germinated in a heated propagating case, 24°C (75°F), with high humidity.

TITANOPSIS Su
Seeds sown during early to mid-spring and germinated in a heated propagating case, 21°C (70°F).

TITHONIA (Mexican sunflower) An
Seeds sown during late winter or early spring and germinated in a heated propagating case, 15.5°C (60°F).

TOLMIEA (Pick-a-back plant) P
Plantlets on leaves. Peg down leaves bearing plantlets into pots of compost during summer and sever from parent when roote
Alternatively, remove leaves containir plantlets complete with stalks and treat leaf cuttings, heated propagating cas 15.5°C (60°F).
Division during mid-spring, potting th portions immediately.

TORENIA (Includes wishbone flower) An/P
Seeds sown during late winter or ear spring and germinated in a heated propaga ing case, 18°C (65°F).

TRACHYMENE (DIDISCUS) An
Seeds sown during early spring and germ nated in a heated propagating case, 15.5 (60°F).

TRADESCANTIA (Wandering Jew) P
Cuttings taken from shoot tips betwe mid-spring and early autumn and rooted a heated propagating case, 15.5°C (60°F Easy to root, even in a jar of water.
Layering in spring or summer simply b pegging down the stems at their nodes. Ma even layer naturally.

TRITONIA Cm
Cormlets removed while repotting in ear autumn and potted separately. They wi reach flowering size within two or thr years.
Seeds sown during spring and germinate in a cool greenhouse.

TROPAEOLUM An/P Cl
Seeds for annuals and perennials, includir *T. peregrinum* (canary creeper), sown durir early spring and germinated in a heate propagating case, 15.5°C (60°F).
Soft basal cuttings for perennial specie taken as soon as available in spring ar rooted in a heated propagating case, 18 (65°F).
Division for perennials during early sprin potting or planting the portio immediately.
Tubers for species which produce ther separating them when repotting or planti in early to mid-spring.

TULBAGHIA P
Division during spring, immediately potting the portions.

JRSINIA An/P
eeds sown during early spring and germi-ated in a heated propagating case, 15.5°C 60°F).

JTRICULARIA (Bladderwort)
a P, inc many Aq
Division during spring or summer, mmediately replacing the portions in the ppropriate growing medium.

'ALLISNERIA (Includes tape grass)
u Aq
Division of tufts during spring or summer, mmediately replanting the portions in 'ater.
lantlets produced on roots of *S. spiralis* ape grass) can be separated and replanted spring or summer.

'ALLOTA (Guernsey or Scarborough lily)

ulblets removed when repotting during te summer, and potted individually. It will e at least five years before they attain owering size.

ANDA O
uttings consisting of stems with aerial ots, summer, rooting them in a heated ropagating case, 21°C (70°F). A leggy plant n be cut back by half, using the top part of e stems as cuttings; when new shoots are roduced use those also as cuttings when ey have developed aerial roots.

ANILLA O
uttings of stem sections taken during mmer and rooted in a heated propagating se, 21°C (70°F). Each cutting should carry to five leaves.

ELTHEIMIA B
ulblets removed and potted individually in te summer while repotting.
eaf cuttings, using entire mature leaves, oting them in a heated propagating case, °C (65°F).
eeds sown as soon as ripe and germinated in a heated propagating case, 15.5–18°C (60–65°F).

× VENIDIO-ARCTOTIS (ARCTOTIS × VENIDIUM) P
Semi-ripe cuttings taken during late summer and rooted in a heated propagating case, 15.5°C (60°F). New plants are best raised annually, discarding old plants in autumn after flowering.

VENIDIUM (Includes monarch of the veldt) An/P
Seeds sown early spring for summer flowering, or in late summer for blooms in early spring, heated propagating case, 15.5°C (60°F). Seeds may also be sown during late spring outdoors where the plants are to flower.

VERBENA (Vervain) An/P
Seeds sown late winter to early spring and germinated in a heated propagating case, 18–21°C (65–70°F). Germination may be slow and unreliable.
Division for perennial species during mid-spring, immediately potting the portions.
Soft cuttings of perennials obtained by forcing stock plants in gentle heat during early spring. Root in a heated propagating case, 15.5°C (60°F). Perennials are best raised afresh each year.

VICTORIA Aq
Seeds sown during mid-winter (after storing in water) by sowing in pots of loam which are then submerged in water at a minimum temperature of 30°C (85°F). Usually grown as annuals in cool temperate climates as it is difficult to overwinter plants.

VRIESIA P
Offsets removed during summer when well developed and rooted, potting them individually and establishing in a heated propagating case.
Seeds sown early to mid-spring and germinated in a heated propagating case, 24°C (75°F), with high humidity.

WASHINGTONIA Pm
Seeds sown during spring and germinated

in a heated propagating case with a minimum temperature of 24°C (75°F).

WATSONIA Cm
Division of established clumps during spring or late summer while repotting.
Seeds sown early to mid-spring and germinated in a heated propagating case, 15.5–18°C (60–65°F).

WITTROCKIA P
Perennials
Offsets removed during summer when well developed and rooted, potting them individually and establishing in a heated propagating case.

WOODWARDIA (Chain fern) F
Ferns
Division during mid-spring, potting the portions immediately.
Bulbils for *W. radicans*, which develop on the ends of the fronds. The ends of the fronds bearing bulbils may be pegged down onto the soil or compost surface, when the bulbils will root and develop into new plants.

XANTHORRHOEA (Grass tree, black boy) P
Offsets removed during spring and immediately replanted or potted.

XANTHOSOMA Tu P
Division during spring, immediately replanting or potting the portions.
Cuttings taken during the spring or summer and rooted in a heated propagating case, 21–24°C (70–75°F). A stem or root can be cut into a number of small sections. Cutting back stems results in shoot production and these young shoots can also be used as cuttings.

YUCCA S/T
Suckers with roots attached removed during mid-spring and immediately potted.
Seeds sown during early or mid-spring and germinated in a heated propagating case, 15.5–18°C (60–65°F).

ZANTEDESCHIA (Arum lily, calla) R P
Offsets removed during late winter while repotting, or mid-autumn for *Z. aethiopic*
and planted in pots of suitable size.
Division of established clumps while repo
ting, as above.
Seeds sown during early to mid-spring an
germinated in a heated propagating cas
15.5–18°C (60–65°F).

ZEA (Ornamental maize) A G
Seeds for *Z. mays* sown during mid-sprin
and germinated in a heated propagatin
case, 15.5–18°C (60–65°F). Sow seeds ind
vidually in small pots to prevent root distu
bance. Seeds may be sown outdoors durin
late spring where they are to grow.

ZEBRINA (Wandering Jew) P
Cuttings taken from shoot tips betwee
mid-spring and early autumn and rooted i
a heated propagating case, 15.5°C (60°F
Easy to root, even in a jar of water.
Layering in spring or summer simply b
pegging down the stems at their nodes. Ma
even layer naturally.

ZEPHYRANTHES (Zephyr lily) B
Bulblets removed during mid-spring whi
repotting and potted individually to grow o
to flowering size.
Division of well-established clumps durin
mid-spring while repotting. Do not distur
clumps until they are becoming congested a
then flowering is better.
Seeds sown during autumn or early to mic
spring and germinated in a cool greenhouse

ZINGIBER (Ginger) Tu P
Division just as growth is starting in sprin
generally late spring, immediately potting o
planting the portions.

ZINNIA An
Seeds sown during early spring and germ
nated in a heated propagating case, 15.5
18°C (60–65°F). If later flowering is accep
able, seeds may also be sown outdoo
during late spring in flowering positions.

ZYGOPETALUM O
Pseudobulbs, mature and without leave
removed while repotting in spring ar
potted separately. Leave at least fo
pseudobulbs on parent plant.

CHAPTER 21

Hardy and Tender Vegetables and Culinary Herbs

EY TO PLANT TYPES

H = Annual herb
(A) = Bulb grown as annual
F = Edible fungi
A = Hardy annual
A H = Hardy annual herb
Bi = Hardy biennial
Bi (A) = Hardy biennial grown as annual
Bi B (A) = Hardy biennial bulb grown as annual
Bi H = Hardy biennial herb
H A = Half-hardy annual
P = Hardy perennial
P (A) = Hardy perennial grown as annual
P Aq = Hardy perennial aquatic
P H = Hardy perennial herb
S (H) = Hardy shrub grown as herb
(A) = Perennial grown as annual
H = Perennial herb
A = Tender annual
A H = Tender annual herb
T (H) = Tree grown as herb
Tu R P = Tuberous-rooted perennial

ease note that some herbs have been cluded in Chapter 19 for their ornamental lue.

NGELICA *(Angelica archangelica)* Bi P H
eeds sown outdoors as soon as ripe in late mmer (seeds quickly lose viability). Thin t or transplant in autumn.
ivision during early spring, replanting the rtions immediately.

RTICHOKES H P
ardy perennials
eeds for globe artichoke *(Cynara scolymus)*, rly spring, heated propagating case, 20°C 8°F). Plant out late spring.
ivision or offsets for globe artichoke, rly to mid-spring, replanting portions mediately.
ubers of Jerusalem artichoke *(Helianthus berosus)* and Chinese artichoke *(Stachys affinis* var. *tubifera)* are planted outdoors during early or mid-spring.

ASPARAGUS *(Asparagus officinalis)* H P
Seeds sown in pots during early spring and germinated in a garden frame; or sow in an outdoor seed bed during mid-spring. Cropping stage reached in three years

ASPARAGUS PEA *(Lotus edulis)* H H A
Seeds sown during late spring outdoors where the plants are to crop. Alternatively sow in pots in a heated greenhouse during mid-spring and plant out late spring.

AUBERGINE *(Solanum melongena)* T A
Seeds sown early spring and germinated in a heated propagating case, 21°C (70°F). Allow up to 12 weeks from sowing to planting (greenhouse or outdoors).

BASIL *(Ocimum basilicum)* T A H
Seeds sown mid-spring and germinated in a heated propagating case, 15.5°C (60°F). Plant outdoors early summer. Alternatively sow outdoors during late spring in cropping position.

BEANS H T A
Seeds sown according to type. French beans *(Phaseolus vulgaris)* sown outdoors in cropping positions between late spring and mid-summer, or mid-spring under cloches; minimum soil temperature of 10°C (50°F) needed. May also be grown to maturity in heated greenhouse, sowing in late winter. Runner beans *(Phaseolus vulgaris* ssp. *vulgaris)* and scarlet runners *(Phaseolus coccineus)* sown in cool greenhouse, 10°C (50°F), in mid-spring for planting out late spring. Sow outdoors in cropping positions in late spring or early summer. Broad beans *(Vicia faba)* sown outdoors in late autumn or early spring in cropping positions.

BEETROOT *(Beta vulgaris)*
H Bi, grown as A
Seeds sown outdoors in cropping positions, late winter under cloches, early spring to mid-summer unprotected, minimum soil temperature 7°C (45°F).

BORAGE *(Borago officinalis)* H A H
Seeds sown outdoors in cropping positions during early autumn or mid-spring.

BROCCOLI, SPROUTING *(Brassica oleracea* var. *botrytis italica)* H Bi
Seeds sown mid- to late spring in an outdoor seed bed, transplanting four to six weeks later.

BRUSSELS SPROUTS *(Brassica oleracea* var. *oleracea gemmifera)* H Bi
Seeds sown in outdoor seed bed during early or mid-spring and transplanted late spring/early summer. Also sow late winter in garden frame or under cloches and plant out mid-spring.

CABBAGES *(Brassica oleracea* vars.) H Bi
Seeds sown in an outdoor seed bed and transplanted when large enough, the sowing time depending on type. Spring cabbage sown late summer. Sow summer and red cabbage between late winter (in a garden frame at that time) and late spring. Winter cabbage, including savoy, is sown during mid-to late spring. Chinese cabbage *(Brassica pekinensis)* is sown during mid-summer in its cropping position.

CALABRESE *(Brassica oleracea* var. *botrytis italica)* H B
Seeds sown in an outdoor seed bed during mid-to late spring and transplanted early summer.

CARAWAY *(Carum carvi)* H Bi H
Seeds sown during mid-spring, outdoors in cropping position.

CARDOON *(Cynara cardunculus)*
H P, grown as A
Hardy perennial, grown as an annual
Seeds sown during mid-spring, outdoors in cropping position.

CARROTS *(Daucus carota* ssp. *sativus)*
H Bi
Seeds are sown outdoors, in garden fram or under cloches where the plants are grow. Early carrots are sown in late winte early spring under cloches or in gard frames. They may also be sown outdoors succession from early spring to late summe Sow maincrop carrots from mid-spring early summer.

CAULIFLOWERS *(Brassica oleracea* va *botrytis botrytis)* H Bi, grown as A
Seeds sown under glass or in an outdo seed bed and transplanted to outdoor cro ping site, timing according to type. Summ cauliflowers may be sown in mid-winter in heated greenhouse, 13°C (55°F). Altern tively sow in early autumn in a garden fran and overwinter seedlings in this. Autum cauliflowers are sown mid-to late spring an outdoor seed bed. Sow winter cauliflo ers in an outdoor seed bed from mid-to la spring.

CELERIAC *(Apium graveolens* va *rapaceum)* H Bi, grown as A
Seeds sown during early spring in a heate greenhouse and germinated in a temper ture of 15.5°C (60°F). Alernatively sow under cold glass in mid-spring. Pla outdoors in cropping position during la spring.

CELERY *(Apium graveolens* var. *dulce)*
H Bi, grown as A
Seeds sown early to mid-spring in a heate greenhouse and germinated in a temper ture of 15.5°C (60°F). Plant outdoors cropping position during late spring.

CELTUCE *(Lactuca sativa* var. *asparagina*
H A
Seeds sown in succession between ear spring and mid-summer, outdoors in cro ping positions.

CHERVIL *(Anthriscus cerefolium)*
H Bi H grown as A
Seeds sown from late winter to mid-autum outdoors in cropping positions.

HICORY *(Cichorium intybus* var. *foliosum)*
[P, grown as A
eeds of forcing chicory sown during late
ɔring or early summer outdoors in crop-
ıng positions. Green chicory is sown in
ırly summer and red chicory (radicchio)
etween late spring and late summer,
ıtdoors in cropping positions.

HIVES *(Allium schoenoprasum)* H P H
eeds sown during mid-spring outdoors in
ermanent positions.
ivision during early or mid-spring,
eplanting the portions immediately.

ORIANDER *(Coriandrum sativum)* H A H
eeds sown during mid-spring outdoors in
ropping positions. Warm soil needed for
ermination. May also be sown during early
ɔring in heated greenhouse and planted out
ı late spring.

ORN SALAD *(Valerianella locusta)* H A
eeds for winter crops sown late summer or
ırly autumn, and for summer crops early to
ıid-spring, outdoors in their cropping posi-
ons.

UCUMBERS *(Cucumis sativus)* H H A
eeds sown under glass during spring,
leally after chitting, singly in small pots,
ıd germinated in a heated propagating
ıse, 21–24°C (70–75°F). Sow greenhouse
ıcumbers in early spring and outdoor
ıltivars during mid-spring. The latter are
lanted outdoors in late spring. Seeds may
so be sown outdoors in cropping positions
ıring late spring/early summer.

ILL *(Anethum graveolens)* H A H
eeds sown outdoors during early to mid-
ɔring, in cropping positions as plants do
ot like to be transplanted. Another sowing
ıay be made during mid-summer for an
ıtumn crop. Dill will also self sow.

NDIVE *(Cichorium endivia)* H A
eeds sown in succession every three or
ɔur weeks from early spring until late
ımmer, outdoors where the plants are to
ıature. Hardier winter cultivars are sown
ıring late summer or early autumn under
oches or in a garden frame.

FENNEL *(Foeniculum vulgare)* H P H
Seeds sown outdoors in permanent positions during mid-spring, thinning the seedlings if necessary. Raise new plants every three or four years to replace old stock.
Division of established plants early spring, every three or four years.

FLORENCE FENNEL *(Foeniculum vulgare* var. *dulce)* H P, grown as A
Seeds sown from mid-spring to early summer in a cool greenhouse and germinated in a heated propagating case, 15.5°C (60°F). Pot seedlings individually. Young plants will be ready for planting outdoors within about one month. Alternatively sow seeds outdoors in cropping position during early or late spring and thin out seedlings. Outdoors seeds may be slow to germinate and seedling growth may also be slow.

FRENCH TARRAGON *(Artemisia dracunculus)* H P H
Division every three or four years in mid-spring, planting the portions immediately. Discard the old declining central portions of the clumps. This keeps the stock young and vigorous.

GARLIC *(Allium sativum)* H P H
Bulblets or 'cloves' are planted annually during late winter, outdoors in their cropping positions. Entire bulbs may also be planted if desired rather than broken down into 'cloves'.

HAMBURG PARSLEY *(Petroselinum crispum* ssp. *tuberosum* H P H
Seeds sown during early spring, outdoors in cropping position. Early sowing is essential to provide roots of suitable size by winter. Seeds germinate extremely slowly. It is important to prevent weed growth during the germination and seedling stages.

HORSERADISH *(Armoracia rusticana)* H P
Root cuttings. Roots of pencil thickness removed during early winter, cut into lengths of 15–20 cm (6–8 in) and stored in slightly moist peat or sand. Plant them during early spring when they are starting into growth, inserting them to their full

length and ensuring they are the right way up (the thickest part is usually the top). The tops should be about 5 cm (2 in) below soil level.

KALE *(Brassica oleracea* var. *acephala sabellica)* H Bi
Seeds sown during mid-to late spring in an outdoor seed bed and the seedlings transplanted to their cropping positions during early summer.

KOHL-RABI *(Brassica oleracea* var. *acephala gongylodes)* H Bi, grown as A
Seeds sown in cropping positions in succession every four weeks from mid-spring to early autumn, to ensure a constant suppy.

LAND OR AMERICAN CRESS *(Barbarea vulgaris)* Bi, grown as A
Seeds sown during early or mid-spring outdoors where the plants are to crop. The soil must be moisture-retentive at all times. Further sowing may be made during summer and into early autumn to ensure a succession. Always cover late sowings with cloches.

LEEKS *(Allium porrum)* H Bi, grown as A
Seeds for early crops sown late winter in a heated greenhouse and germinated in a temperature of 13°C (55°F), planting the young plants outdoors in cropping positions during mid-spring. Sow maincrop leeks in outdoor seed bed during early or mid-spring (using cloches if desired) and transplant to cropping position during late spring or early summer. Seeds sown during early summer and transplanted in mid-summer will produce crops in spring.

LETTUCES *(Lactuca sativa* vars.)
H A/H H A
Seeds sown according to types. Year-round production is possible. For summer and autumn crops outdoors sow in succession outdoors from early spring to late summer, in cropping positions. Cloches can be used for early and late sowings. For spring lettuce sow winter-hardy cultivars outdoors or under cloches where they are to crop during late summer and early autumn. Winter lettuces are grown in heated or unheated greenhouses and sown between late summ and mid-winter, depending on cultiva They are grown in small pots and plant out when sufficiently large. With all of the sowings always use the correct cultivars.

LOVAGE *(Levisticum officinale)* H P H
Seeds sown when ripe in late summer early autumn, in outdoor seed bed, tran planting the seedlings the following spri to cropping positions.
Root cuttings, each containing a bud 'eye', taken during spring and inserti them 5 cm (2 in) below soil level.

MARJORAM *(Origanum* spp) A/P H
Seeds for sweet marjoram *(Origanu majorana)*, a half-hardy annual. Sc outdoors in cropping position during la spring. May also be sown in a co greenhouse during mid-spring, in heat propagating case, 15.5°C (60°F). Pot seec ings and plant outdoors in late spring.
Cuttings with a heel for pot marjora *(Origanum onites)*, a hardy shrubby pere nial, taken during late spring or ear summer and rooted in a garden frame.

MARROWS *(Cucurbita pepo ovifera)*
H H A
Seeds sown during late spring outdoo where the plants are to mature, ideal under cloches. Sow two seeds in each po tion and thin seedlings to leave t strongest. Alternatively sow seeds in heated greenhouse during mid-to la spring. Sow two seeds in small pots an germinate in a heated propagating cas 15.5–18°C (60–65°F). Remove weake seedlings from each pot. Plant young plan outdoors in cropping positions during la spring or early summer.

MINT, COMMON AND APPLE *(Ment spicata* and *M. rotundifolia)* H P H
Division during autumn or early/mi spring.
Soft basal cuttings taken during mid-to la spring and rooted in a garden frame.
Cuttings of side shoots taken duri summer when available and rooted as abov

MUSHROOMS *(Agaricus bisporus)* E F
Spawn obtained from specialist suppliers is planted on a specially prepared bed of compost indoors (shed, cellar, etc) at any time of year. The bed temperature should be 24°C (75°F) when planting spawn. The growing room should be kept dark and at a temperature of 10–13°C (50–55°F). Mushroom growing is a fairly complex procedure which should be studied in depth.

MUSTARD AND CRESS *(Sinapsis alba and Lepidium sativum)*
H A which are grown together
Seeds are sown at any time of year in containers of compost, or on pads of flannel or similar material which are kept constantly moist. Sow thickly and do not cover with soil, but cover with black polythene sheeting until germination occurs. Raise in a greenhouse with a minimum temperature of 7°C (45°F). Cress should be sown three days before mustard to ensure the two are ready at the same time. Harvest in the seedling stage.

NEW ZEALAND SPINACH *(Tetragonia tetragonoides)* T A
Seeds sown in a heated greenhouse during early spring and germinated in a heated propagating case, 15.5°C (60°F), planting the young plants outdoors in cropping positions during late spring or early summer. Soak seeds in water for half a day before sowing. Seeds may also be sown outdoors in their cropping positions during late spring.

OKRA OR LADY'S FINGER *(Abelmoschus esculentus)* T A
Seeds sown during early spring in a heated greenhouse and germinated in a heated propagating case, 18–21°C (65–70°F). Sow seeds individually in small pots. Plant young plants in a heated greenhouse during mid-spring and in an unheated greenhouse during late spring.

ONIONS *(Allium cepa)* H Bi B, grown as A
Seeds sown in heated greenhouse during mid-winter, planting out young plants in cropping positions in mid-spring. Alternatively sow outdoors in cropping position during late winter (cover with cloches) or early spring. Overwintering onions are sown outdoors during late summer or early autumn in cropping position. Sow salad onions in succession during spring and summer outdoors in cropping positions.

PARSLEY *(Petroselinum crispum)*
H Bi, grown as A
Seeds sown early to mid-spring outdoors where the plants are to crop. Sow only in warm soil. Germination is normally slow. Seeds may also be sown during mid-summer for winter crops. Cover plants with cloches over winter.

PARSNIPS *(Pastinaca sativa)* H Bi
Seeds sown during early or mid-spring outdoors in cropping positions. Do not sow into cold soil – a soil temperature of 7°C (45°F) is needed for germination.

PEAS *(Pisum sativum)* H A Cl
Seeds sown between late winter and mid-spring, using first early, second early and maincrop cultivars in that order. Early sowings, using the hardier round-seeded cultivars, may be started under cloches. Sowings may be made until early summer for succession. A sowing of a first early may be made in mid-summer – cover with cloches in autumn. A sowing of a round-seeded cultivar may be made in late autumn outdoors under cloches in mild areas with well-drained soil. Sugar peas (mangetout) are sown early to mid-spring.

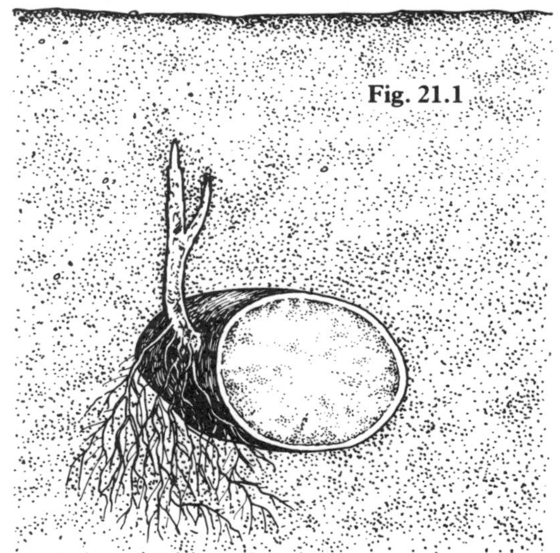

Fig. 21.1 Potatoes may be propagated by cutting large 'seed' tubers into portions, each containing growth buds or 'eyes'. When planted the buds produce shoots, which then form roots, so forming a new plant.

POTATOES *(Solanum tuberosum)*
Tu P, grown as A
Tubers. Small tubers known as 'seed' obtained annually from specialist producers. 'Seed' should be certified virus free. Large tubers may be cut into smaller portions each containing growth buds or 'eyes' (Fig.21.1). Plant early potatoes in early spring, second early and maincrop cultivars in mid-spring. Tubers may be 'chitted' or sprouted in gentle warmth and good light before planting to ensure rapid establishment.

PUMPKINS *(Cucurbita* spp) H H A
Seeds sown during mid- to late spring in a heated greenhouse, providing a germination temperature of 15.5–18°C (60–65°F). Sow two seeds in each small pot and remove weakest seedling. Plant outdoors in cropping positions late spring or early summer. Alternatively sow outdoors in cropping position during late spring, ideally covering with cloches. Sow two seeds in each position and remove the weakest seedling.

RADISHES *(Raphanus sativus)*
H Bi, grown as A
Seeds sown in cropping positions every two weeks throughout spring and summer. Earliest sowings during late winter in garden frame. Start sowing outdoors in early spring and continue until the end of summer. Sow winter radishes between early and late summer.

RHUBARB *(Rheum rhaponticum)* H P
Division of established clumps every five years or so during mid-autumn or early spring, replanting the portions immediately. Each division must have at least one bud. A knife will be needed to cut through the thick rootstock.

ROSEMARY *(Rosmarinus officinalis)*
H S, grown as H
Semi-ripe cuttings taken early or mid-autumn and rooted in a garden frame or under a low polythene tunnel. Alternatively take cuttings mid-to late summer.

SAGE *(Salvia officinalis)* H S, grown as H
Seeds sown during early spring and germinated in a cool greenhouse or garden frame. Alternatively sow in an outdoor seed be[d] during mid-or late spring.
Semi-ripe cuttings with a heel taken durin[g] late summer or early autumn and rooted in [a] garden frame. Overwinter young plants i[n] the frame.

SALSIFY *(Tragopogon porrifolius)*
H Bi, grown as A
Hardy biennial grown as an annual
Seeds sown during mid-spring outdoors i[n] cropping position.

SAVORY *(Satureja* species) A/P H
Seeds for the annual summer savor[y] *(Satureja hortensis)* and perennial winte[r] savory *(S. montana)*, sown during mid[-] spring outdoors in their cropping positions.
Division for winter savory during early o[r] mid-spring, immediately replanting th[e] portions.
Cuttings of side shoots for winter savor[y] during late spring, rooting them in a garde[n] frame and overwintering the young plants i[n] this.

SCORZONERA *(Scorzonera hispanica)*
H P
Seeds sown during mid-or late sprin[g] outdoors in cropping position.

SEAKALE *(Crambe maritima)* H P
Seeds sown during early or mid-spring in a[n] outdoor seed bed. Transplant young plant[s] to permanent site the following early spring. Plants should not be forced or blanche[d] until they are two or three years old.
Root cuttings taken during autumn o[r] winter while lifting plants to be force[d] indoors. Cuttings are the thickness of [a] pencil and 15 cm (6 in) in length. Bundl[e] them, heel in under a garden frame an[d] plant in permanent positions during earl[y] spring.

SEAKALE BEET OR SWISS CHAR[D]
(Beta vulgaris variety) H Bi
Seeds sown during mid-spring outdoors i[n] cropping position. Sow three seeds at eac[h] station and thin seedlings to leave th[e] strongest. A further sowing may be mad[e] during mid-summer, protecting overwinter[-] ing plants with cloches.

SHALLOTS *(Allium cepa* variety*)*
H B, grown as A
Bulblets saved from previous crop, planted outdoors during late winter or early spring.

SPINACH *(Spinacia oleracea)* H A
Seeds sown outdoors in cropping positions. Summer spinach is sown in succession from early spring to mid-summer, at three-weekly intervals. Sow winter spinach in uccession from mid-summer to early utumn, every three weeks. Protect over-wintering plants with cloches.

PINACH BEET OR PERPETUAL SPINACH *(Beta vulgaris* variety*)* H Bi
Seeds sown outdoors in cropping positions during mid-spring and again in mid-ummer. Sow three seeds at each station and thin out seedlings to leave the strongest.

QUASHES *(Cucurbita* spp*)* H H A
Seeds sown during mid-to late spring in a eated greenhouse, providing a germination emperature of 15.5–18°C (60–65°F). Sow wo seeds in each small pot and remove weakest seedling. Plant outdoors in crop-ing positions late spring or early summer. Alternatively sow outdoors in cropping posi-on during late spring, ideally covered with cloches. Sow two seeds in each position and emove the weakest seedling.

WEDE *(Brassica napus)* H Bi
eeds sown late spring or early summer utdoors in cropping position.

WEET BAY *(Laurus nobilis)*
', grown as H
emi-ripe cuttings taken early or mid-utumn and rooted in a garden frame or nder a low polythene tunnel.
eeds sown during early or mid-spring in an utdoor seed bed.

WEET CICELY *(Myrrhis odorata)* H P H
eeds sown late winter or early spring, or te summer to mid-autumn, outdoors in ermanent position. Cold period required to nsure germination so before sowing in pring place seeds in refrigerator, mixed ith sand, for six to eight weeks. Tempera-ure 0–5°C (32–40°F).

Root cuttings each with a dormant bud taken during autumn or winter and planted outdoors 5 cm (2 in) deep.

SWEETCORN *(Zea mays)* H H A
Seeds sown in a heated greenhouse during mid-spring and germinated in a heated propagating case, 15.5–18°C (60–65°F). Sow one seed per small pot. Plant outdoors in cropping position during late spring/early summer. Seeds may also be sown outdoors in cropping position during late spring, ideally covered with cloches.

SWEET PEPPERS *(Capsicum annuum)*
P, grown as A
Seeds sown during early spring in a heated greenhouse and germinated in a heated propagating case, 18°C (65°F). Plant in cropping positions when 15 cm (6 in) high, in heated or unheated greenhouse, or outdoors during late spring/early summer.

TARRAGON *(Artemisia dracunculus)*
H P H
Division during early or mid-spring, replanting the portions immediately in their permanent positions. Best to lift and divide every four years.

THYME *(Thymus vulgaris)* H P H
Division early to mid-spring, replanting the portions immediately in permanent posi-tions.
Cuttings taken with a heel during early summer and rooted in a garden frame.
Seeds sown during early spring in an outdoor seed bed.

TOMATOES *(Lycopersicon lycopersicum)*
H H A
Seeds sown during early spring in a heated greenhouse and germinated in a heated propagating case, 18°C (65°F). Plant in cropping positions when 15 cm (6 in) high, in heated or unheated greenhouse. Outdoor cultivars are planted late spring/early summer.
Grafting during spring, using disease- and eelworm-resistant F_1 hybrid rootstocks, on bench of heated greenhouse, under polythene 'tent' for humidity.

TURNIPS *(Brassica rapa)*
H Bi, grown as A
Seeds sown outdoors in cropping positions. For early supplies, sow late winter/early spring, using early cultivars, and cover with cloches. Further sowings for summer crops are made during mid- and late spring. Sowing in mid- and late summer using maincrop cultivars will provide crops in autumn and winter. For turnips tops sow during early autumn and do not thin o seedlings.

WATERCRESS *(Rorippa nasturium aquaticum)* H P Aq
Cuttings taken during early spring an rooted in jars of clean water. When rooted rapid process) plant in permanent positio in running water or constantly wet soil.

Appendix

USEFUL ADDRESSES

Mail-order Seed Companies

General
J.W. Boyce, 67 Station Road, Soham, Ely, Cambridgeshire CB7 5ED
D.T. Brown and Co. Ltd., Station Road, Poulton-le-Fylde, Blackpool, Lancashire FY6 7HX
Chelsea Choice Seeds Ltd., Notley Road, Braintree, Essex CM7 7HA
Samuel Dobie and Son Ltd., PO Box 90, Paignton, South Devon TQ3 1XY
Mr Fothergill's Seeds, Gazeley Road, Kentford, Newmarket, Suffolk CB8 7QB
S.E. Marshall and Co. Ltd., Regal Road, Wisbech, Cambridgeshire PE13 2RF
Sutton Seeds Ltd., Hele Road, Torquay, South Devon TQ2 7QJ
Thompson and Morgan (Ipswich) Ltd., London Road, Ipswich, Suffolk IP2 OBA
Unwins Seeds Ltd., Histon, Cambridge CB4 4LE

Specialist seedsmen

Wildflowers
John Chambers, 15 Westleigh Road, Barton Seagrave, Kettering, Northamptonshire NN15 5AJ
Seeds-by-Post, Suffolk Herbs, Sawyers Farm, Little Cornard, Sudbury, Suffolk CO1O ONT

Vegetables
Chase Organics Ltd., Terminal Hous Station Approach, Shepperton, Middles TW17 8AS
Country Gardens, 69/71 Main Street, Ea Leake, Leicestershire LE12 6PF (includi the relatively rare tomato rootstock KNVF
W. Robinson and Sons Ltd., Sunny Ban Forton, nr Preston, Lancashire PR34 OBN

Unusual plants
Thomas Butcher, 60 Wickham Road, Sh ley, Croydon, Surrey CR9 8AG
Chiltern Seeds, Bortree Stile, Ulversto Cumbria LA12 7PB

Makers of Propagation Equipment

General
Autogrow Products Ltd., North Walsha Norfolk NR28 OAW
George H. Elt Ltd., Eltex Work Bromyard Road, Worcester
Humex, Gore Road Industrial Estate, Ne Milton, Hampshire
SoTemp Equipment, Basildon Ro Gardens, Burnt Mills Road, Basildon, Ess SS13 1EA
Thermoforce Ltd., Heybridge Work Maldon, Essex
Two Wests and Elliot, Unit 4, Carrwo

Road, Sheepbridge Industrial Estate, Chesterfield, Derbyshire S41 9RH

Greenhouses

Baco Aluminium Products Ltd., Middlemore Road, Smethwick, Warley, West Midlands B66 2EE

Banbury Homes and Gardens Ltd., PO Box 17, Banbury, Oxfordshire OX17 3NS

NHM Ltd., Chiswick Avenue, Mildenhall, Suffolk IP28 7AZ

D.W. Pound Garden Buildings, Rock, Kidderminster, Worcestershire DY12 2UX

Robinsons of Winchester Ltd., Chilcomb Lane, Chilcomb, Winchester, Hampshire SO21 1HU

Micropropagation

Flow Laboratories Ltd., Woodcock Hill, Harefield Road, Rickmansworth, Hertfordshire WD3 1PQ

Imperial Laboratories Ltd., West Portway, Andover, Hampshire SP10 3LF

Composts and Fertilizers

Boots, Nottingham NG2 3AA

Camland Products Ltd., 36 Regent Street, Cambridge CB2 1DB

ICI Garden Products, Woolmead House East, Woolmead Walk, Farnham, Surrey GU9 7UB

Maskell and Sons Ltd., Dirleton Works, Stephenson Street, London E16

PBI Ltd., Britannica House, Waltham Cross, Hertfordshire EN8 7DY

Silvaperl Products Ltd., PO Box 8, Harrogate, North Yorks HG2 8JW

J. Arthur Bowers, Wigford House, Brayford Pool, Lincoln LN5 7BL

Chempak Products, Gedding Road, Hoddesden, Hertfordshire EN11 OLR

Fisons Ltd., Paper Mill Lane, Bramford, Ipswich, Suffolk IP8 4B2

Societies to Join

The following groups run schemes for seed distribution amongst members:

Alpine Garden Society, Lye End Link, St Johns, Woking, Surrey GU21 1SW

Hardy Plant Society, 10 St Barnabas Street, Emmer Green, Caversham, Reading, Berkshire RG4 8RA

National Society of Allotment and Leisure Gardeners, Hunters Road, Corby, Northamptonshire NN17 1JE

Northern Horticultural Society, Harlow Carr Gardens, Crag Lane, Harrogate, North Yorkshire HG3 1QB

Royal Horticultural Society, 80 Vincent Square, London SW1P 2PE

INDEX

Acer, 108
Acidanthera, 47
Acorus, 85
Agapanthus, 37
Ajuga, 84, 99
Alchemilla mollis, 37
Allium, 103, 104
Alpines, 64
  division, 83–84
  outdoor sowing, 42–44
  protecting in cloches, 17
Alstroemeria, 87, 105
Amaranthus caudatus, 44
Androsace sarmentosa, 83
Anemone × *hybrida*, 87
Annuals
  half-hardy, 54–56, 60
  hardy, 44–46
Antirrhinums, 31
Aphelandra squarrosa, 81
Aphids, 30, 32
Aquatic and water plants, 84–85
Aquilegia alpina, 37
Arabis, 83
Arenaria, 83
Arum, 105
Asparagus, 39
Aster, 32, 83
Astrantia, 84
Atriplex hortensis rubra, 44
Aubergine, 37, 39, 57, 58
Aubrieta, 83
Auricula, 87–88

Basal cuttings, 66–67
Beans, 39
  bean seed fly, 28
  bean weevil, 30
  broad, 57, 59
  damping off, 32
  dwarf, 57
  French, 49, 57, 58, 59
  millepedes, 30–31
  runner, 49, 57, 58, 59
  seed dressings, 28
Bedding plants, 32, 54
Beetroot, 30–31, 33, 39
Begonia, 55, 76, 79–80, 105–6, 121
Bell glass, 10–11, 17
Bellis, 46
Berberis, 67
Bergenia, 48, 82, 88, 90
Berrying plants
  seed collection, 38
  stratification, 39–40
Betula, 120
Biennials
  hardy, 46–47
Birds
  seedlings damaged by, 30
Blackberry, 81, 91, 94
Blackcurrant, 32, 95
Blackfly, 30
Border plants *see* Hardy
  perennials
Bottle gardens, 9–10
Bottom heat, 10, 66, 70
Brassicas
  cabbage root fly, 31–32
  damping off, 32
  flea beetles, 28
  seed collection, 37
  seed dressings, 28
Broccoli, 39
Bromeliads, 101
Brussels sprouts, 39, 57, 58, 59
Bryophyllum tubiflorum, 101
Budding, 8–9, 23-24, 116–19
  budding knives, 23–24, 117
  budding patches, 24, 118
  budsticks, 116
  chip budding, 118–19
  equipment and tools, 23–24
  hygiene, 25
  rootstock, 116
  shield budding, 117–18
  trees, 119
Buddleia, 32, 127
Bulbs, 47, 101, 102–7
  bulbils, 47, 48, 103–4
    ferns, 101
    outdoor sowing, 47–48
  bulblets, 102–3
    pests and diseases, 32–34
  diseases, 33
  division, 106–7
  hygiene, 26
  outdoor sowing, 47–48
  scales, 104–5
  scoring and scooping, 107
  seed collection, 37
  stems and bulb eelworm, 32
Cabbage, 39, 54, 57
  cabbage root fly, 31–32
Cacti, 54, 56, 101
  grafting, 115
Caladium, 105
Calluna, 43, 54, 67, 96
Camellia, 65, 80–81
Campanula, 32
Cardiocrinum, 103
Carnation, 33
Carrot, 20, 39, 49, 59
  carrot fly, 31–32
Cato, 8
Cats
  seedlings damaged by, 30
Cauliflower, 20, 39, 54, 57, 59
Ceanothus, 75
Celeriac, 39
Celery, 39, 49, 58
Cerei, 57
Cheiranthus, 46
Chicory, 39
Chimonanthus, 91
Chinese
  propagation by, 7–9
Chinese cabbage, 39
Chionodoxa, 103
Chipping seeds, 6
Chitted seed, 49–50
Chlorophytum, 85, 99
Choisya ternata, 120
Chrysanthemum, 33
  chrysanthemum eelworm, 32
Cineraria, 54
Clarkia elegans, 44
Clay soil
  drainage, 25
Clematis, 76, 81, 93
Climbers, 51–52
Cloches, 16–18, 54, 58, 59
  soil-warming cables, 19–20
Cockroach, 28–29
Codiaeum, 97
Colchicum, 103
Compost, 21–22, 54–55
  cuttings, 66
  formulated, 21
  soilless, 21–22
  sowing spores, 62
Compost heap, 25
Conifers, 38, 51–52, 67, 112

onvallaria, 82
ordylines, 71
orms, 47
cormlets, 103
diseases, 33
hygiene, 26
outdoor sowing, 47–48
ornus, 127
otinus coggygria, 95
otoneaster, 127
rataegus, 109
rocus, 103
ucumber, 20, 37, 39, 49, 54, 57, 58
uttings, 6, 7, 10–11, 18
compost, 21–22
equipment and tools, 23
hormone rooting powder, 23, 65, 69
hygiene, 25
'Irishman's cuttings', 7
leaf, 76–80
leaf-bud, 80–81
mist propagation, 22–23, 66
pegged down leaf, 79–80
pests and diseases, 32–34
root, 7, 73–75
potting up, 75
preparation, 73–74
thick-rooted plants, 74–75
thin-rooted plants, 75
stem, 64–72
basal cuttings, 66–67
eye-cuttings, 71–72
hardwood cuttings, 67
mallet cuttings, 67
pipings, 72
preparation, 64
selecting the parent, 64
semi-ripe cuttings, 67–69
softwood cuttings, 64–66
stem sections, 71
unheated propagators, 18
utworm, 29
yclamen, 37, 47, 105
yperus, 85

aboecia, 96
ahlia, 32, 33, 47, 105, 106
amping off, 32
ead wood, 25
elphinium, 32
eutzia, 95, 127
ianthus, 46, 72, 83, 87, 91
layering, 96–97
ibbers, 14
icentra eximia, 120
ieffenbachia, 71, 97
igitalis, 46

Diseases
basal rot, 33
bulbs, tubers and corms, 33
choosing healthy plants, 26
damping off, 32
dead wood, 25
diseased leaves and plants, 25
dry rot, 33
feeding and conditioning, 25–26
galls, 33
greenhouses, 26
grey bulb rot, 33
hard rot, 33
hygiene *see* Hygiene
micropropagation, 121
neck rot, 28, 33
overcrowding, 25
potato blight, 33
seed dressings, 28
smoulder disease, 33
storage rot, 33
storing seeds, 39
tulip fire, 33
vegetative propagation, 32–34
viruses, 28, 33–34, 121
white mould, 33
Division, 24, 82–90
Alpine and rock plants, 83–84
aquatic and water plants, 84–85
bulbous plants, 106–7
greenhouse and house plants, 85–87
rhizomes, 82, 88
simple, 82–83
suckers, 82
tools and equipment, 24
trees and shrubs, 90–91
tubers, 105–6
Dormancy, seeds, 35–36
Doronicum, 87
Dracaena, 71, 81, 97
Drainage
hygiene, 25
Dropping, 96
Dry rot, 33

Echeverias, 101
Eelworm, 32–33, 73
Epipremnum, 81
Equipment and tools, 13
artificial lighting, 20–21
budding knives, 23–24, 117
budding patches, 24
compost, 21–22
cuttings, knives for, 23
dibbers, 14
division, 24
'electronic leaf', 22–23
forks, 13, 24
frames and cloches, 16–17, 16–18
grafting knives, 23–24
hoes, 13
hormone rooting powders, 23
misting equipment, 22–23
pots, 17–18
presser, 18
propagators, 18–19
rakes, 13
secateurs, 23–24
seedtrays, 17–18
sieves, 22
soil blocks, 22
soil-warming cables, 19–20
spades, 13
trowels, 13–14, 24
watering cans, 14
Erica, 96
Erinus alpinus, 43
Erythronium, 103
Euphorbia, 37, 44
Eye-cuttings, 71–72

F_1 hybrids, 40
Fagus, 91
Fatsia, 97
Feeding and conditioning
plant hygiene, 25–26
Fennel, 39
Ferns, 85, 87, 101
fern case, 11
raising from spores, 61–63
Fertilizer, 42
Ficus, 70, 81, 97
Flea beetle, 28, 29
Fluid sowing, 50
Forks, 13
Forsythia, 91, 127
Frames, 16–18, 54, 57–58, 67
soil-warming cables, 19–20
Freesia, 47
Fritillaria, 103, 107
Fruit trees, 108
grafting, 109
training, 122–5
Fuchsia, 64

Galanthus, 32, 47, 102, 103, 106
Galls, 33
Gentiana sino-ornata, 83
Geranium, 33
Germination, 6
Geum, 99
Gladiolus, 33, 47, 103
Gloxinia, 54, 57, 77
Gooseberry, 69, 95

Grafting, 8–9, 9, 23–24, 108–15
  equipment and tools, 23–24
  hygiene, 25
  rootstock, 108, 109–10
  saddle, 112
  scion, 108
  spliced side, 112–13
  spliced side veneer, 113-14
  veneer, 113
  whip and tongue, 110, 112
Grape vine, 71
Grasses, 48, 87, 88
Greeks
  propagation by, 6–7, 8
Greenfly, 30
Greenhouses, 14–16, 54
  artificial lighting, 20–21
  cleaning, 26–27
  cockroaches, 28–29
  condensation, 15
  division of greenhouse plants, 85–87
  glazing, 15–16
  heating, 15
  hygiene, 25, 26–27
  plastic, 15–16
  seed propagation, 56–57
  smoke pesticide or fumigant, 27
  soil-warming cables, 19–20
  structure, 14–15
  vegetables, 54, 57, 59
Grey bulb rot, 33

Haberlandt, 11
Haberlea, 76
Half-hardy annuals, 54–56, 60
Half-ripe cuttings, 67
Hamamelis, 91, 97, 109
Hardwood cuttings, 69–71
Hardy annuals, 44–46
Hardy biennials, 46–47
Hardy perennials
  division, 83, 87–90
  outdoor sowing, 48–49
  root cuttings, 73
Heating
  artificial, 54
Hebe, 64
Hedera, 76, 81
Helleborus, 48, 87
Hemerocallis, 105
Herbaceous plants *see* Hardy perennials
Hill, Thomas
  The Gardener's Labyrinth, 9
History of propagation, 6–12
Hoes, 13
Hollyhock, 33
Hormone rooting powder, 23, 65, 69
House plants, 54
  division, 85–87
  growing from seed, 56–57
  hygiene, 26
Hyacinthus, 32, 47, 103, 107
Hydrangea, 32, 95
Hygiene, 51
  budding, 25
  choosing healthy plants, 26
  compost heap, 25
  cuttings, 25
  dead wood, 25
  diseased leaves and plants, 25
  drainage, 25
  feeding and conditioning, 25–26
  garden, 25
  grafting, 25
  greenhouses, 25, 26–27
  layering, 25
  overcrowding, 25
  pots and trays, 27
  rubbish, 25
  seeds and seedlings, 25, 28–31
  smoke pesticide or fumigant, 27
  vegetative propagation, 32–34
  viruses, 33–34
  see also Diseases; Pests

Iberis, 44
Ilex, 65, 108
Imported species, 9–10
Iris, 32, 82, 87, 88
'Irishman's cuttings', 7

Jasminum, 93
John Innes composts, 21
Juglans, 108
Juncus, 85
Juniperus, 91

Kalanchoe tubiflora, 101
Kale, 39
Kohl rabi, 39

Labels, 14
Laburnum, 30
Laurel, 67
Lawns, 52–53
Layering, 24, 91–98
  air (Chinese), 7–8, 24, 97–98
  continuous (etiolation; French layering), 95–96
  dropping, 96
  hygiene, 25
  mound, 95
  plantlets, 99–100
  serpentine, 93–94
  simple, 91–93
  tip, 94–95
  tools and equipment, 24
Leaf and bud eelworm, 32
Leaf cuttings, 76–80
Leaf-bud cuttings, 80–81
Leatherjacket, 29
Leek, 39, 54, 57
Leontopodium alpinum, 43
Lettuce, 39, 49, 54, 57, 58, 59
  damping off, 32
  mosaic virus, 28
  seed collection, 37
  soil-warming cables, 20
Leucojum, 103
Lewisia, 76
Lighting
  artificial, 20–21, 54, 57
Lilium
  bulbils, 104
  bulblets, 103
  scales, 104–5
Lime, 42
Linaria, 43
Liquidambar styaciflua, 91
Liverwort, 83
Lobelia, 55
Loganberry, 81, 91, 94
Lonicera, 93
Lupinus, 33, 44, 87
  pea and bean weevils, 30

Magnolia, 91, 97
Mahonia, 76, 81
Mallet cuttings, 67
Maranta, 85, 121
Marrow, 20, 33, 37, 39, 49, 54, 57, 58
Meconopsis, 43
Melon, 20, 39, 49, 54, 57, 58
  forcing, 17, 20
Meristem, 120
Micropropagation, 11–12, 120–1
Millepedes, 30–31
Mimulus, 84
Mist propagation, 22–23, 66
Monstera deliciosa, 81
Muscari, 103
Myosotis, 46

Narcissus, 32, 33, 47, 103, 106
Neck rot, 28, 33
Nephrolepis exaltata, 121
Nerine, 103
Nigella, 44
Nitrogen, 42
Nursery beds, 41, 46–47, 49

ffsets, 101
nion, 37, 39, 54, 57, 104
neck rot fungus, 28
onion fly, 31–32
salad onion, 49
stem and bulb eelworm, 32
puntia, 57
rchids, 54, 57, 85, 86–87
vercrowding, 25
verheating, 22

aeonia, 87
apaver orientale, 74, 87
arsley, 39, 49, 57
arsnip, 39, 49
assiflora, 81, 90
eas, 39, 59
bean seed fly, 28
damping off, 32
millepedes, 30–31
pea weevil, 30
seed dressings, 28
eat, 42
elargonium, 54
elleted seeds, 57
eony, 32
eperomia, 76
eppers, 37, 39, 57, 58
erennials
hardy, 48–49, 83, 87–90
ernettya, 96
ests
aphids, 30, 32
bean seed fly, 28
blackfly, 30
cabbage root fly, 31–32
carrot fly, 31–32
choosing healthy plants, 26
cockroach, 28–29
cutworm, 29
eelworm, 32–33
flea beetle, 28, 29
greenfly, 30
greenhouses, 26
growing seeds under cover, 54
leatherjacket, 29
millepedes, 30–31
onion fly, 31–32
pea and bean weevils, 30
root fly, 31–32
seed dressings, 28
seedlings, 28–32
slugs, 28
smoke pesticide or fumigant, 27
snails, 28
springtail, 29–30
storing seeds, 39
turnip gall weevil, 30
vegetative propagation, 32–34
viral infections transmitted by, 33–34
wireworm, 29
woodlice, 28
Petunia, 31
Philadelphus, 64, 95
Philodendron, 81, 97, 127
Phlox, 32–33, 73
Phosphate, 42
Phototropism, 54
Pipings, 72
Plantlets
layering, 99–100
offsets, 101
pests and diseases, 32–34
rooting, 100–1
Polyanthus, 87–88
Potash, 42
Potato, 47, 105
potato blight, 33
Pots, 17–18
cleaning, 27
Presser, 18
Pricking out, 21–22, 59–60
Primula, 46, 83, 87, 87–88
Propagators, 66, 70
electric, 18–19
unheated, 18
Pulmonaria, 87
Pulsatilla vulgaris, 37
Pumpkin, 39

Radish, 20, 39, 59
Rakes, 13
Ramonda, 76
Raolia, 83
Raspberry, 82, 90
Redcurrant, 69
Reseda odorata, 44
Rhizomes, 48, 103
division, 82, 88
outdoor sowing, 47–48
Rhododendron, 43, 54, 65, 91, 109, 112, 121
Rhoicissus, 81
Rhubarb, 32
Rhus typhina, 74, 82, 90
Ribes, 127
Rock plants, 83–84
Romans
propagation by, 7, 8
Romneya coulteri, 75
Root cuttings, 73–75
Root fly, 31–32
Rosa, 39, 82, 90, 109, 116
training, 126–7
Rubbish, garden, 25
Rudbeckia, 83
Runners, 100
pests and diseases, 32–34

Sagittaria, 105
Saintpaulia, 32, 57, 76, 77, 78–79, 85, 121
Salsify, 39
Sansevieria, 76–77, 85
Saxifraga stolonifera, 99
Scilla, 103
Scorzonera, 39
Seakale, 39
Secateurs, 23–24
Seed sowing and propagation, 6, 13–22
broadcasting, 44, 58
chipping seeds, 6
chitted seed, 49–50
compost, 21–22
depth, 50–51
dibbers, 14
dormancy, 35–36
drills, 13, 44, 46, 49, 58
drying and storing seeds, 38–39
F_1 Hybrids, 40
fluid sowing, 50
frames and cloches, 16–18
freezing seeds, 6
greenhouses, 14–16
hygiene, 25, 28–31, 51
labels, 14
nursery beds, 46–47, 49
outdoor sowing, 13–14, 41–53
alpines, 42–44
autumn digging, 41–42
bulbs, corms and tubers, 47–48
fertilizers, 42
hardy annuals, 44–46
hardy biennials, 46–47
hardy perennials, 48–49
lawns, 52–53
nursery beds, 41
soil preparation, 41–42
spring digging, 42
woody plants, 51–52
pelleted seeds, 57
pots, 17–18
presser, 18
pricking out, 21–22, 46–47, 59–60
seed collecting, 35–40
seed dressings, 28
seeds treated by suppliers, 28
seedtrays, 17–18
soaking seeds, 6
soil preparation, 51

sowing too early, 51
space sowing, 49
stratification, 39–40
thinning out, 25, 46
tools and equipment, 13–14
under cover, 14–22, 54–60
compost, 54–55, 56–57
house and greenhouse plants, 56–57
lighting, 54, 57
pricking out, 59–60
soil-warming cables, 19–20
vegetables, 57–60
watering, 55
viability, 39
watering, 14, 45–46, 49, 55
see also Seedlings
Seedlings
damping off, 32
hygiene, 28–31
leaves and growing points eaten away, 28–9
leaves pitted, 29–30
pests, 28–32
pitted and holed, 29
pricking out, 59–60
scratched or pulled from soil, 30
stems eaten away at soil level, 29
see also Seed sowing and propagation
Seedtrays, 17–18
hygiene, 27
Semi-ripe cuttings, 67–69
Sempervivum, 83, 101
Shrubs
dead wood, 25
outdoor sowing, 51–52
root cuttings, 73
semi-ripe cuttings, 67
suckers, 90–91
training, 127
Sieves, 22
Sinningia, 76, 105
Slugs, 28
Snails, 28
Softwood cuttings, 64–66
Soil blocks, 22
Soil preparation, 41–42, 51
autumn digging, 41–42
fertilizers, 42
outdoor sowing, 41–42
spring digging, 42
tilth, 42
Soil types
alkalinity and acidity, 42
clay, 25
Soil-warming cables, 19–20, 66, 70
Soilless composts, 212
Spades, 13
Sphagnum moss, 24
Spinach, 39
Spores, raising plants from, 61–63
collecting spores, 62
germination, 63
potting up, 63
sowing spores, 62–63
Sports, 120
Springtail, 29–30
Stem and bulb eelworm, 32
Stem cuttings, 64–72
Stem sections, 71
Sternbergia, 103
Stock, 31
Stooling, 95
Stratification, 39–40
Strawberry, 32, 33, 91, 100
forcing, 17
stem and bulb eelworm, 32
Streptocarpus, 76–78
Succulents, 54, 56, 57, 76, 101
Suckers, 69, 73, 90–91
division, 82
Sweet corn, 39, 49, 57, 58
Sweet pea, 33
Symphoricapos, 82
Syringa, 73, 82, 90, 108

Tagetes, 31
Theophrastus
Enquiry into Plants, 6-7
Thinning out, 46
plant hygiene, 25
Thymus, 83
Tilia, 91, 97
Tilth, 42
Tissue culture, 11-12
Tolmiea menziesii, 100–1
Tomato, 33, 39, 49, 54, 57, 58
damping off, 32
grafting, 114–5
mosaic virus, 28
protecting in cloches, 17
seed collection, 37
Tools *see* Equipment and tools
Totipotency, 120
Training, 122–7
cordons, 123–4
espaliers, 124–5
roses, 126–7
shrubs, 127
staking, 123
Trees
budding, 119
bush fruit trees, 122–3
cordon training, 123–4
dead wood, 25
espalier training, 124–5
grafting, 108–14
root cuttings, 73
seeds, 37, 51–52
staking, 123
standard and half-standard, 123
suckers, 90–91
training, 122–5
Tropaeolum, 44
Trowels, 13–14, 24
Tubers
diseases, 33
division, 105–6
hygiene, 26
outdoor sowing, 47–48
Tulipa, 32, 33, 37, 47, 103, 107
tulip fire, 33
Tunnels, 54
Turnip, 39, 59
turnip gall weevil, 30
Typha, 85

Vaccinium, 96
Vegetables
growing in greenhouses, 59
outdoor sowing, 49–51
seed viability, 39
sowing under cover, 54, 57–60
Vegetative propagation
bulbs, 102–7
cuttings *see* Cuttings
hygiene, 32–34
plantlets, 99–101
runners, 100
Veronica, 83
Viburnum, 91, 127
Viola odorata, 99
Viruses, 28, 33–34, 121
vegetative propagation, 32–34
Ward, Dr. Nathaniel
Wardian Case, 9–10, 11
Water melon, 39
Water lily, 85
Watercress, 39
Watering
damping off, 32
seeds, 14, 45-46, 49, 55
Watering cans, 14
Weevils, 30
Whitecurrant, 69
Wireworm, 29
Wisteria, 93
Woodlice, 28
Woody plants
outdoor sowing, 51–52